REED'S APF
FOR ENGINEERS

REED'S APPLIED HEAT FOR ENGINEERS

WILLIAM EMBLETON OBE
CEng, FIMarE, MIMechE
Extra First Class Engineers' Certificate

Revised by
LESLIE JACKSON
BSc, MA, CEng, FIMarE, MIMechE
Extra First Class Engineers' Certificate

ADLARD COLES NAUTICAL
London

Published by Adlard Coles Nautical
an imprint of Bloomsbury Publishing Plc
50 Bedford Square, London WC1B 3DP
www.adlardcoles.com

First edition published by Thomas Reed Publications 1963
Reprinted 1966, 1969
Second edition 1971
Reprinted 1974, 1976, 1979
Third edition 1982
Reprinted 1986
Fourth edition 1991
Reprinted 1997, 1998, 1999 (twice), 2000
Reprinted by Adlard Coles Nautical 2003, 2006, 2008, 2009,
2011, 2012 (twice), 2013, 2014 and 2015

ISBN 978-0-7136-6733-2

A CIP catalogue record for this book is available from the British Library.

This book is produced using paper that is made from wood grown in managed, sustainable forests. It is natural, renewable and recyclable. The logging and manufacturing processes conform to the environmental regulations of the country of origin.

Printed and bound in Great Britain by
CPI Group (UK) Ltd, Croydon CR0 4YY

PREFACE

This book covers the syllabuses in Applied Heat for all classes of the Marine Engineers' Certificates of Competency of the Department of Transport (DTp). The examinations are now administered by the Scottish Vocational Educational Council (SCOTVEC). It is a useful aid to students on Business and Technician Education Council (BTEC) and SCOTVEC engineering courses.

Basic principles are dealt with commencing at a fairly elementary stage. Each chapter has fully worked examples interwoven into the text, test examples are set at the end of each chapter for the student to work out, and finally there are some typical examination questions included. The prefix '*f*' is used to indicate those parts of the text, and some test examples, of Class 1 standard.

The author has gone beyond the normal practice of merely supplying bare answers to the test examples and examination questions by providing fully worked step by step solutions leading to the final answers.

This latest revision is a major update in the subject so taking the material for study through the 1990s.

CONTENTS

CHAPTER 1

UNITS AND COMMON TERMS

MASS is the quantity of matter possessed by a body and is proportional to the volume and the density of the body. It is a constant quantity, that is, the mass of a body can only be changed by adding more matter to it or taking matter away from it.

The abbreviation for mass is m and the unit is the kilogramme [kg]. For very large or small quantities, multiples or submultiples of the gramme [g] are used. Large masses are common in engineering and these are measured in megagrammes [Mg]. One megagramme is equal to 10^3 kilogrammes and called a tonne [t].

Mass is proportionally accelerated or retarded by an applied force. To maintain a coherent system of units, a unit of force is chosen which will given unit acceleration to unit mass. This unit of force is called the *newton* [N]. Hence, one newton of force acting on one kilogramme of mass will give it an acceleration of one metre per second per second, therefore:

Accelerating force [N] = mass [kg] × acceleration [m/s^2]. In symbols:

$$F = ma$$

FORCE OF GRAVITY. All bodies are attracted towards each other, the force of attraction depending upon the masses of the bodies and their distances apart. Newton's law of gravitation states that this force of attraction is proportional to the product of the masses of the bodies and inversely proportional to the square of the distance apart.

An important example of this is the mass of the earth which attracts all comparatively smaller bodies towards its, the attractive force by which a body tends to be drawn towards the centre of the earth is the force of gravity and is called the *weight* of the body.

If a body is allowed to fall freely, it will fall with an acceleration of 9·81 m/s^2, this is termed gravitational acceleration and is represented by g. Since one newton is the force which will give one kilogramme of mass an acceleration of one m/s^2, then the force in newtons to give m kg of mass an acceleration of 9·81 m/s^2 is $m \times 9{\cdot}81$. Hence, at the earth's surface, the gravitational force on a mass of m kg is mg newtons, or in other words:

$$\text{Weight [N]} = \text{mass [kg]} \times g\ [\text{m/s}^2].$$

The further the distance between the centre of gravity of the mass and the centre of gravity of the earth, the less is the attractive force between them. Thus, the weight of a mass measured by a spring balance (not a pair of scales which is merely a means of comparing the weight of one mass with another) will vary slightly at different parts of the earth's surface due to the earth not being a perfect sphere.

If a body is projected in a space-rocket, the attractive force of the earth on the body becomes less as its distance from the earth increases until, in complete outer-space, it becomes nil, that is, it is then weightless. The mass of the body of course remains unchanged.

WORK is done when a force applied on a body causes it to move and is measured by the product of the force and the distance through which the force moves.

The unit of work is the *joule* [J] which is defined as the work done when the point of application of a force of one newton moves through a distance of one metre in the direction in which the force is applied. Hence, one joule is equal to one newton-metre. In symbols, J = Nm.

$$\text{Work done [J]} = \text{force [N]} \times \text{distance moved [m]}.$$

The joule is a small unit. Moderate quantities of work may be expressed in kilojoules [1 kJ = 10^3 J] and larger quantities in megajoules [1 MJ = 10^6 J].

POWER is the rate of doing work, that is, the quantity of work done in a given time. The unit of power is the *watt* [W] which is equal to the rate of one joule of work being done every second. In symbols, W = J/s = N m/s.

$$\text{Power [W]} = \frac{\text{work done [J]}}{\text{time [s]}}$$

The watt is a small unit and only suitable for small powers. For normal powers in engineering, the kilowatt [1 kW = 10^3 W] and megawatt [1 MW = 10^6 W] are usually more convenient units.

ENERGY is the capacity for doing work and it is measured by the amount of work done. Energy is therefore expressed in the same units as work, that is, joules, kilojoules and megajoules.

Another useful unit of energy is the *kilowatt-hour* [kW h]. This, as its name implies, represents the energy used or the work done when one kilowatt of power is exerted continually for 3600 seconds, *i.e.* one hour.

$$\begin{aligned} \text{Energy} &= \text{power} \times \text{time} \\ 1\ \text{kWh} &= 1000\ \text{watts} \times 3600\ \text{seconds} \\ &= 1000\ [\text{J/s}] \times 3600\ [\text{s}] \\ &= 3{\cdot}6 \times 10^6\ \text{J} \\ &= 3{\cdot}6\ \text{MJ} \end{aligned}$$

EFFICIENCY is the ratio of the work got out of a machine to the work put into it, and, as this is done in the same time, it is also the ratio of the output power to the input power. Since no machine is perfect, the output is always less than the input, due to frictional and other losses, therefore the efficiency is always less than unity.

The symbol for efficiency is η and it may be expressed as a fraction or as a percentage.

$$\eta = \frac{\text{output power}}{\text{input power}}$$

Example. A mass of 1600 kg is lifted by a winch through a height of 25 m in 30 seconds. Calculate (i) the work done, and, if the efficiency of the winch is 60%, find (ii) the input power in kW and (iii) the energy consumed in kWh.

Force [N] to lift mass against gravity

$$\begin{aligned} &= \text{weight of mass} = mg \\ &= 1600 \times 9{\cdot}81\ \text{newtons} \end{aligned}$$

$$\begin{aligned} \text{Work done}\ [\text{J} = \text{Nm}] &= \text{force}\ [\text{N}] \times \text{distance}\ [\text{m}] \\ &= 1600 \times 9{\cdot}81 \times 25 \\ &= 392\,400\ \text{J} = 392{\cdot}4\ \text{kJ} \quad \text{Ans. (i)} \end{aligned}$$

$$\begin{aligned} \text{Output power}\ [\text{kW} = \text{kJ/s}] &= \frac{\text{work done [kJ]}}{\text{time [s]}} \\ &= \frac{392{\cdot}4}{30} = 13{\cdot}08\ \text{kW} \end{aligned}$$

$$\text{Efficiency} = \frac{\text{output power}}{\text{input power}}$$

$$\text{Input power} = \frac{13{\cdot}08}{0{\cdot}6} = 21{\cdot}8\ \text{kW} \quad \text{Ans. (ii)}$$

$$\text{Energy consumed} = \frac{392{\cdot}4}{0{\cdot}6} = 654 \text{ kJ}$$

1 kWh = 3·6 MJ

$$654 \text{ kJ} = \frac{0{\cdot}654}{3{\cdot}6} = 0{\cdot}1817 \text{ kW h} \quad \text{Ans. (iii)}$$

Alternatively,

$$\begin{aligned} \text{Energy [kW h]} &= \text{power [kW]} \times \text{time [h]} \\ &= 21{\cdot}8 \times \frac{30}{3600} \\ &= 0{\cdot}1817 \text{ kW h.} \end{aligned}$$

PRESSURE is expressed as the intensity of force, that is, the force acting on unit area. The unit of force is the newton [N] and the unit of area is the square metre [m^2], therefore the fundamental unit of pressure is newton per square metre [N/m^2]. The symbol representing pressure is usually p.

Pressures of liquids and gases reach high values which are expressed in multiples of the basic unit of force. For example, the steam pressure in low pressure boilers is often in the region of 8×10^5 N/m² and in high pressure boilers it could be 6×10^6 N/m², the former can be conveniently written 800 k/Nm² and the latter 600 kN/m² or 6 MN/m². Another very convenient unit of pressure commonly used is the *bar*, this has the advantage of being easy to "think" in these units since one bar is approximately equal to one atmosphere of pressure (1 atm = 1·013 bar). One bar is 10^5 N/m², which is 100 kN/m², hence the working pressure of the boilers given as an example above would be stated as 8 bar and 60 bar, respectively.

Pressures in internal combustion engines vary from a little above or below one bar during the air charging period to about 100 bar (= 100×10^5 N/m² = 10 000 kN/m²) during combustion.

Low pressures and vacua are usually measured in millimetres of mercury [mmHg], very small pressures in millimetres of water [mm water]. The instrument used is the *manometer*. This is a glass U-tube partially filled with mercury or water, one end is connected to the source of pressure and the other end is open to the atmosphere. The difference between the levels of liquid in the two legs indicates the difference in pressure between the source and the atmosphere.

The difference between the levels of liquid in the two legs indicates the difference in pressure between the source and the atmosphere.

Considering the manometer containing mercury, if we take the density of mercury as $13{\cdot}6 \times 10^3$ kg/m^3 and the force of gravity on a mass of one kilogramme as 9·80665 newtons (a more accurate figure for the standard value of gravitational acceleration than 9·81 which is usually acceptable in engineering), then the weight of one cubic metre of mercury is $13{\cdot}6 \times 10^3 \times 9{\cdot}80665$ newtons = 133·3 kN. Hence a column of mercury one metre high exerts a pressure of 133·3 kN on one square metre, or a column of mercury one millimetre high is equivalent to a pressure of 133·3 N/m^2.

Similarly, each millimetre of water pressure is equal to 9·80665 N/m^2 which is usually taken as 9·81 N/m^2.

Small pressures may also be expressed in millibars [mbar]: One mbar = $1 \text{ bar} \times 10^{-3} = 10^5 \times 10^{-3}$ N/m^2 = 100 N/m^2.

The mercurial *barometer* works on the principle of the atmospheric pressure supporting a column of mercury. The vertical column of mercury left standing up the tube, perfect vacuum above, is supported by the outside atmospheric pressure and is therefore a measure of the pressure of the atmosphere. As the atmospheric pressure rises and falls, the level of the supported column of mercury rises and falls accordingly.

For example, if the column of mercury supported by the atmospheric pressure is 760 mm, then the atmospheric pressure will be:

$760 \times 133{\cdot}3 = 1{\cdot}013 \times 10^5$ N/m^2 = 1·013 bar or 101·3 kN/m^2.

GAUGE PRESSURE AND ABSOLUTE PRESSURE. Most pressure recording instruments, including the ordinary pressure gauge and the open-ended manometer, measure the pressure from the level of atmospheric pressure. The pressure so recorded is termed the *gauge pressure* and the word "gauge" should follow the units of pressure. Thus, if a pressure gauge reads 2000 kN/m^2 the pressure should be stated as 2000 kN/m^2 gauge, meaning that this is the pressure over and above the atmospheric pressure.

The true pressure is measured above a perfect vacuum and called the *absolute pressure* and this is the value which is used in calculations. The absolute pressure is therefore obtained by adding the atmospheric pressure to the gauge pressure, the gauge pressure being read from the pressure gauge and the atmospheric pressure obtained from the barometric reading.

As an example, if the pressure of a fluid is 550 kN/m^2 gauge, and the barometer stands at 758 mmHg then:

Atmospheric pressure $= 758 \times 133{\cdot}3$
$= 1{\cdot}01 \times 10^5 \text{ N/m}^2 = 101 \text{ kN/m}^2$
Absolute pressure = gauge pressure + atmospheric pressure
$= 550 + 101$
$= 651 \text{ kN/m}^2$

This could be written 651 kN/m^2 absolute. However it is usual to omit the word absolute and take it for granted that if the word gauge does not follow the value of the pressure then it means that it is an absolute pressure. This will be the practice throughout this book.

Pressure gauges are not always perfectly accurate and, in any case, it is difficult to read to an accuracy of one or two kN/m^2. It is therefore quite common when exact accuracy is not essential to assume the atmospheric pressure to be 100 kN/m^2.

In the above example, if the barometer reading was not known, the absolute pressure would be taken as:

$$550 + 100 = 650 \text{ kN/m}^2$$

with very little difference in the final result of a calculation.

If the manometer was to be used as a vacuum gauge, say for a steam condenser, the level of the mercury in the leg connected to the condenser will be higher than the level in the leg open to atmosphere. The difference in level indicates the pressure *below* atmospheric and is written "mmHg vacuum".

For example, if the gauge reads 600 mmHg on vacuum and the barometer stands at 758 mmHg then:

Pressure below atmospheric
$= 600 \times 133{\cdot}3 = 8 \times 10^4 \text{ N/m}^2$

Atmospheric pressure
$= 758 \times 133{\cdot}3 = 1{\cdot}01 \times 10^5 \text{ N/m}^2 = 101 \text{ kN/m}^2$

Therefore absolute pressure in condenser is 80 kN/m^2 *below* 101 kN/m^2 which is 21 kN/m^2 or, more simply calculated:

Abs. press. = (barometer mmHg – vacuum gauge mmHg) $\times$ 133·3
$= (758 - 600) \times 133{\cdot}3$
$= 158 \times 133{\cdot}3$
$= 2{\cdot}1 \times 10^4 \text{ kN/m}^2 = 21 \text{ kN/m}^2$

VOLUME has the basic unit of the cubic metre [m^3]. A common submultiple is the litre (l) and this is equal in volume to one cubic decimetre and is used only for fluid measure.

$$1 \text{ m}^3 = 10^3 \text{ dm}^3 \text{ therefore } 10^3 \text{ litres} = 1 \text{ m}^3$$

The millilitre [ml] is 1×10^{-3} litre and therefore equal in volume to one cubic centimetre. Whereas the basic unit of density is kilogramme per cubic metre [kg/m^3], densities of liquids are sometimes expressed in grammes per millilitre [g/ml] and densities of solids in grammes per cubic centimetre [g/cm^3].

SPECIFIC VOLUME is the volume occupied by unit mass and the basic unit is cubic metre per kilogramme [m^3/kg] thus, the specific volume is the reciprocal of density. In certain cases, specific volume may be expressed in cubic metres per tonne [m^3/t] and litres per kilogramme [l/kg].

TEMPERATURE is an indication of hotness or coldness and therefore is a measure of the intensity of heat.

The most common temperature measuring instrument is the mercurial thermometer. This consists of a glass tube of very fine bore with a bulb at its lower end, the bulb and tube are exhausted of air, partially filled with mercury and hermetically sealed at the top end. When the thermometer is placed in a substance whose temperature is to be measured, the mercury takes up the same temperature and expands (if heated) or contracts (if cooled) and the level, which rises or falls in consequence indicates on the termometer scale the degree of heat intensity.

The Celsius scale (formerly known as Centigrade) is used in the SI system of measuring and specifying temperatures. The point at which pure water freezes into ice is marked zero on the Celsius scale and that point at which pure water boils into steam at atmospheric pressure is assigned the number 100. The former is sometimes referred to as the *lower fixed point* or *ice point*, the latter as the *upper fixed point* or *steam point*. The unit representing a temperature reading on the Celsius scale is °C and the symbol for temperature is θ.

The advantages of mercury are that it does not wet the bore of the glass and therefore none sticks to the glass as the temperature falls. It can be used over a wide useful range of temperature as its freezing point is low, about –38°C, and boiling point high, about 358°C.

Alcohol is sometimes used but requires to be coloured. Although its freezing point is in the region of –115°C and can be useful for measuring very low temperature, its boiling point is also low, about 78°C, and therefore has a limited use.

The Farenheit scale was used in the past and to convert a *temperature interval* from Farenheit to Celsius:

$$\text{Interval on Celsius scale} = \frac{5}{9} \times \text{interval on Farenheit scale.}$$

For example, if a body is heated through 153°F, this is

$$153 \times \frac{5}{9} = 85°\text{C}$$

To convert a *temperature reading* from Farenheit to Celsius:

$$\text{Reading on the Celsius scale} = (\text{F} - 32) \times \frac{5}{9}$$

For example, a temperature of 77 °F is equivalent to

$$(77 - 32) \times \frac{5}{9} = 25°\text{C}$$

ABSOLUTE TEMPERATURE. All gases expand at practically the same rate when heated through the same range of temperature, and contract at the same rate when cooled.

The rate of expansion or contraction of a perfect gas is (very nearly) 1/273 of its volume at 0°C when heated or cooled at constant pressure through one degree Celsius. Hence, if a gas initially at 0°C could be cooled at constant pressure until its temperature is 273 Celsius degrees below 0°C, the volume would contract until there was nothing left and no further reduction of temperature would be possible, that is, the gas would then have reached its *absolute zero of temperature*. In practice of course, it is not possible to cool a gas down to the absolute zero. As the absolute zero of temperature is approached the gas will change into a liquid and the laws of gases are then no longer applicable.

From the above, temperatures can be expressed as *absolute* quantities, that is, stating the degrees of temperature above the level of Absolute Zero, by adding 273 to the ordinary Celsius thermometer reading.

Absolute temperature is often referred to as *thermodynamic temperature*, the symbol for this is T and the unit is the kelvin which is represented by K (often °K), thus:

Thermodynamic temperature = Celsius temperature + 273.

In symbols,

$$T[\text{K}] = \theta\,[°\text{C}] + 273$$

VOLUME FLOW is the volume of a fluid flowing past a given point in unit time. The basic unit is cubic metres per second [m^3/s], other convenient units are cubic metres per hour [m^3/h], cubic metres per minute [m^3/min] and litres per hour [l/h].

For example, if a fluid is flowing at a velocity of v metres per second full bore through a pipe of internal diameter d metres, the quantity flowing in cubic metres per second is:

$$\text{Volume flow } [m^3/s] = \text{area } [m^2] \times \text{velocity } [m/s]$$
$$= 0{\cdot}7854\, d^2 \times v$$

$\dot{V}$ is used to indicate volume flow rate

MASS FLOW is the mass of fluid flowing past a given point in unit time, the basic unit being kilogrammes per second [kg/s]. Since density is the mass per unit volume, then:

$$\text{Mass flow } [kg/s] = \text{volume flow } [m^3/s] \times \text{density } [kg/m^3].$$

Mass flow may also be expressed in other convenient units such as tonnes per hour [t/h] and kilogrammes per hour [kg/h]; $\dot{m}$ is used to indicate mass flow rate.

SWEPT VOLUME OR STROKE VOLUME is the volume swept through by a piston in the cylinder of a reciprocating engine, pump, compressor, etc., it is the product of the piston area [m^2] and the stroke of the piston [m].

The space left between the piston at its inner dead centre (top of its stroke) and the cylinder head is termed the *clearance volume*. This may be expressed as the actual volume of the clearance space, or as a fraction of the stroke volume.

SYSTEM. A *system* is the term given to the collection of matter under consideration enclosed within a *boundary*, the region outside the boundary is termed the *surroundings*. The boundary may be an imaginary enclosure, or it may be real such as the cylinder wall, cylinder head and piston of an internal combustion engine which encloses the mixture of gases within. Energy is transferred across a boundary from one system to another.

If there is no transfer of matter across the boundary, that is, if no substance can enter the system or leave it during investigation, it is a *closed system*, energy only being transferred across the boundary, and whatever changes take place to the substance, is termed a **non-flow** process because no matter flows into or out of it.

If there is a flow of matter through the boundary, it is an *open system*. If the mass flow entering the system is equal to the mass flow leaving so that at any time the quantity of matter within the system is constant, the series of changes to the matter is referred to as a **steady-flow** process.

An example of a closed system within which a non-flow process takes place is in the cylinder of an air compressor and an IC engine.

A turbine is an example of an open system in which a steady-flow process takes place, as is a nozzle and a uniformly cycling reciprocating unit.

A CYCLE is a recurrent period of a complete set of a series of connected processes which a system undergoes, the final state of the system at the end of the cycle being exactly as it was at the beginning of the cycle.

TEST EXAMPLES 1

1. A pump discharges 50 tonne of water per hour to a height of 8 m the overall efficiency of the pumping system being 69%. Calculate the output power and the input power. Calculate also the energy consumed by the pump in 2 hours, expressed in kWh and in MJ.

2. (a) Express a pressure of 20 mm water in N/m^2 and mbars.
   (b) Express a pressure of 750 mmHg in kN/m^2 and bars.

3. A condenser vacuum gauge reads 715 mmHg when the barometer stands at 757 mmHg. State the absolute pressure in the condenser in kN/m^2 and bars.

4. Convert the following temperature readings from °F to °C:
   140 °F   5 °F   −31 °F   −40 °F

5. Oil flows full bore at a velocity of 2 m/s through a nest of 16 tubes in a single pass cooler. The internal diameter of the tubes is 30 mm and the density of the oil is 0·85 g/ml. Find the volume flow in litres per second and the mass flow in kilogrammes per minute.

CHAPTER 2

HEAT

Heat is a form of energy associated with the movement of the molecules which constitute the heated body. Heat is transferred from one substance to another by temperature difference between the two substances, it is interchangeable with other forms of energy and can be made available for doing work and producing mechanical and electrical power.

The basic unit of all energy, including heat, is the *joule* [J]. Thus, units of heat are expressed in joules or multiples of the joule, the commonest being kilojoules [kJ] and megajoules [MJ]. The symbol representing quantity of heat is Q.

SPECIFIC HEAT (or specific heat capacity) of a substance is the quantity of heat required to raise the temperature of unit mass of the substance by one degree. The units of specific heat are therefore heat units per unit mass per unit temperature. The symbol for specific heat is c. In most cases involving specific heat the kilojoule is the most convenient size of heat unit for unit mass of one kilogramme, hence the specific heat is usually expressed in kilojoules per kilogramme per kelvin, in symbols this is kJ/kg K. Note also that one Celsius degree of temperature interval on the thermometer scale is the same as one kelvin degree of temperature interval on the absolute scale, the above units could therefore be written kJ/kg°C, but rarely are.

It follows from the above definition of specific heat that the quantity of heat energy transferred to a substance to raise its temperature is the product of the mass of the substance, its specific heat, and its rise in temperature.

In symbols:

$$Q[\text{kJ}] = m\ [\text{kg}] \times c\ [\text{kJ/kg K}] \times (T_2 - T_1)\ [\text{K}]$$

Different substances have different specific heat values. Also, the specific heat of any one particular substance is not always a constant value over a large range of temperature, the variation in specific heat however, is small, and an average value over the temperature range under consideration may be taken for most practical purposes. For example, the specific heat of water decreases from 4·21 kJ/kg K at 0°C to 4·178 kJ/kg K at 35°C and increases thereafter with rise in temperature, being 4·219 kJ/kg K at 100°C.

Between 0°C and 100°C a mean value is usually taken as 4·2.

Example. Calculate the quantity of heat to be transferred to 2·25 kg of brass to raise its temperature from 20°C to 240°C, taking the specific heat of the brass as 0·394 kJ/kg K.

$$\begin{aligned} \text{Temperature increase} &= 240 - 20 = 220\text{°C} = 220 \text{ K} \\ Q \text{ [kJ]} &= m \text{ [kg]} \times c \text{ [kJ/kg K]} \times (T_2 - T_1) \text{ [K]} \\ &= 2{\cdot}25 \times 0{\cdot}394 \times 220 \\ &= 195 \text{ kJ Ans.} \end{aligned}$$

The characteristics of gases vary considerably at different temperatures and pressures, and heat may be transferred under an infinite number of different conditions, consequently the specific heat can have an infinite number of different values. Two important conditions are transferring heat to or from a gas which is at constant pressure, and transferring heat while its volume is constant. The specific heat of a gas at constant pressure is represented by c_p and at constant volume by c_v. This is dealt with in detail later.

Note: $(T_2 - T_1)$ [K] = $(\theta_2 - \theta_1)$ [°C].

MECHANICAL EQUIVALENT OF HEAT is the relationship between mechanical energy and heat energy. This was determined by Joule, one of the first scientists to demonstrate that heat was a form of energy, using apparatus which generated heat by the expenditure of mechanical work. When work is done in overcoming friction, the mechanical energy expended is converted into heat energy. The force required to overcome sliding friction between two bodies is the product of the coefficient of friction (μ) and the normal force between the surfaces of the bodies.

Example. A shaft runs at a rotational speed of 50 rev/s in oil-cooled bearings 178 mm diameter. The force between the surfaces of the shaft journals and bearings is 2·67 kN and the coefficient of friction is 0·04. Find (i) the friction force at the surface of the journals, (ii) the mechanical energy expended in friction per revolution, (iii) the power loss due to friction, (iv) the temperature rise of the oil if the volume flow through the bearings is 18 litre/min, the specific heat of the oil being 2 kJ/kg K and its density 0·9 g/ml.

$$\begin{aligned} \text{Friction force} &= \mu \times \text{normal force between surfaces} \\ &= 0{\cdot}04 \times 2{\cdot}67 \times 10^3 \\ &= 106{\cdot}8 \text{ N} \quad \text{Ans. (i)} \end{aligned}$$

Work done to overcome friction per revolution [J = N m]

$$= \text{friction force [N]} \times \text{circumference of journal [m]}$$
$$= 106{\cdot}8 \times \pi \times 0{\cdot}178$$
$$= 59{\cdot}7 \text{ J} \quad \text{Ans. (ii)}$$

Power expended [W = J/s]

$$= \text{energy per revolution} \times \text{rev/s}$$
$$= 59{\cdot}7 \times 50$$
$$= 2985 \text{ W} = 2{\cdot}985 \text{ kW} \quad \text{Ans. (iii)}$$

Density of oil $= 0{\cdot}9$ g/ml $= 0{\cdot}9$ kg/litre

Mass flow of oil [kg/s] = volume flow [l/s] × density [kg/l]

$$= \frac{18}{60} \times 0{\cdot}9 = 0{\cdot}27 \text{ kg/s}$$

$$Q \text{ [kJ/s]} = m \text{ [kg/s]} \times c \text{ [kJ/kgK]} \times (T_2 - T_1) \text{ [K]}$$
$$2{\cdot}985 = 0{\cdot}27 \times 2 \times \text{temp. rise}$$
$$\text{Temp. rise} = 5{\cdot}527 \text{ K or } 5{\cdot}527^\circ\text{C} \quad \text{Ans. (iv)}$$

WATER EQUIVALENT of a mass of a substance is the mass of water that would require the same heat transfer as the mass of that substance to cause the same change of temperature.

For example, taking an aluminium vessel of mass 2 kg, and the specific heat of aluminium as 0·912 kJ/kgK:

$$Q = \text{mass} \times \text{specific heat} \times \text{temperature change}$$

$$\text{For water, } Q_W = m_W \times c_W \times (T_2 - T_1)_W$$
$$\text{For aluminium, } Q_A = m_A \times c_A \times (T_2 - T_1)_A$$

Since Q is to be the same quantity of heat,

$$Q_W = Q_A$$
$$m_W \times c_W \times (T_2 - T_1)_W = m_A \times c_A \times (T_2 - T_1)_A$$

and the temperature change is to be the same,

$$m_W \times c_W = m_A \times c_A$$

$$\therefore m_W = m_A \times \frac{c_A}{c_W}$$

Taking the specific heat of water as 4·2 kJ/kgK, the water equivalent of this mass of aluminium is

$$m_W = 2 \times \frac{0{\cdot}912}{4{\cdot}2}$$
$$= 0{\cdot}4343 \text{ kg}$$

That is to say, 0·4343 kg of water would require the same amount of heat transferred to it as the 2 kg of aluminium to raise it through the same range of temperature.

It is useful to know the water equivalent of laboratory calorimeters. When water is contained in a vessel, the temperature of the vessel is the same as that of the water inside, and when the temperature of the water is changed, the temperature of the vessel changes with it. The vessel can therefore be considered as an extra mass of water equal to the water equivalent of the vessel.

Example. The mass of a copper calorimeter is 0·28 kg and it contains 0·4 kg of water at 15°C. Taking the specific heat of copper as 0·39 kJ/kg K, calculate the heat required to raise the temperature to 20°C.

Water equivalent of calorimeter

$$= 0{\cdot}28 \times \frac{0{\cdot}39}{4{\cdot}2} = 0{\cdot}026\ \text{kg}$$

Heat received by water and calorimeter,

$$\begin{aligned} Q\ [\text{kJ}] &= m\ [\text{kg}] \times c\ [\text{kJ/kg K}] \times (T_2 - T_1)\ [\text{K}] \\ &= (0{\cdot}4 + 0{\cdot}026) \times 4{\cdot}2 \times (20 - 15) \\ &= 0{\cdot}426 \times 4{\cdot}2 \times 5 \\ &= 8{\cdot}946\ \text{kJ Ans.} \end{aligned}$$

When two substances at different temperatures are mixed together, heat will transfer from the hotter substance to the colder until both become the same temperature. Unless otherwise is stated, it is assumed that no heat is transferred to or from an outside source during the mixing process and therefore the quantity of heat absorbed by the colder substance is all at the expense of the loss of heat by the hotter substance.

Example. In an experiment to find the specific heat of lead, 0·5 kg of lead shot at a temperature of 51°C is poured into an insulated calorimeter containing 0·25 kg of water at 13·5°C and the resultant temperature of the mixture is 15·5°C. If the water equivalent of the calorimeter is 0·02 kg, find the specific heat of the lead.

Heat received by water and calorimeter when their temperature is raised from 13·5°C to 15·5°C:

$$Q = m \times c \times (T_2 - T_1)$$
$$= (0{\cdot}25 + 0{\cdot}02) \times 4{\cdot}2 \times (15{\cdot}5 - 13{\cdot}5)$$
$$= 0{\cdot}27 \times 4{\cdot}2 \times 2 \text{ kJ}$$

Heat lost by lead in cooling from 51°C to 15·5°C:

$$Q = m \times c \times (T_3 - T_2)$$
$$= 0{\cdot}5 \times c \times (51 - 15{\cdot}5)$$
$$= 0{\cdot}5 \times c \times 35{\cdot}5 \text{ kJ}$$

Heat transferred from the lead is equal to the heat received by the water and calorimeter:

$$0.5 \times c \times 35{\cdot}5 = 0{\cdot}27 \times 4{\cdot}2 \times 2$$
$$c = \frac{0{\cdot}27 \times 4{\cdot}2 \times 2}{0{\cdot}5 \times 35{\cdot}5}$$
$$= 0{\cdot}1278 \text{ kJ/kg K} \quad \text{Ans.}$$

LATENT HEAT is the heat which supplies the energy necessary to overcome some of the binding forces of attraction between the molecules of a substance and is responsible for it changing its physical state from a solid into a liquid, or from a liquid into a vapour, the change taking place without any change of temperature.

The process of changing the physical state from a solid into a liquid is called *melting* or *fusion*, and the quantity of heat required to change unit mass of the substance from solid to liquid at the same temperature is the *latent heat of fusion.*

For example, the latent heat of fusion for ice is 335 kJ/kg at 0°C. This means that one kilogramme of ice at 0°C would require 335 kilojoules of heat transferred to it to completely melt it into one kilogramme of water at 0°C. Also, one kilogramme of water at 0°C would require to lose 335 kilojoules of heat to completely freeze it into ice at 0°C.

The process of changing the physical state of a substance from a liquid into a vapour is called *boiling* or *evaporation* and the quantity of heat to bring about this change at constant temperature to unit mass is the *latent heat of evaporation.*

The latent heat of evaporation of water at atmospheric pressure is 2256·7 kJ/kg. This means that one kilogramme of water at 100°C would require 2256·7 kilojoules of heat to completely boil it into one kilogramme of steam at 100°C. Also, one kilogramme of steam at 100°C would require to lose 2256·7 kilojoules of heat to completely condense it into one kilogramme of water at 100°C.

The temperature at which a liquid boils and the latent heat of evaporation depend strictly upon the pressure, the higher the pressure, the higher the boiling point and the smaller the amount of latent heat required to evaporate it. For example, at atmospheric pressure, the temperature at which water boils is 100°C and the latent heat of evaporation is 2256·7 kJ/kg, at a pressure of 15 bar (1500 kN/m^2) the boiling point is 198·3°C and the latent heat 1947 kJ/kg, at 30 bar (3000 kN/m^2) the boiling point is 233·8°C and the latent heat 1795 kJ/kg. These values are obtained from steam tables which are described later.

When heat is transferred to or from a substance which changes only its temperature, and there is no physical change of state, it is sometimes referred to as *sensible heat*. This distinguishes it from latent heat which changes the physical state of the substance without change of temperature.

We shall see later that, for a constant pressure process, the heat energy transferred to a substance is termed *enthalpy*, then, latent heat of fusion is termed *enthalpy of fusion*, and latent heat of evaporation is termed *enthalpy of evaporation*, and so on.

Example. Calculate the heat required to be given to 2 kg of ice at –15°C to change it into steam at atmospheric pressure, taking the values:

$$\begin{aligned} \text{Specific heat of ice} &= 2{\cdot}04 \text{ kJ/kg K} \\ \text{Latent heat of fusion} &= 335 \text{ kJ/kg} \\ \text{Specific heat of water} &= 4{\cdot}2 \text{ kJ/kg K} \\ \text{Latent heat of evaporation} &= 2256{\cdot}7 \text{ kJ/kg} \end{aligned}$$

Heat to raise the temperature of the ice from –15°C to its melting point of 0°C, *i.e.* a temperature rise of 15°C:

$$\begin{aligned} \text{Sensible heat} &= m \times c \times \text{temp. rise} \\ &= 2 \times 2{\cdot}04 \times 15 \\ &= 61{\cdot}2 \text{ kJ} \end{aligned}$$

Heat to change the ice at 0°C into water at 0°C:

$$\begin{aligned} \text{Latent heat} &= 2 \times 335 \\ &= 670 \text{ kJ} \end{aligned}$$

Heat to raise the temperature of the water from 0°C to its boiling point of 100°C, *i.e.* a temperature rise of 100°C:

$$\begin{aligned} \text{Sensible heat} &= 2 \times 4{\cdot}2 \times 100 \\ &= 840 \text{ kJ} \end{aligned}$$

Heat to evaporate the water at 100°C into steam at 100°C:

$$\text{Latent heat} = 2 \times 2256{\cdot}7$$
$$= 4513{\cdot}4 \text{ kJ}$$
$$\text{Total heat} = 61{\cdot}2 + 670 + 840 + 4513{\cdot}4$$
$$= 6084{\cdot}6 \text{ kJ Ans.}$$

Working is simplified by finding the total heat transfer required per unit mass and then finally multiplying by the total mass, thus,

$$Q = 2(2{\cdot}04 \times 15 + 335 + 4{\cdot}2 \times 100 + 2256{\cdot}7)$$
$$= 2(30{\cdot}6 + 335 + 420 + 2256{\cdot}7)$$
$$= 2 \times 3042{\cdot}3$$
$$= 6084{\cdot}6 \text{ kJ}$$

Change of temperature, change of quantity of heat energy etc., are often written $\Delta\theta$, ΔQ, etc.

TEST EXAMPLES 2

1. A water brake coupled to an engine on test absorbs 70 kW of power. Find the heat generated at the brake per minute and the mass flow of fresh water through the brake, in kg/min if the temperature increase of the water is 10°C. Assume all the heat generated is carried away by the cooling water.

2. The effective radius of the pads in a single collar thrust block is 230 mm and the total load on the thrust block is 240 kN when the shaft is running at 93 rev/min. Taking the coefficient of friction between thrust collar and pads as 0·025, find (i) the power lost due to friction, (ii) the heat generated per hour, (iii) the mass flow of oil in kilogrammes per hour through the block assuming all the heat is carried away by the oil, allowing an oil temperature rise of 20°C and taking the specific heat of the oil as 2 kJ/kg K.

3. To ascertain the temperature of flue gases, 1·8 kg of copper of specific heat 0·395 kJ/kg K was suspended in the flue until it attained the temperature of the gases, and then dropped into 2·27 kg of water at 20°C. If the resultant temperature of the copper and water was 37·2°C, find the temperature of the flue gases.

4. In an experiment to find the specific heat of iron, 2·15 kg of iron cuttings at 100°C are dropped into a vessel containing 2·3 litre of water at 17°C and the resultant temperature of the mixture is 24·4°C. If the water equivalent of the vessel is 0·18 kg, determine the specific heat of the iron.

5. 0·5 kg of ice at –5°C is put into a vessel containing 1·8 kg of water at 17°C and mixed together, the result being a mixture of ice and water at 0°C. Calculate the final masses of ice and water, taking the water equivalent of the vessel as 0·148 kg, specific heat of ice 2·04 kJ/kg K, latent heat of fusion 335 kJ/kg.

CHAPTER 3

THERMAL EXPANSION

EXPANSION OF METALS

The effect of increasing the temperature of metals is generally to cause their dimensions to increase. Most metals expand when they are heated and contract when they are cooled, the amount of expansion per degree rise of temperature differs with different metals. Some alloys are manufactured to have a minimum amount of expansion over a considerable working temperature range, these are usually for special purposes such as measuring instruments and gauges.

Although the expansion is in all directions so that there is an increase in all dimensions, it is sometimes only relevant to consider the expansion in one direction.

LINEAR EXPANSION. When a linear dimension is under consideration, the amount that a metal will expand lengthwise is expressed by its *coefficient of linear expansion*. This is the increase in length per unit length per degree increase in temperature. For instance, if the coefficient of linear expansion of copper is given as $1{\cdot}7 \times 10^{-5}/°C$ (which is 0·000017 per degree Celsius) it means that each metre of length will increase in length by $1{\cdot}7 \times 10^{-5}$ metre when heated through one degree Celsius. This coefficient may be represented by α. Hence, representing the original length by l, and temperature rise by $(\theta_2 - \theta_1)$,

$$\text{Increase in length} = \alpha \times l \times (\theta_2 - \theta_1).$$

The new length of the metal will then be:

$$\begin{aligned} \text{new length} &= \text{original length} + \text{increase in length} \\ &= l + \alpha l\,(\theta_2 - \theta_1), \\ &= l\{1 + \alpha(\theta_2 - \theta_1)\} \end{aligned}$$

Example. A main steam pipe is 6·5 m long when fitted at a temperature of 15°C. Calculate how much allowance should be made for its increase in length if it is subjected to a steam temperature of 300°C, taking the coefficient of linear expansion of the material as $1{\cdot}2 \times 10^{-5}/°C$.

$$\begin{aligned}\text{Increase in length} &= \alpha \times l \times (\theta_2 - \theta_1)\\ &= 1{\cdot}2 \times 10^{-5} \times 6{\cdot}5 \times (300 - 15)\\ &= 0{\cdot}02223 \text{ m}\\ &= 22{\cdot}23 \text{ mm Ans.}\end{aligned}$$

Example. A brass liner is 270 mm diameter when the temperature is 17°C. Take the coefficient of linear expansion of the brass as $1{\cdot}9 \times 10^{-5}$/°C and find the temperature to which the liner should be heated in order to increase the diameter by 2 mm.

Diameter is a linear dimension and therefore the same rule can be applied as for length. Note that the diameter and increase in diameter must be expressed in the same units.

$$\begin{aligned}\text{Increase in diameter} &= \alpha \times d \times (\theta_2 - \theta_1)\\ 2 &= 1{\cdot}9 \times 10^{-5} \times 270 \times (\theta_2 - \theta_1)\end{aligned}$$

$$\text{Increase in temperature} = \frac{2 \times 10^5}{1{\cdot}9 \times 270} = 389{\cdot}7\text{°C}$$

$$\begin{aligned}\therefore \text{Required temperature} &= 17 + 389{\cdot}7\\ &= 406{\cdot}7\text{°C Ans.}\end{aligned}$$

SUPERFICIAL EXPANSION refers to increase in area. The coefficient of superficial expansion is the increase in area per unit area per degree increase in temperature. Therefore, if A represents original area, and $(\theta_2 - \theta_1)$ the increase in temperature, then,

Increase in area = coeff. of superficial expansion $\times A \times (\theta_2 - \theta_1)$.

Consider an area of metal of unit length and unit breadth (Fig. 1) and let this be heated through one degree.

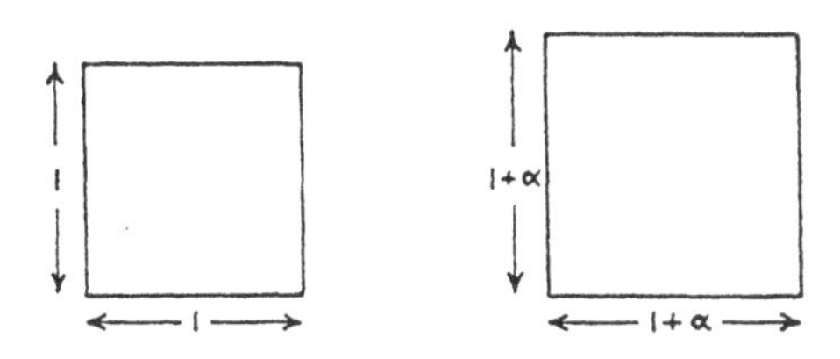

Fig.1

The length and breadth will each increase by an amount equal to α, the coefficient of linear expansion.

$$\begin{aligned}\text{original area} &= 1 \times 1 = 1\\ \text{new length and new breadth} &= 1 + \alpha\\ \text{new area} &= (1 + \alpha)^2 = 1 + 2\alpha + \alpha^2\end{aligned}$$

$$\begin{aligned}\text{Increase in area} &= \text{new area} - \text{original area}\\ &= 1 + 2\alpha + \alpha^2 - 1\\ &= 2\alpha + \alpha^2\end{aligned}$$

α is a very small quantity for any metal (such as about $1{\cdot}2 \times 10^{-5}$ for steel), therefore α^2 being the second order of smallness is a very small quantity indeed and is completely negligible as a quantity to be added for all practical purposes. We can therefore take the increase to be 2α. As this is the increase in area per unit area for one degree increase in temperature, it is the value of the coefficient of superficial expansion, hence,

Coeff. of superficial expansion = 2 × coeff. of linear expansion therefore,

$$\text{Increase in area} = 2\alpha \times A \times (\theta_2 - \theta_1).$$

CUBICAL EXPANSION refers to the increase in volume. The *coefficient of cubical* (or *volumetric*) *expansion* is the increase in volume per unit volume per degree increase in temperature. Therefore, if V represents the original volume, and $(\theta_2 - \theta_1)$ increase in temperature, then:

Increase in volume = coeff. of cubical expansion $\times V \times (\theta_2 - \theta_1)$.

Consider a block of metal of unit length, unit breadth, and unit thickness (Fig. 2) and let this be heated through one degree.

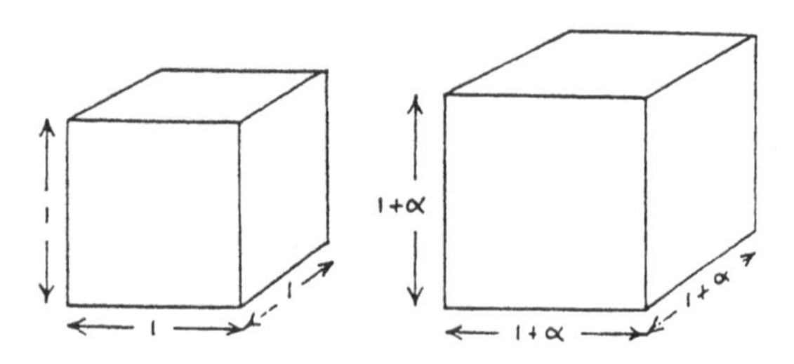

Fig. 2

The length, breadth and thickness will each increase by an amount equal to α, the coefficient of linear expansion.

$$\begin{aligned}\text{Original volume} &= 1 \times 1 \times 1 = 1\\ \text{New volume} &= (1 + \alpha)^3\\ &= 1 + 3\alpha + 3\alpha^2 + \alpha^3\end{aligned}$$

$$\begin{aligned} \text{Increase in volume} &= \text{new volume} - \text{original volume} \\ &= 1 + 3\alpha + 3\alpha^2 + \alpha^3 - 1 \\ &= 3\alpha + 3\alpha^2 + \alpha^3 \end{aligned}$$

α^2 and α^3 being the second and third order of smallness respectively, are negligible quantities for addition, hence the increase may be taken as 3α. As this is the increase in volume per unit volume per degree increase in temperature, it is the value of the coefficient of cubical expansion.

Coeff. of cubical expansion = 3 × coeff. of linear expansion.

Therefore,

$$\text{Increase in volume} = 3\alpha \times V \times (\theta_2 - \theta_1).$$

Note that *linear* refers to any linear dimension such as length, breadth, thickness, diameter, circumference, and so on, and the expression for linear expansion can be applied to any of these dimensions. It is true for internal dimensions as well as external.

The expression for superficial expansion covers any area of the solid, cross-sectional area, surface area, etc., and holds good for internal areas as well as external.

The expression for cubical expansion is also applicable for internal volumes of a hollow vessel.

Example. A metal sphere is exactly 25 mm diameter at 20°C. Find the increase in diameter, increase in surface area, and increase in volume, when heated to 260°C, if the coefficient of linear expansion of the metal is $1{\cdot}8 \times 10^{-5}$/°C.

$$\begin{aligned} \text{Increase in temperature} &= 260 - 20 = 240^\circ\text{C} \\ \text{Increase in diameter} &= \alpha \times d \times (\theta_2 - \theta_1) \\ &= 1{\cdot}8 \times 10^{-5} \times 25 \times 240 \\ &= 0{\cdot}108 \text{ mm} \quad \text{Ans. (i)} \\ \text{Surface area of sphere} &= \pi d^2 \\ \text{Increase in area} &= 2\alpha \times A \times (\theta_2 - \theta_1) \\ &= 2 \times 1{\cdot}8 \times 10^{-5} \times \pi \times 25^2 \times 240 \\ &= 16{\cdot}96 \text{ mm}^2 \quad \text{Ans (ii)} \end{aligned}$$

$$\text{Volume of sphere} = \frac{\pi}{6} d^2$$

$$\begin{aligned} \text{Increase in volume} &= 3\alpha \times V \times (\theta_2 - \theta_1) \\ &= 3 \times 1{\cdot}8 \times 10^{-5} \times \frac{\pi}{6} \times 25^3 \times 240 \\ &= 106 \text{ mm}^3 \quad \text{Ans. (iii)} \end{aligned}$$

EXPANSION OF LIQUIDS

Liquids have no definite shape of their own, therefore no linear dimensions, hence the coefficient of cubical expansion of a liquid is an independent quantity. The coefficient of cubical expansion is usually represented by β.

$$\text{Increase in volume} = \beta \times V \times (\theta_2 - \theta_1)$$

Example. 2500 litres of oil are heated through 50°C. If the coefficient of cubical expansion of this oil is 0·0008/°C, find the increase in volume in cubic metres.

$$\begin{aligned} 2500 \text{ litres} &= 2{\cdot}5 \text{ m}^3 \\ \text{Increase in volume} &= \beta \times V \times (\theta_2 - \theta_1) \\ &= 0{\cdot}0008 \times 2{\cdot}5 \times 50 \\ &= 0{\cdot}1 \text{ m}^3 \quad \text{Ans.} \end{aligned}$$

APPARENT CUBICAL EXPANSION. The tank or vessel which contains a liquid will also expand when heated. It is therefore useful to know the expansion of the liquid relative to its container so that the correct allowance can be made for changes of temperature.

The apparent or relative increase in volume of a liquid is the difference between the volumetric expansion of the liquid and the volumetric expansion of its container. If both have the same initial volume and are raised through the same range of temperature, then, letting suffix L represent the liquid and suffix C the container:

Apparent increase in volume of the liquid

$$\begin{aligned} &= \text{vol. increase of liquid} - \text{vol. increase of container} \\ &= \beta_L \times V \times (\theta_2 - \theta_1) - \beta_C \times V \times (\theta_2 - \theta_1) \\ &= (\beta_L - \beta_C) \times V \times (\theta_2 - \theta_1) \end{aligned}$$

The difference between the coefficients of cubical expansion of the liquid and its container can therefore be termed the *apparent* coefficient of cubical expansion of the liquid.

RESTRICTED THERMAL EXPANSION

If the natural thermal expansion of a metal is restricted, the metal will be strained from its natural length and the material will be stressed. To enable the effects of restricted expansion to be seen here, it is necessary to revise the relationships between stress, strain and modulus of elasticity.

STRESS is the internal resistance set up in a material when an external force is applied. It is expressed as the force carried by the material per unit area of its cross-section.

$$\text{Stress [N/m}^2\text{]} = \frac{\text{force [N]}}{\text{area [m}^2\text{]}}$$

High values of stress are expressed in multiples of the force unit, such as kN/m^2 (= 10^3 N/m^2) and MN/m^2 (= 10^6 N/m^2). In some industries, stress may be expressed in hectobars [hbar], one hectobar = 10^7 N/m^2.

STRAIN is the change of shape that takes place in a material due to it being stressed. Linear strain is the change of length per unit length, thus,

$$\text{Linear strain} = \frac{\text{change of length}}{\text{original length}}$$

MODULUS OF ELASTICITY of a material is the constant obtained by dividing the stress set up in it by the strain it endures under that stress, provided the elastic limit is not exceeded.

$$\text{Modulus of elasticity } (E) = \frac{\text{stress}}{\text{strain}}$$

Example. A solid steel stay 50 mm diameter and 300 mm long at a temperature of 25°C is firmly secured at each end so that expansion is fully restricted. Find the stress set up in the stay and the equivalent total axial force when it is heated to 150°C, taking the coefficient of linear expansion of steel as $1{\cdot}2 \times 10^{-5}$/°C and its modulus of elasticity as 206 GN/m^2.

If the stay was perfectly free to expand without restriction,

$$\begin{aligned}\text{Free expansion} &= \alpha \times l \times (\theta_2 - \theta_1)\\ &= 1{\cdot}2 \times 10^{-5} \times 300 \times (150 - 25)\\ &= 0{\cdot}45 \text{ mm}\end{aligned}$$

Free and unrestrained length would then be

$$300 + 0{\cdot}45 = 300{\cdot}45 \text{ mm}$$

If prevented from expanding the effect is that the stay is compressed from its natural unstrained length of 300·45 mm to 300 mm,

$$\text{Strain} = \frac{\text{change of length}}{\text{original length}}$$

$$= \frac{0{\cdot}45}{300{\cdot}45} = 0{\cdot}001498$$

It is usual, however, to make a slight approximation at this stage by taking the original cold length of 300 mm instead of the heated length of 300·45 mm. The difference in the final result is negligible and it proves much more convenient in solving more complicated problems. It will also be seen that by making this slight approximation the length of the material will not be required because the strain can be obtained direct from:

$$\text{Strain} = \frac{\text{change of length}}{\text{original length}}$$

$$= \frac{\alpha \times l \times (\theta_2 - \theta_1)}{l} = \alpha(\theta_2 - \theta_1)$$

Hence the strain for this stay can be taken as:

$$\text{Strain} = \alpha(\theta_2 - \theta_1)$$
$$= 1{\cdot}2 \times 10^{-5} \times 125 = 0{\cdot}0015$$

which is the same as $\frac{0{\cdot}45}{300} = 0{\cdot}0015$

$$E\ [\text{N/m}^2] = \frac{\text{stress [N/m}^2]}{\text{strain}}$$

$$\therefore \text{Stress} = \text{strain} \times E$$
$$= 0{\cdot}0015 \times 206 \times 10^9$$
$$= 3{\cdot}09 \times 10^8\ \text{N/m}^2$$
$$= 309\ \text{MN/m}^2 \text{ or } 30{\cdot}9 \text{ hbar} \quad \text{Ans. (i)}$$

$$\text{Stress [N/m}^2] = \frac{\text{force [N]}}{\text{area [m}^2]}$$

$$\therefore \text{Total axial force} = \text{stress} \times \text{area}$$
$$= 3{\cdot}09 \times 10^8 \times 0{\cdot}7854 \times 0{\cdot}05^2$$
$$= 6{\cdot}068 \times 10^5\ \text{N}$$
$$= 606{\cdot}8\ \text{kN} \quad \text{Ans. (ii)}$$

TEST EXAMPLES 3

1. A steam pipe is 3·85 m long when fitted at a temperature of 18°C. Find the increase in length if free to expand, when carrying steam at a temperature of 260°C, taking the coefficient of linear expansion of the pipe material as $1{\cdot}25 \times 10^{-5}/°C$.

2. A solid cast iron sphere is 150 mm diameter. If 2110 kJ of heat energy is transferred to it, find the increase in diameter, taking the following values for cast iron: density = 7·21 g/cm^3, specific heat = 0·54 kJ/kg K, coefficient of linear expansion = $1{\cdot}12 \times 10^{-5}/°C$.

3. A bi-metal control device is made up of a thin flat strip of aluminium and a thin flat strip of steel of the same dimensions, connected together in parallel and separated from each other by two brass distance pieces 2·5 mm long, their centres being 50 mm apart. Find the radius of curvature of the strips when heated through 200°C, taking the following values for the coefficients of linear expansion:

$$\text{Aluminium } \alpha = 2{\cdot}5 \times 10^{-5}/°C$$
$$\text{Steel } \alpha = 1{\cdot}2 \times 10^{-5}/°C$$
$$\text{Brass } \alpha = 2{\cdot}0 \times 10^{-5}/°C$$

4. The pipe line of a hydraulic system consists of a total length of steel pipe of 13·7 m and internal diameter 30 mm. If the coefficient of linear expansion of the steel is $1{\cdot}2 \times 10^{-5}/°C$ and coefficient of cubical expansion of the oil in the pipe is $9 \times 10^{-4}/°C$, calculate the volumetric allowance in litres to be made for oil overflow from the pipe when the temperature rises by 27°C.

5. A straight length of steam pipe is to be fitted between two fixed points with no allowance for expansion. If the compressive stress in the pipe is to be limited to 35 hbar (= 350 MN/m^2) calculate the initial tensile stress to be exerted on the pipe when fitted cold at 17°C to allow for a steam temperature of 220°C. Take the coefficient of linear expansion of the pipe material as $1{\cdot}12 \times 10^{-5}/°C$ and the modulus of elasticity as 206 GN/m^2.

CHAPTER 4

HEAT TRANSFER

Heat is transferred from one system to another by one of the three methods known as *Conduction*, *Convection*, and *Radiation*, or by a combination of these.

CONDUCTION

Conduction is the flow of heat energy through a body, or from one body to another in contact with each other, due to difference in temperature. The natural flow of heat takes place from a region of high temperature to a region of lower temperature.

Generally speaking, metals are good conductors of heat. Air, and some materials such as asbestos, cork, glass wool, are very bad conductors, these are called insulators and are used to minimise heat loss. Special plastic base compositions are used to lag boilers, pipes, casings, etc. to reduce loss of heat energy to the colder outside surroundings. Cork and fibre glass are common insulating materials to pack the hollow walls of refrigerating chambers to reduce heat flow into the cold chambers from the warmer outside surroundings.

The quantity of heat conducted through a material in a given time depends upon the thermal conductivity of the material, is proportional to the surface area exposed to the source of heat, is proportional to the temperature difference between the hot and cold ends, and is inversely proportional to the distance or thickness through which the heat is conducted, thus,

$$\text{Quantity of heat varies as } \frac{\text{area} \times \text{time} \times \text{temp. difference}}{\text{thickness}}$$

The thermal conductivity depends upon the nature of the material and its ability to conduct heat. This varies for different materials and sometimes varies slightly for the same material depending upon the temperature range.

The *thermal conductivity (k)* of a material expresses the quantity of heat energy conducted through unit area in a unit time for unit temperature difference between two opposite faces of a material of unit distance apart.

Considering the heat flow through a flat wall, taking the following symbols and basic units, and referring to Fig. 3:

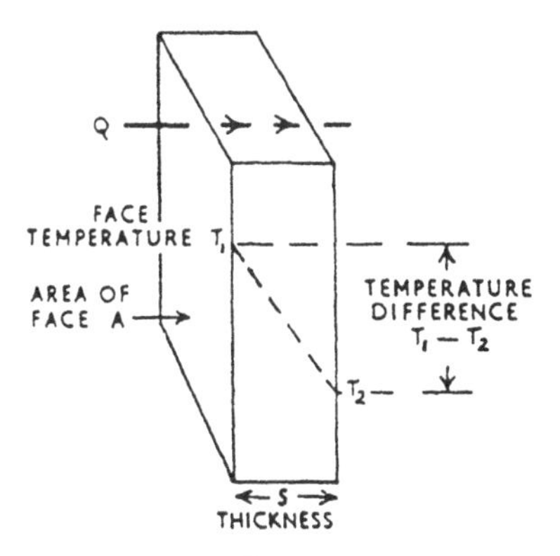

Fig. 3

Q = quantity of heat energy conducted, in joules [J].
A = area through which heat flows, in square metres [m^2].
t = time of heat flow, in seconds [s].
$T_1 - T_2$ = temperature difference between the two faces [K].
S = thickness of wall, in metres [m].

Then, the units of k the thermal conductivity are Jm/m^{2}s K. For convenience this is usually shortened by cancelling m into m^2 and substituting W watts in the place of J/s.

$$\frac{\text{Jm}}{\text{m}^2\text{s K}} = \frac{\text{J}}{\text{ms K}} = \frac{\text{W}}{\text{mK}} = \text{W/m K}$$

Hence, the quantity of heat energy transferred by conduction is:

$$Q = \frac{kAt(T_1 - T_2)}{S}$$

Example. Calculate the heat transfer per hour through a solid brick wall 6 m long, 2·9 m high, and 225 mm thick, when the outer surface is at 5°C and the inner surface 17°C, the thermal conductivity of the brick being 0·6 W/m K.

$$Q[\text{J}] = \frac{k[\text{W/m K}] \times A[\text{m}^2] \times t[\text{s}] \times (T_1 - T_2)[\text{K}]}{S[\text{m}]}$$

$$= \frac{0{\cdot}6 \times 6 \times 2{\cdot}9 \times 3600 \times (17 - 5)}{0{\cdot}225}$$

$$= 2{\cdot}004 \times 10^6 \text{ J}$$

$$= 2{\cdot}004 \text{ MJ or } 2004 \text{ kJ} \quad \text{Ans.}$$

Note: that T_1 is 17 + 273 = 290 K and T_2 is 5 + 273 = 278 K, their difference, $T_1 - T_2$, is the same as 17°C − 5°C.

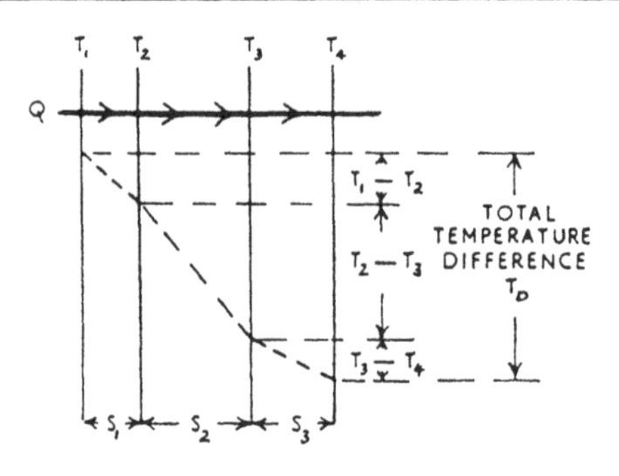

Fig. 4

COMPOSITE WALL. Consider the transfer of heat energy by conduction through a wall made up of a number of layers of different materials, take as an example three slabs of different thicknesses as shown in Fig. 4.

For each thickness:

$$Q = \frac{kAt \times \text{temp. diff.}}{S} \quad \therefore \text{ temp. diff.} = \frac{QS}{kAt}$$

Total drop in temperature across three thicknesses:

$$T_1 - T_4 = \frac{Q_1 S_1}{k_1 A_1 t_1} + \frac{Q_2 S_2}{k_2 A_2 t_2} + \frac{Q_3 S_3}{k_3 A_3 t_3}$$

The same quantity of heat energy is transferred across each layer through the same area in the same time, therefore Q, A and t are common:

$$T_1 - T_4 = \frac{Q}{At}\left\{\frac{S_1}{k_1} + \frac{S_2}{k_2} + \frac{S_3}{k_3}\right\}$$

For any number of layers, let T_D represent the total temperature drop, that is,

$$T_D = (T_1 - T_2) + (T_2 - T_3) + (T_3 - T_4) + \text{etc.}$$

and using the summation sign Σ for the sum of the quantities inside the brackets, that is,

$$\Sigma\left\{\frac{S}{k}\right\} = \frac{S_1}{k_1} + \frac{S_2}{k_2} + \frac{S_3}{k_3} + \text{etc.}$$

then,

$$T_D = \frac{Q}{At} \Sigma \left\{ \frac{S}{k} \right\} \quad \text{or} \quad Q = \frac{AtT_D}{\Sigma \left\{ \frac{S}{k} \right\}}$$

Example. One insulated wall of a cold-storage compartment is 8 m long by 2·5 m high and consists of an outer steel plate 18 mm thick, an inner wood wall 22·5 mm thick, the steel and wood are 90 mm apart to form a cavity which is filled with cork. If the temperature drop across the extreme faces of the composite wall is 15°C, calculate the heat transfer per hour through the wall and the temperature drop across the thickness of the cork. Take the thermal conductivity for steel, cork and wood as 45, 0·045, and 0·18 W/mK respectively.

For the composite wall,

$$T_1 - T_4 = \frac{Q}{At} \left\{ \frac{S_1}{k_1} + \frac{S_2}{k_2} + \frac{S_3}{k_3} \right\}$$

where

$$\begin{aligned} T_1 - T_4 &= 15 \text{ K} \\ A &= 8 \times 2{\cdot}5 = 20 \text{ m}^2 \\ t &= 3600 \text{ seconds} \end{aligned}$$

$$\begin{aligned} \Sigma \left\{ \frac{S}{k} \right\} &= \frac{0{\cdot}018}{45} + \frac{0{\cdot}09}{0{\cdot}045} + \frac{0{\cdot}0225}{0{\cdot}18} \\ &= 0{\cdot}0004 + 2 + 0{\cdot}125 \\ &= 2{\cdot}1254 \end{aligned}$$

$$\therefore \; 15 = \frac{Q}{20 \times 3600} \times 2{\cdot}1254$$

$$\begin{aligned} Q &= \frac{15 \times 20 \times 3600}{2{\cdot}1254} \\ &= 5{\cdot}082 \times 10^5 \text{ J} \\ &= 508{\cdot}2 \text{ kJ} \quad \text{Ans. (i)} \end{aligned}$$

Temperature drop across the cork:

$$\begin{aligned} &= \frac{QS}{Atk} \\ &= \frac{5{\cdot}082 \times 10^5 \times 0{\cdot}09}{20 \times 3600 \times 0{\cdot}045} = 14{\cdot}11 \text{ K} \\ &= 14{\cdot}11\text{°C} \quad \text{Ans. (ii)} \end{aligned}$$

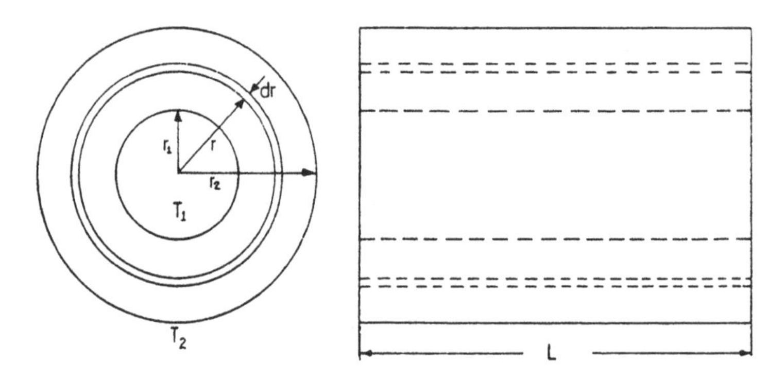

Fig. 5

f CYLINDRICAL WALL. A cylindrical wall, *e.g.* a pipe, could be considered as being made up of many thin cylindrical elements all with the same coefficient of thermal conductivity.

In Fig. 5. If dT is the small temperature difference radially across the cylindrical element whose thickness is dr, length L and area A.

$$\text{Then } Q = kAt\left(-\frac{dT}{dr}\right)$$

$$A = 2\pi rL$$

$$\text{Hence } Q = k2\pi rL\left(-\frac{dT}{dr}\right)$$

Using integration by the use of the calculus to account for the complete cylinder thickness:

$$Q\int_{r_1}^{r_2}\frac{dr}{r} = -k \times 2\pi Lt\int_{T_1}^{T_2} dT$$

$$Q \ln\left(\frac{r_2}{r_1}\right) = -2\pi kLt\,(T_2 - T_1)$$

$$Q = \frac{2\pi kLt\,(T_1 - T_2)}{\ln\left(\dfrac{r_2}{r_1}\right)}$$

(ln x, the natural logarithm of x, is often written $\log_e x$)

f Example. A steam pipe is 85 mm external diameter and 25 m long. It carries 1250 kg of steam per hour at a pressure of 20 bar. The steam enters the pipe 0·98 dry and it is to leave the outlet end of the pipe with a minimum dryness fraction of 0·96. The pipe is to be lagged with a material having a thermal conductivity of 0·2 W/m K and outer surface temperature of 26°C. Any temperature drop across the wall of the pipe is to be neglected and the rate of heat transfer through the wall of a thick cylinder is given by:

$$Q = \frac{2\pi k\theta}{\ln\left(\frac{D_2}{D_1}\right)} \quad \text{per metre length}$$

where:

k = the thermal conductivity of cylinder material
θ = temperature difference
D_2 = external diameter
D_1 = internal diameter

Determine the thickness of the lagging.

$$\text{Heat loss} = m \times hfg\,(x_i - x_o)$$

$$= \frac{1250 \times 1890 \times (0{\cdot}98 - 0{\cdot}96)}{3600}$$

$$= 13{\cdot}125 \text{ kW}$$

$$Q = \frac{2\pi k\theta}{\ln\left(\frac{D_2}{D_1}\right)}$$

$$\frac{13125}{25} = \frac{2\pi \times 0{\cdot}2 \times (212{\cdot}4 - 26)}{\ln\left(\frac{D_2}{0{\cdot}085}\right)}$$

$$\ln\left(\frac{D_2}{0{\cdot}085}\right) = \frac{0{\cdot}4\pi \times 186{\cdot}4}{525}$$

$$\lg\left(\frac{D_2}{0{\cdot}085}\right) = \frac{0{\cdot}4462}{2{\cdot}3026}$$

$$D_2 = 0{\cdot}085 \times 1{\cdot}563$$

$$= 0{\cdot}1329 \text{ m}$$

$$\text{Thickness} = \frac{132{\cdot}9 - 85}{2}$$

$$= 23{\cdot}9 \text{ mm} \quad \text{Ans.}$$

Notes: (i) at 20 bar $t_s = 212{\cdot}4°C$

(ii) lg x, the common (base 10) logarithm of x, is often written log x

(iii) ratio of diameters is the same as ratio of radii

CONVECTION

Convection is the method of transferring heat through a fluid by the movement of heated particles of the fluid.

Fig. 6 shows a vessel with an inclined tube connected at the bottom. When this contains water, and heat is applied to the tube, the heated particles of the water become less dense and rise, denser particles move to take their place and thus convection currents are set moving resulting in all the water in the vessel and tube becoming heated almost uniformly due to the continuous circulation of the water. This is the principle of the water-tube boiler.

Fig. 6 also shows air in a room heated by convection, the fire, radiator or other heat source being placed at the bottom of the room.

Fig. 6 also illustrates the air in the room cooled by convection, the coolers (such as refrigerator pipes) being situated near the top of the room.

These above are examples of natural circulation of the fluid and is referred to as *free convection*. When the motion of the fluid is produced mechanically, such as by means of a pump or a fan, it is referred to as *forced convection*.

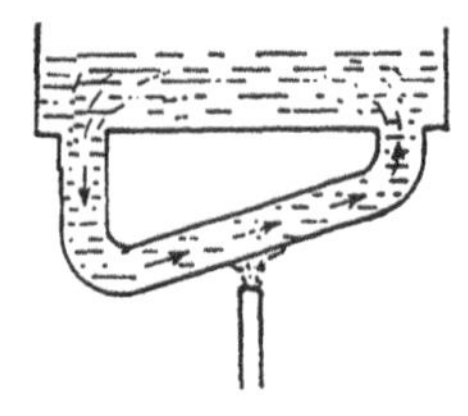

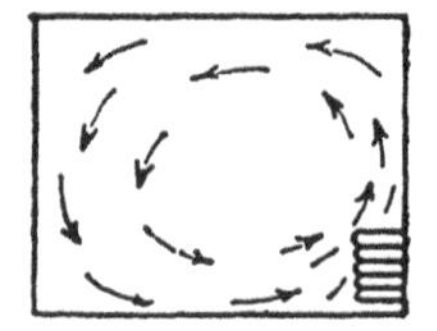

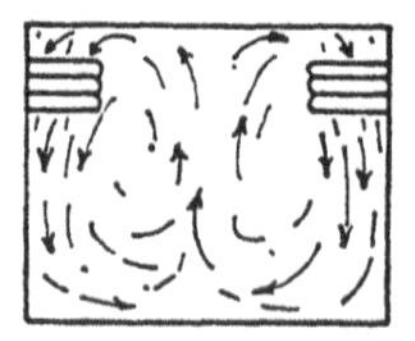

Fig. 6

RADIATION

Radiation is the transfer of heat energy from one body to another through space by rays of electro-magnetic waves. The rays of heat travel in straight lines in all directions at the same velocity as light, that is, at nearly 300 000 kilometres per second.

Some of the radiant heat falling upon a body is reflected in the same manner as light is reflected, the remainder is absorbed. Dark and rough surfaces are good absorbers of radiant heat whereas bright and polished surfaces reflect most of the heat and therefore the amount of absorption is small. A body which is a poor absorber is also a poor radiator, and a good absorber is a good radiator. A perfect absorber and radiator of heat energy is termed a "perfect" black body.

The *emissivity* of a radiating body is the ratio of the heat emitted by that body compared with the heat emitted by a perfect black body of the same surface area and temperature in the same time. Emissivity may be represented by ε, and its value for the ideal radiator is therefore unity.

The STEFAN-BOLTZMANN LAW states that the heat energy radiated by a perfect radiator is proportional to the fourth power of its absolute temperature. Hence Q = quantity of heat energy radiated, A = surface area radiating heat, t = time of radiation, T = absolute temperature, then:

$$Q = AtT^4 \times \text{a constant}$$

The value of the constant depends upon the units employed, and for the basic units of area in square metres, time in seconds, and absolute temperature in degrees Kelvin, the value of the constant to give kilojoules of heat energy is $5{\cdot}67 \times 10^{-11}$ kJ/m²s K⁴.

Thus, the quantity of heat energy radiated from a hot body of absolute temperature T_1 to its surroundings at absolute temperature T_2 is therefore:

$$Q = 5{\cdot}67 \times 10^{-11} \times \varepsilon At\,(T_1^4 - T_2^4)$$

Example. The temperature of the flame in a furnace is 1277°C and the temperature of its surrounds is 277°C. Calculate the maximum theoretical quantity of heat energy radiated per minute per square metre to the surrounding surface area.

$T_1 = 1550$ K
$T_2 = 550$ K
$Q = 5{\cdot}67 \times 10^{-11} \times At(T_1{}^4 - T_2{}^4)$
$= 5{\cdot}67 \times 10^{-11} \times 1 \times 60 \times (1550^4 - 550^4)$
$= 5{\cdot}67 \times 10^{-11} \times 1 \times 60 \times 2{\cdot}705 \times 2{\cdot}1 \times 10^{12}$
$= 1{\cdot}933 \times 10^4$ kJ or 19·33 MJ Ans.

COMBINED MODES

TRANSFER OF HEAT FROM ONE FLUID TO ANOTHER THROUGH A DIVIDING WALL. This is a practical application in many engineering appliances. Consider transfer of heat from a fluid to a flat plate, through the plate, and from the plate to another fluid, as illustrated in Fig. 7. On each side of the plate, a thin film of almost stagnant fluid clings to the surface, the heat transfer through the film is by conduction, convection and radiation.

The quantity of heat transfer through the film of fluid per unit area of surface, in unit time, for unit temperature drop across the thickness of the film, is expressed by the *surface heat transfer coefficient (h)*, and this depends to a large extent upon the velocity of the fluid and the condition of the surfaces.

Since the quantity of heat conducted in a given time is proportional to the surface area and the temperature drop, then the units of h are joules per metre² of area per second per degree = J/m² s K = W/m² K, hence:

$Q[\text{J}] = h[\text{W/m}^2\,\text{K}] \times A[\text{m}^2] \times t[\text{s}] \times (T_1 - T_2)[\text{K}]$

Referring to Fig. 7,

Heat transfer through film of fluid A:

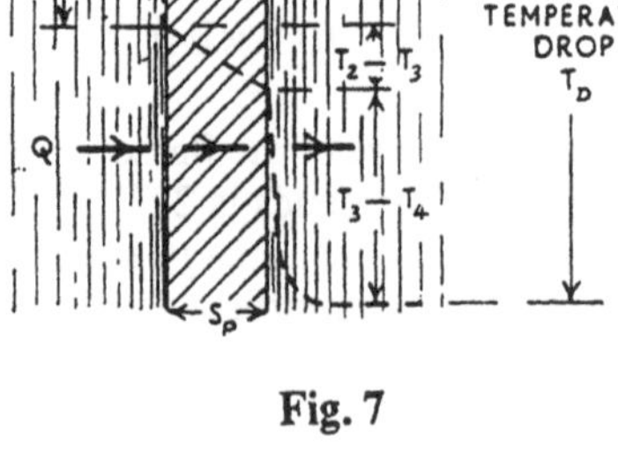

Fig. 7

$$Q = h_A At(T_1 - T_2) \qquad \therefore T_1 - T_2 = \frac{Q}{h_A At}$$

Heat transfer through solid plate:

$$Q = \frac{k_P At(T_2 - T_3)}{S_P} \qquad \therefore T_2 - T_3 = \frac{QS_P}{k_P At}$$

Heat transfer through film of fluid B:

$$Q = h_B At(T_3 - T_4) \qquad \therefore T_3 - T_4 = \frac{Q}{h_B At}$$

Total drop in temperature:

$$T_D = (T_1 - T_2) + (T_2 - T_3) + (T_3 - T_4)$$

$$= \frac{Q}{At}\left\{\frac{1}{h_A} + \frac{S_P}{k_P} + \frac{1}{h_B}\right\}$$

The quantity inside the brackets may be represented by $1/U$, where U is called the *overall heat transfer coefficient*, then,

$$\frac{1}{U} = \frac{1}{h_A} + \frac{S_P}{k_P} + \frac{1}{h_B}$$

hence,

$$T_D = \frac{Q}{UAt} \qquad \text{or,} \quad Q = UAtT_D$$

Example. A cubical tank of 2 m sides is constructed of metal plate 12 mm thick and contains water at 75°C. The surrounding air temperature is 16°C. Calculate (i) the overall heat transfer coefficient from water to air, and (ii) the heat loss through each side of the tank per minute. Take the thermal conductivity of the metal as 48 W/m K, the heat transfer coefficient of the water 2·5 kW/m² K, and the heat transfer coefficient of the air 16 W/m² K.

$$\frac{1}{U} = \frac{1}{h_W} + \frac{S_P}{k_P} + \frac{1}{h_A}$$

$$= \frac{1}{2{\cdot}5 \times 10^3} + \frac{0{\cdot}125}{48} + \frac{1}{16}$$

$$= 0{\cdot}0004 + 0{\cdot}00025 + 0{\cdot}0625$$

$$= 0{\cdot}06315$$

$$U = 1/0{\cdot}06315 = 15{\cdot}84 \text{ W/m}^2\text{K} \quad \text{Ans. (i)}$$

$$Q = UAtT_D$$

$$= 15{\cdot}84 \times 2^2 \times 60 \times (75 - 16)$$

$$= 2{\cdot}243 \times 10^5$$

$$= 224{\cdot}3 \text{ kJ} \quad \text{Ans. (ii)}$$

f HEAT EXCHANGERS. Common examples of heat exchangers in which heat is transferred from one fluid to another are: feed heaters, lubricating oil coolers, gas/air heaters, etc.

To determine the amount of heat transferred in a heat exchanger it is necessary to know the overall heat transfer coefficient U for the wall and fluid boundary layers and also the *logarithmic mean temperature difference*.

$$\text{Heat transferred } Q = UAt\,\theta_m \text{ [J]}$$

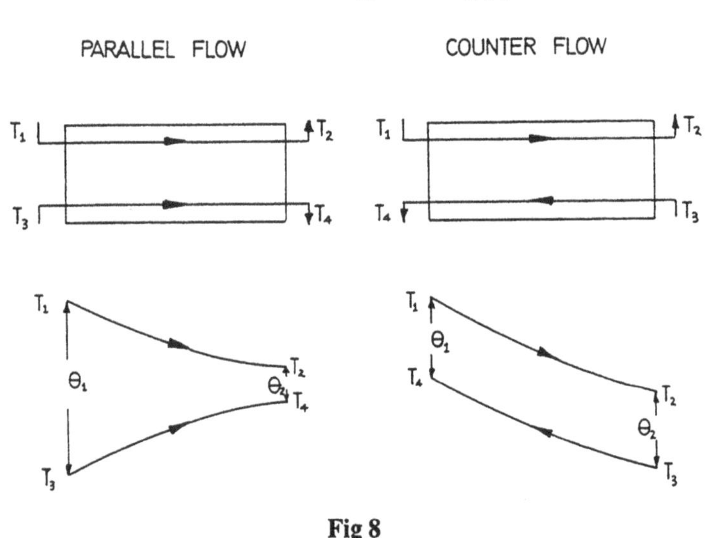

Fig 8

For both heat exchangers shown in Fig. 8:

$$\theta_m = \frac{\theta_1 - \theta_2}{\ln\left(\frac{\theta_1}{\theta_2}\right)}$$

Where,

θ_1 = the temperature difference between hot and cold fluid at inlet

θ_2 = the temperature difference between hot and cold fluid at outlet.

θ_m = mean temperature difference between the fluids, usually, called the *logarithmic mean temperature difference*.

f Example. A counter-flow cooler is required to cool 0·5 kg/s of oil from 50°C to 20°C with a cooling water inlet temperature at

10°C and mass flow rate of 0·375 kg/s. There are 200 thin walled tubes of diameter 12·5 mm. Calculate the length of the tube required.

Overall heat transfer coefficient: 70 W/m²K

Specific heat capacities: oil 1400 J/kg K, water 4200 J/kg K

Let T_4 be water outlet temperature.

$$Q = 0{\cdot}5 \times 1400\,(50 - 20) = 0{\cdot}375 \times 4200\,(T_4 - 10) = 21\,000 \text{ W}$$

$$T_4 = 13{\cdot}33\text{°C}$$

$$= \frac{\theta_1 - \theta_2}{\ln\left(\frac{\theta_1}{\theta_2}\right)}$$

$$= \frac{(T_1 - T_4) - (T_2 - T_3)}{\ln\left(\frac{T_1 - T_4}{T_2 - T_3}\right)}$$

$$= \frac{(50 - 13{\cdot}33) - (20 - 10)}{\ln\left(\frac{50 - 13{\cdot}33}{20 - 10}\right)}$$

$$= \frac{26{\cdot}67}{1{\cdot}299}$$

$$= 20{\cdot}53\text{°C}$$

$$Q = UAt\,\theta_m$$

$$21\,000 = 70(\pi \times 0{\cdot}0125 \times l \times 200) \times 1 \times 20{\cdot}53$$

$$l = 1{\cdot}86 \text{ m} \quad \text{Ans.}$$

TEST EXAMPLES 4

1. Calculate the quantity of heat conducted per minute through a duralumin circular disc 127 mm diameter and 19 mm thick when the temperature drop across the thickness of the plate is 5°C. Take the thermal conductivity of duralumin as 150 W/m K.

2. A cold storage compartment is 4·5 m long by 4 m wide by 2·5 m high. The four walls, ceiling and floor are covered to a thickness of 150 mm with insulating material which has a thermal conductivity of $5{\cdot}8 \times 10^{-2}$ W/m K. Calculate the quantity of heat leaking through the insulation per hour when the outside and inside face temperatures of the material is 15°C and –5°C respectively.

3. One side of a refrigerated cold chamber is 6 m long by 3·7 m high and consists of 168 mm thickness of cork between outer and inner walls of wood. The outer wood wall is 30 mm thick and its outside face temperature is 20°C, the inner wood wall is 35 mm thick and its inside face temperature is –3°C. Taking the thermal conductivity of cork and wood as 0·042 and 0·2 W/m K respectively, calculate (i) the heat transfer per second per square metre of surface area, (ii) the total heat transfer through the chamber side per hour (iii) the interface temperatures.

4. A flat circular plate is 500 mm diameter. Calculate the theoretical quantity of heat radiated per hour when its temperature is 215°C and the temperature of its surrounds is 45°C. Take the value of the radiation constant as $5{\cdot}67 \times 10^{-11}$ kJ/m^{2}s K^4.

5. Hot gases at 280°C flow on one side of a metal plate of 10 mm thickness and air at 35°C flows on the other side. The surface heat transfer coefficient of the gases is 31·5 W/m^2 K and that of air is 32 W/m^2 K. The thermal conductivity of the metal plate is 50 W/m K. Calculate (i) the overall heat transfer coefficient, and (ii) the heat transfer from gases to air per minute per square metre of plate area.

6. The wall of a cold room consists of a layer of cork sandwiched between outer and inner walls of wood, the wood walls being each 30 mm thick. The inside atmosphere of the room is maintained at –20°C when the external atmospheric temperature is 25°C, and the heat loss through the wall is 42 W/m^2. Taking the thermal conductivity of wood and cork as 0·2 W/m K and 0·05 W/m K respectively, and the surface heat transfer coefficient between each exposed wood surface and their respective atmospheres as 15 W/m^2 K, calculate (i) the temperatures of the exposed surfaces, (ii) the

temperatures of the interfaces, and (iii) the thickness of the cork.

f 7. The steam drum of a water-tube boiler has hemispherical ends, the diameter is 1·22 m and the overall length is 6 m. Under steaming conditions the temperature of the shell before lagging was 230°C and the temperature of the surrounds was 51°C. The temperature of the cleading after lagging was 69°C and the surrounds 27°C. Assuming 75% of the total shell area to be lagged and taking the radiation constant as $5{\cdot}67 \times 10^{-11}$ kJ/m²s K⁴, estimate the saving in heat energy per hour due to lagging.

f 8. A pipe 200 mm outside diameter and 20 m length is covered with a layer, 70 mm thick, of insulation having a thermal conductivity of 0·05 W/m K and a surface heat transfer coefficient of 10 W/m² K at the outer surface.

If the temperature of the pipe is 350°C and the ambient temperature is 15°C calculate:

(i) the external surface temperature of the lagging

(ii) the heat flow rate from the pipe

Heat flow rate through the lagging per metre length of pipe is

$$\frac{2\pi kT}{\ln (D/d)}$$

where k is the thermal conductivity of the insulation
T is the temperature difference across the insulation
D and d are the external and internal diameters of the insulation respectively.

f 9. In an inert gas system the boiler exhaust is cooled from 410°C to 130°C in a parallel flow heat exchanger. Gas flow rate is 0·4 kg/s, cooling water flow rate is 0·5 kg/s, cooling water inlet temperature 10°C. Overall heat transfer coefficient from the gas to the water is 140 W/m² K. Determine the cooling surface area required.

Take c_p for exhaust gas as 1130 J/kg K and c_p for water as 4190 J/kg K.

Note: logarithmic mean temperature difference $\theta_m = \dfrac{\theta_1 - \theta_2}{\ln\left(\dfrac{\theta_1}{\theta_2}\right)}$

Where θ_1 = temperature difference between hot and cold fluid at inlet.

Where θ_2= temperature difference between hot and cold fluid at outlet.

f 10. Steam at 0·07 bar condenses on the outside of a thin tube 0·025 m diameter and 2·75 m long. Cooling water of density 1000 kg/m^3, c_p 4·18 kJ/kg K, flows through the tube at 0·6 m/s rising in temperature from 12°C to 24°C. The surface heat transfer coefficient between steam and tube is 17000 $W/m^2 K$. Determine:

(a) the overall heat transfer coefficient

(b) the surface heat transfer coefficient between water and tube.

CHAPTER 5

LAWS OF PERFECT GASES

When a substance has been evaporated it can exist as a gas or vapour and one of its most important characteristics is its elastic property. For instance, if a certain volume of a liquid is put into a vessel of large volume, the liquid will only partially fill the vessel, taking up no more nor less volume than it did before, but when a gas enters a vessel it immediately fills up every part of that vessel no matter how large it is. Practically speaking, liquids cannot be compressed nor expanded, but gases can be compressed into smaller volumes and expanded to larger volumes.

A perfect gas is a theoretically ideal gas which strictly follows Boyle's and Charles' laws of gases.

Consider a given mass of a perfect gas enclosed in a cylinder by a gas-tight movable piston. When the piston is pushed inward, the gas is compressed to a small volume, when pulled outward the gas is expanded to a larger volume. However, not only is there a change in volume but the pressure and temperature also change. These three quantities, pressure, volume and temperature, are related to each other, and to determine their relationship it is usual to perform experiments with each one of these quantities in turn kept constant while observing the relationship between the other two.

In such basic laws, the pressure, the volume and temperature must all be the absolute values, that is, measured from absolute zero, and not measured from some artificial level. Absolute values were explained in Chapter 1, a brief reminder will suffice:

ABSOLUTE PRESSURE (p) is the pressure measured above a perfect vacuum, obtained by adding the atmospheric pressure to the gauge pressure

"Absolute" need not follow the given pressure, it is to be taken as such unless the pressure is distinctly marked "gauge" to indicate that it is a pressure gauge reading.

VOLUME (V). The volume of the gas is equal to the full volume of the vessel containing it. Note that in the case of the cylinder of an air compressor or reciprocating engine, the total volume of air or gas includes the volume of the clearance space between the piston at its top dead centre and the cylinder cover, as well as the piston swept volume.

ABSOLUTE TEMPERATURE *(T)*. This is the temperature in degrees kelvin measured above the absolute zero of temperature.

$$T[\text{K}] = \theta[^\circ\text{C}] + 273$$

BOYLE'S LAW

Boyle's law states that the absolute pressure of fixed mass of a perfect gas varies inversely as its volume if the temperature remains unchanged.

$$p \propto \frac{1}{V} \qquad \text{therefore } p \times V = \text{ a constant.}$$

Hence, $p_1 \times V_1 = p_2 \times V_2$

To illustrate this, imagine 2 m³ of gas at a pressure of 100 kN/m² (= 10^5 N/m² = 1 bar) contained in a cylinder with a gas-tight movable piston as illustrated in Fig 9. When the piston is pushed inward the pressure will increase as the gas is compressed to a smaller volume and, provided the temperature remains unchanged, the product of pressure and volume will be a constant quantity for all positions of the piston. From the known initial conditions the constant is calculated:

$$p_1 \times V_1 = \text{constant}$$
$$100 \times 2 = 200$$

and the pressure at any other volume can be determined:

When the volume is 1·5 m³,

$$p_2 \times 1{\cdot}5 = 200$$
$$p_2 = 133{\cdot}3 \text{ kN/m}^2$$

When the volume is 1 m³,

$$p_3 \times 1 = 200$$
$$p_3 = 200 \text{ kN/m}^2$$

When the volume is 0·5 m³,

$$p_4 \times 0{\cdot}5 = 200$$
$$p_4 = 400 \text{ kN/m}^2$$

And so on.

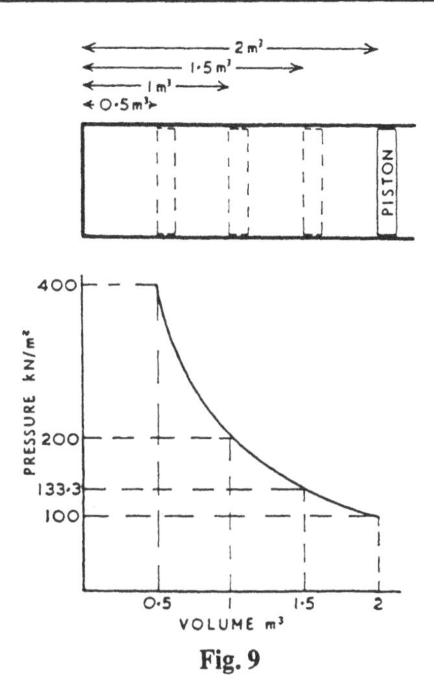

Fig. 9

The variation of pressure with change of volume is shown in the graph below the cylinder in Fig. 9. The graph produced by joining up the plotted points is a rectangular hyperbola, consequently we refer to compression or expansion where pV = constant as hyperbolic compression or hyperbolic expansion. When the temperature is constant as in this example, the operation may also be termed "isothermal".

Note that as the ordinates (vertical measurements) represent pressure, and the abscissae (horizontal measurements) represent volume, and since the product of pressure and volume is constant, then all rectangles drawn from the axes with their corners touching the curve, will be of equal area.

Example. 3·5 m³ of air at a pressure of 20 kN/m² gauge is compressed at constant temperature to a pressure of 425 kN/m² gauge. Taking the atmospheric pressure as 100 kN/m² calculate the final volume of air.

Initial absolute pressure = 20 + 100 = 120 kN/m²
Final absolute pressure = 425 + 100 = 525 kN/m²

$$p_1V_1 = p_2V_2$$
$$120 \times 3{\cdot}5 = 525 \times V_2$$
$$V_2 = \frac{120 \times 3{\cdot}5}{525} = 0{\cdot}8 \text{ m}^3 \qquad \text{Ans.}$$

CHARLES' LAW

Charles' law states that the volume of a fixed mass of a perfect gas varies directly as its absolute temperature if the pressure remains unchanged, also, the absolute pressure varies directly as the absolute temperature if the volume remains unchanged.

From the above statement we have:

For constant pressure, $V \propto T \qquad \therefore \frac{V}{T} = \text{constant}$

hence, $\frac{V_1}{T_1} = \frac{V_2}{T_2}$ or $\frac{V_1}{V_2} = \frac{T_1}{T_2}$

For constant volume, $p \propto T \qquad \therefore \frac{p}{T} = \text{constant}$

hence, $\frac{p_1}{T_1} = \frac{p_2}{T_2}$ or $\frac{p_1}{p_2} = \frac{T_1}{T_2}$

Example. The pressure of the air in a starting air vessel is 40 bar (= 40×10^5 N/m^2 or 4 MN/m^2) and the temperature is 24°C. If a fire in the vicinity causes the temperature to rise to 65°C, find the pressure of the air. Neglect any increase in volume of the vessel.

As the term "gauge" does not follow the given pressure, it is assumed that this is the initial absolute pressure.

Initial absolute temperature = 24°C + 273 = 297 K
Final absolute temperature = 65°C + 273 = 338 K

$$\frac{p_1}{T_1} = \frac{p_2}{T_2} \qquad \therefore p_2 = \frac{p_1T_2}{T_1}$$

$$p_2 = \frac{40 \times 338}{297} = 45{\cdot}52 \text{ bar}$$

$$= 45{\cdot}52 \times 10^5 \text{ N/m}^2 \text{ or } 4{\cdot}552 \text{ MN/m}^2 \qquad \text{Ans.}$$

COMBINATION OF BOYLE'S AND CHARLES' LAWS

Each one of these laws states how one quantity varies with another if the third quantity remains unchanged, but if the three quantities change simultaneously, it is necessary to combine these laws in order to determine the final conditions of the gas.

Referring to Fig. 10 which again represents a cylinder with a piston, gas-tight so that the mass of gas within the cylinder is always the same. Let the gas be compressed from its initial state of pressure p_1 volume V_1 and temperature T_1 to its final state of p_2V_2 and T_2, but to arrive at the final state let it pass through two stages, the first to satisfy Boyle's law and the second to satisfy Charles' law.

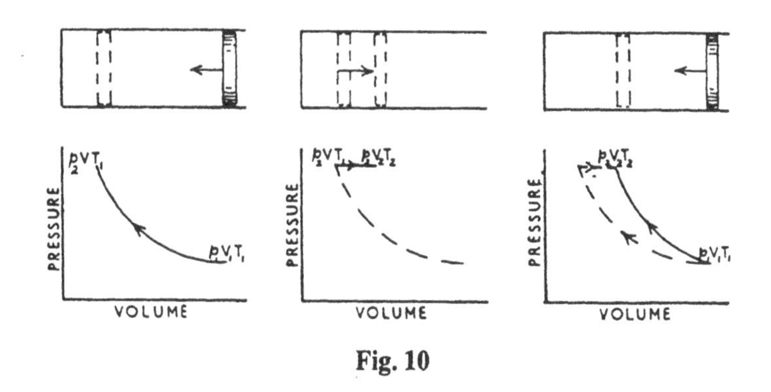

Fig. 10

Imagine the piston pushed inward to compress the gas until it reaches the final pressure of p_2 and let its volume then be represented by V. Normally the temperature would tend to increase due to the work done in compressing the gas, but any heat so generated must be taken away from it during compression so that its temperature remains unchanged at T_1 hence following Boyle's law:

$$p_1V_1 = p_2V \qquad \text{(i)}$$

Now apply heat to raise the temperature from T_1 to T_2 and at the same time draw the piston outward to prevent a rise of pressure and keep it constant at p_2. The volume will increase in direct proportion to the increase in absolute temperature according to Charles' law:

$$\frac{V_2}{V} = \frac{T_2}{T_1} \qquad \text{(ii)}$$

By substituting the value of V from (ii) into (i) this quantity will be eliminated:

From (ii) $$V = \frac{V_2 T_1}{T_2}$$

Substituting into (i)

$$p_1 V_1 = p_2 \times \frac{V_2 T_1}{T_2}$$

$$\therefore \frac{p_1 V_1}{T_1} = \frac{p_2 V_2}{T_2}$$

This combined law of Boyle's and Charles' is true for a given mass of any perfect gas subjected to any form of compression or expansion.

Example. 0·5 m^3 of a perfect gas at a pressure of 0·95 bar (= 95 kN/m^2) and the temperature 17°C are compressed to a volume of 0·125 m^3 and the final pressure is 5·6 bar (= 560 kN/m^2). Calculate the final temperature.

Initial absolute temperature = 290 K

$$\frac{p_1 V_1}{T_1} = \frac{p_2 V_2}{T_2}$$

$$\frac{95 \times 0{\cdot}5}{290} = \frac{560 \times 0{\cdot}125}{T_2}$$

$$T_2 = \frac{560 \times 0{\cdot}125 \times 290}{95 \times 0{\cdot}5}$$

$$= 427{\cdot}4 \text{ K}$$

$$= 154{\cdot}4°\text{C} \quad \text{Ans.}$$

CHARACTERISTIC EQUATION

Since pV/T is a constant, its value can be determined for a given mass of any perfect gas. To form a basis on which to work, the constant is calculated on the specific volume (v) of the gas, that is, the volume in cubic metres occupied by a mass of one kilogramme. The constant so obtained is termed the *gas constant*, it is represented by R and is different for all different gases:

$$\frac{pv}{T} = R \quad \text{or} \quad pv = RT$$

The volume occupied by m kg of mass being represented by V then:

$$pV = mRT$$

This expression is called the *characteristic equation of a perfect gas*.

Taking air as an example, experiments show that at standard atmospheric pressure and temperature, that is, at 101·325 kN/m^2 and 0°C, the specific volume is 0·7734 m^3/kg. Inserting these values to find R:

$$R = \frac{pv}{T}$$

$$= \frac{101{\cdot}325\ [\mathrm{kN/m^2}] \times 0{\cdot}7734\ [\mathrm{m^3/kg}]}{273\ [\mathrm{K}]}$$

$$= 0{\cdot}287\ \mathrm{kJ/kg\,K}$$

Note the units for R. Repeating the above with the units only:

$$R = \frac{\mathrm{kN/m^2 \times m^3/kg}}{\mathrm{K}} = \frac{\mathrm{kN}}{\mathrm{m^2}} \times \frac{\mathrm{m^3}}{\mathrm{kg}} \times \frac{1}{\mathrm{K}}$$

$$= \frac{\mathrm{kN\,m}}{\mathrm{kg\,K}} = \mathrm{kN\,m/kg\,K} = \mathrm{kJ/kg\,K}$$

Example. An air compressor delivers 0·2 m^3 of air at a pressure of 850 kN/m^2 and 31°C into an air reservoir. Taking the gas constant for air as 0·287 kJ/kg K, calculate the mass of air delivered.

$$pV = mRT$$

$$m = \frac{pV}{RT}$$

$$= \frac{850 \times 0{\cdot}2}{0{\cdot}287 \times 304} = 1{\cdot}948\ \mathrm{kg} \quad \text{Ans.}$$

f UNIVERSAL GAS CONSTANT

The *kilogramme-mol* of a substance is a mass of that substance numerically equal to its molecular weight. The kilogramme-mol may simply be abbreviated to *mol.*

As examples, the molecular weight of oxygen is 32, therefore the mass of one mol of oxygen is 32 kg. The molecular weight of hydrogen is 2 hence one mol of hydrogen is a mass of 2 kg. And so on.

f AVOGADRO'S LAW states that under equal conditions of temperature and pressure, equal volumes of all gases contain the same number of molecules.

Consider two gases, one of mass m_1 of molecular weight M_1 and containing n_1 molecules, the other of mass m_2, of molecular weight M_2 and containing n_2 molecules:

$$\text{Ratio of masses} = \frac{m_1}{m_2} = \frac{n_1 M_1}{n_2 M_2}$$

For equal volumes of the gases at the same temperature and pressure, the gases contain an equal number of molecules, therefore $n_1 = n_2$ and cancel:

$$\frac{m_1}{m_2} = \frac{M_1}{M_2}$$

From the characteristic gas equation $pV = mRT$, substituting for $m = pV/RT$:

$$\frac{p_1V_1/R_1T_1}{p_2V_2/R_2T_2} = \frac{M_1}{M_2} \qquad \therefore \frac{p_1V_1R_2T_2}{p_2V_2R_1T_1} = \frac{M_1}{M_2}$$

pV and T cancel because they are equal, therefore,

$$\frac{R_2}{R_1} = \frac{M_1}{M_2} \qquad \text{or} \quad R_1M_1 = R_2M_2$$

Hence the product of the gas constant R and the molecular weight M is the same for all gases. This product is termed the *Universal Gas Constant*, it is represented by R_0 and its value has been found by experiment to be 8·314 kJ/mol K.

$$R_0 = RM$$

Any particular gas constant R can therefore be found if its molecular weight is known.

Thus, the molecular weight of nitrogen is 28, therefore the gas constant for nitrogen is:

$$R = \frac{R_0}{M} = \frac{8 \cdot 314}{28} = 0 \cdot 2969 \text{ kJ/kg K}$$

f DALTON'S LAW OF PARTIAL PRESSURES

This law states that the pressure exerted in a vessel by a mixture of gases is equal to the sum of the pressures that each separate gas would exert if it alone occupied the whole volume of the vessel.

The pressure exerted by each gas is termed a partial pressure:

Total pressure of the mixture

$$= \begin{matrix}\text{partial press.}\\ \text{due to gas}_1\end{matrix} + \begin{matrix}\text{partial press.}\\ \text{due to gas}_2\end{matrix} + \begin{matrix}\text{partial press.}\\ \text{due to gas}_3\end{matrix} + \text{etc.}$$

$$p = p_1 + p_2 + p_3 + \text{etc.}$$

From the characteristic gas equation $pV = mRT$, substituting for the pressure of each gas, and also for the mixture:

$$\frac{mRT}{V} = \frac{m_1R_1T_1}{V_1} + \frac{m_2R_2T_2}{V_2} + \frac{m_3R_3T_3}{V_3} + \text{etc.}$$

Since it can be considered as though each gas on its own occupied the whole space then the volume V is common throughout. The temperature is also common, therefore V and T being common to all terms will cancel:

$$\frac{mRT}{V} = \frac{T}{V}\{m_1R_1 + m_2R_2 + m_3R_3 + \text{etc.}\}$$

$$mR = m_1R_1 + m_2R_2 + m_3R_3 + \text{etc.}$$

where m is the total mass of the mixture and R is the gas constant of the mixture.

f Example. The analysis by mass of a sample of air is 23·14% oxygen, 75·53% nitrogen, 1·28% argon and 0·05% carbon dioxide. Estimate the gas constant for air (to the nearest four figures) taking the molecular weights of O_2, N_2, Ar and CO_2 as 32, 28, 40 and 44 respectively, and the universal gas constant $R_0 = 8 \cdot 314$ kJ/mol K.

$$R = \frac{R_0}{M} \qquad \text{Considering 1 kg of air:}$$

$$\text{Oxygen} \quad R_1 = \frac{8{\cdot}314}{32} = 0{\cdot}2598$$

$$m_1R_1 = 0{\cdot}2314 \times 0{\cdot}2598 = 0{\cdot}06012$$

$$\text{Nitrogen} \quad R_2 = \frac{8{\cdot}314}{28} = 0{\cdot}2969$$

$$m_2R_2 = 0{\cdot}7553 \times 0{\cdot}2969 = 0{\cdot}2242$$

$$\text{Argon} \quad R_3 = \frac{8{\cdot}314}{40} = 0{\cdot}2078$$

$$m_3R_3 = 0{\cdot}0128 \times 0{\cdot}2078 = 0{\cdot}002661$$

$$\text{Carbon dioxide} \quad R_4 = \frac{8{\cdot}314}{44} = 0{\cdot}1889$$

$$m_4R_4 = 0{\cdot}0005 \times 0{\cdot}1889 = 0{\cdot}00009445$$

$$\Sigma mR = 0{\cdot}28707545$$

$$R \text{ air} = \frac{0{\cdot}2871}{1} = 0{\cdot}2871 \text{ kJ/kg K} \quad \text{Ans.}$$

f PARTIAL VOLUMES. If each gas of a mixture in a closed vessel at the full volume of V and partial pressure p_1(p_2, etc.), is considered as being compressed until its pressure is the full pressure p and its volume is its partial volume V_1 (V_2, etc.), then, since its temperature remains the same:

$$p_1 \times V = p \times V_1 \qquad \text{and} \qquad p_2 \times V = p \times V_2 \text{ etc.}$$

$$\therefore (p_1 + p_2 + \text{etc.}) \times V = p \times (V_1 + V_2 + \text{etc.})$$

hence, the total volume of the mixture is equal to the sum of the partial volumes of the gases at the same total pressure and temperature.

$$\text{Also, } \frac{p_1}{p_2} = \frac{V_1}{V_2} \text{ therefore:}$$

the ratio of the partial volumes is equal to the ratio of the partial pressures.

Dalton's law is also applied for the determination of the quantity of air-leakage into steam condensers. An example is given in Chapter 10 after the study of properties of steam.

SPECIFIC HEATS OF GASES

The specific heat (c) of a substance is defined as the quantity of heat energy required to be transferred to unit mass of that substance to raise its temperature by one degree. Hence, the quantity (Q) of heat in kilojoules required to be given to a mass of m kilogrammes of the substance to raise its temperature from T_1 to T_2 is:

$$Q[\text{kJ}] = m[\text{kg}] \times c[\text{kJ/kgK}] \times (T_2 - T_1)\ [K]$$

Since the characteristics of gases vary considerably at different temperatures and pressures, heat energy may be transferred under an infinite number of conditions, therefore the specific heat can have an infinite number of different values. Consider, however, two important conditions under which heat may be transferred, (i) while the *volume* of the gas remains constant, (ii) while the *pressure* of the gas remains constant. The specific heat of a gas at constant volume is represented by c_V.

The specific heat of a gas at constant pressure is represented by c_P. This is a higher value than c_V because, when the gas is receiving heat it must be allowed to expand in volume to prevent a rise in pressure and, whilst expanding, the gas is expending energy in doing external work, hence extra heat energy must be supplied equivalent to the external work done.

Example. A quantity of air of mass 0·23 kg, pressure 100 kN/m^2, volume 0·1934 m^3 and temperature 20°C is enclosed in a cylinder with a gas-tight movable piston, and heat energy is transferred to the air to raise its temperature to 142°C.

(a) If the piston is prevented from moving during heat transfer so that the volume of the air remains unchanged, calculate (i) the heat supplied, taking the specific heat at constant volume $c_V = 0{\cdot}718$ kJ/kg K, and (ii) the final pressure.

(b) If the piston moves to allow the air to expand in volume at such a rate as to keep the pressure constant, calculate (i) the heat supplied, taking the specific heat at constant pressure $c_P = 1{\cdot}005$ kJ/kg K, and (ii) the final volume.

Initial absolute temperature = 293 K
Final absolute temperature = 415 K

Temperature rise = 142 − 20 = 122°C = 122 K

(a)
$$Q = m \times c_V \times (T_2 - T_1)$$
$$= 0{\cdot}23 \times 0{\cdot}718 \times 122$$

$$= 20{\cdot}15 \text{ kJ} \quad \text{Ans. (a)(i)}$$

$$\frac{p_1V_1}{T_1} = \frac{p_2V_2}{T_2}$$

The volume is constant $\therefore$ $V_2 = V_1$ and cancels,

$$p_2 = \frac{p_1T_2}{T_1} \text{ (which is Charles' law)}$$

$$= \frac{100 \times 415}{293} = 141{\cdot}6 \text{ kN/m}^2 \quad \text{Ans. (a)(ii)}$$

(b)

$$Q = m \times c_p \times (T_2 - T_1)$$
$$= 0{\cdot}23 \times 1{\cdot}005 \times 122$$
$$= 28{\cdot}19 \text{ kJ} \quad \text{Ans. (b)(i)}$$

$$\frac{p_1V_1}{T_1} = \frac{p_2V_2}{T_2}$$

The pressure is constant $\therefore$ $p_1 = p_2$ and cancels

$$V_2 = \frac{V_1T_2}{T_1} \text{ (which is Charles' law)}$$

$$= \frac{0{\cdot}1934 \times 415}{293} = 0{\cdot}2738 \text{ m}^3 \quad \text{Ans. (b)(ii)}$$

Note the difference between the quantity of heat energy supplied to the air in the two cases:

$$28{\cdot}19 - 20{\cdot}15 = 8{\cdot}04 \text{ kJ}$$

This amount of heat energy was expended in moving the piston:

Referring to Fig. 11, the piston is pushed forward by the gas at a constant pressure of, say p[kN/m^2]. Let A[m^2] represent the area of the piston, then the total force on the piston is $p \times A$[kN]. If the piston moves S[metres], the work done, being the product of force and distance, is $p \times A \times S$[kN m = kJ]. The product of the area A and

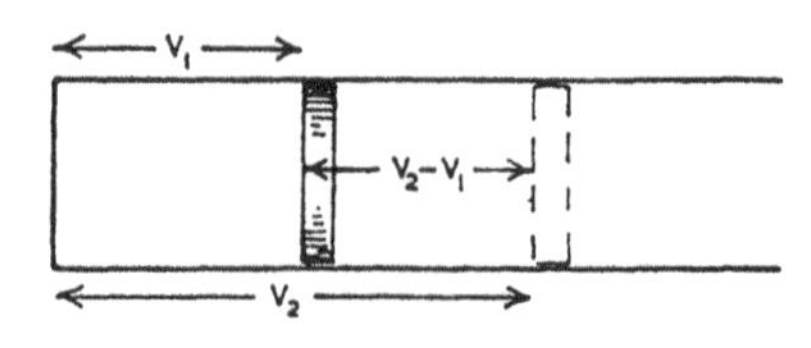

Fig 11

the distance moved S is the volume swept through by the piston, hence,

$$\text{Work done} = p(V_2 - V_1)$$

In the last example, when the pressure was constant at 100 kN/m^2 the volume of the air increased from 0·1934 m^3 to 0·2738 m^3 so that the piston swept volume is the difference between these.

$$\begin{aligned} \text{Work done} &= p(V_2 - V_1) \\ &= 100(0{\cdot}2738 - 0{\cdot}1934) \\ &= 100 \times 0{\cdot}0804 \\ &= 8{\cdot}04 \text{ kJ} \end{aligned}$$

Showing that the extra heat energy given to the air at constant pressure compared with that at constant volume is the heat energy expended in doing external work by pushing the piston forward.

ENERGY EQUATION (CLOSED SYSTEMS)

The *internal energy* (U) of a gas is the energy contained in it as stored up work by virtue of the movement of its molecules.

JOULE'S LAW states that the internal energy of a gas depends only upon its temperature and is independent of changes in pressure and volume.

By the principle of the conservation of energy, that is, that energy can neither be created nor destroyed, it follows that the total heat energy (Q) transferred to a gas will be the sum of the increase in its internal energy ($U_2 - U_1$) and any work (W) that is done by the gas during the transfer of heat energy to it, thus:

Heat energy transferred to the gas	=	Increase in internal energy of the gas	+	External work done by the gas

$$Q = (U_2 - U_1) + W$$

This is known as the *First Law of Thermodynamics*.

Any of these three terms may be negative i.e. heat rejected (extracted), decrease of internal energy, work done on the gas (compression).

These are non-flow (reversible) processes for perfect gases.

RELATIONSHIP BETWEEN SPECIFIC HEATS. Referring again to the last example, with Fig. 11, and applying the energy equation:

In the first case, heat energy is supplied to the gas to raise its temperature from T_1 to T_2 at *constant volume*, the heat supplied is $mc_V(T_2 - T_1)$ but no work is done because, as the volume is constant, the piston does not move.

Heat supplied = Increase in internal energy + External work done

$$mc_V(T_2 - T_1) = (U_2 - U_1) + 0 \qquad \text{(i)}$$

In the second case, heat energy is supplied to raise the temperature through the same range, from T_1 to T_2 at *constant pressure*. The heat supplied is $mc_P(T_2 - T_1)$ and external work is done equal to $p(V_2 - V_1)$. Expressing the work done in terms of temperature by substituting the value of V from the characteristic gas equation, and inserting into the energy equation,

$$p(V_2 - V_1) = p\left\{\frac{mRT_2}{p} - \frac{mRT_1}{p}\right\} = mR(T_2 - T_1)$$

Heat supplied = Increase in internal energy + External work done

$$mc_P(T_2 - T_1) = (U_2 - U_1) + mR(T_2 - T_1)$$

$$U_2 - U_1 = mc_P(T_2 - T_1) - mR(T_2 - T_1) \qquad \text{(ii)}$$

Since the temperature change is the same in each case then the change in internal energy is the same, hence, from (i) and (ii):

$$mc_V(T_2 - T_1) = mc_P(T_2 - T_1) - mR(T_2 - T_1)$$

m and $(T_2 - T_1)$ are common to all terms and cancel:

$$c_V = c_P - R$$

$$\therefore R = c_P - c_V$$

Inserting the values for air as previously given,

$$R = 1{\cdot}005 - 0{\cdot}718 = 0{\cdot}287 \text{ kJ/kg K}$$

RATIO OF SPECIFIC HEATS. Another important relationship between the specific heats is the ratio c_P/c_V, the symbol denoting this ratio is γ:

$$\gamma = \frac{c_P}{c_V}$$

and for air this ratio is:

$$\gamma = \frac{1{\cdot}005}{0{\cdot}718} = 1{\cdot}4$$

Taking this further:

$$\gamma = \frac{c_P}{c_V} \qquad \therefore\ c_P = \gamma c_V$$

substituting this into the relationship:

$$\begin{aligned} R &= c_P - c_V \\ &= \gamma c_V - c_V \\ \therefore\ R &= c_V(\gamma - 1) \\ \text{or } c_V &= \frac{R}{\gamma - 1} \end{aligned}$$

These expressions will be found useful in later calculations.

ENTHALPY

The enthalpy (H) of any fluid (liquid, vapour, gas) is a convenient grouping of terms representing (in open systems for reversible steady-flow processes) the sum of two kinds of energy transfer, i.e. internal energy (U) and work transfer (pV).

$$H = U + pV \ \text{(J)}$$

$$h = u + pv \ \text{(J/kg) specific enthalpy}$$

ENERGY EQUATION (OPEN SYSTEMS).

Consider mass of fluid m_1, volume V_1, entering system.
Work done against resistance $= p_1V_1$.
Consider mass of fluid m_2, volume V_2, leaving system.
Work done against resistance $= p_2V_2$.
If $m_1 = m_2$ (steady flow process, mass of fluid within system constant)
$\therefore p_1V_1 = p_2V_2$.
This is a work, or energy, transfer through the system.
Referring to Fig 12, the mass of fluid within the boundary is constant, *i.e.* the mass entering equals the mass leaving.

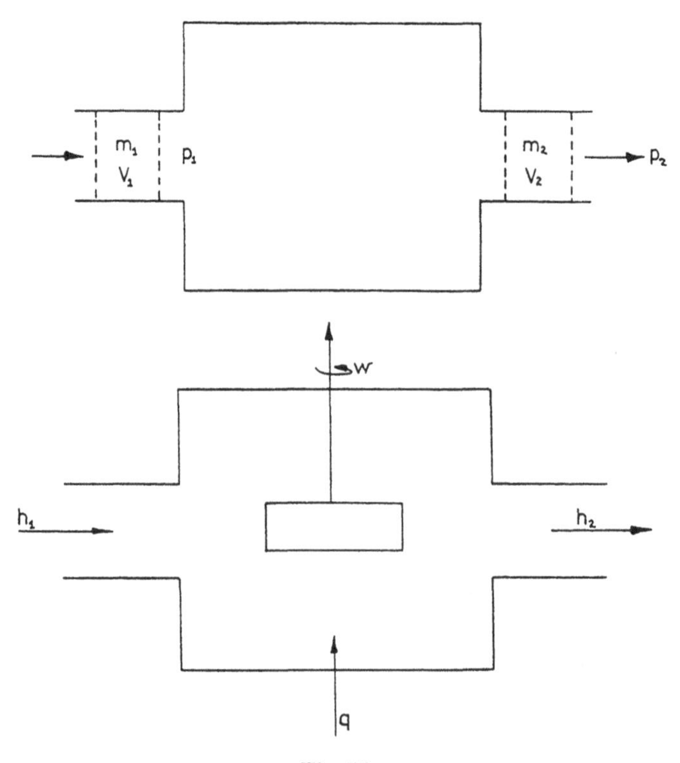

Fig. 12

Let h_1 and h_2 = specific enthalpy of fluid entering and leaving the system respectively (kJ/kg).

Let c_1 and c_2 = velocity of fluid entering and leaving the system respectively (m/s)

Let q = quantity of heat per unit mass crossing the boundary into the system (kJ/kg)

Let w = external work done per unit mass by the fluid (kJ/kg)

Assume no change in potential energy, then:

$$h_1 + \text{Kinetic Energy}_1 + q = h_2 + \text{Kinetic Energy}_2 + w.$$

i.e. $h_1 + \frac{1}{2}c_1^2 + q = h_2 + \frac{1}{2}c_2^2 + w$

This is known as the Steady Flow Energy Equation.

Example. The working fluid passes through a gas turbine at a steady rate of 10 kg/s. It enters with a velocity of 100 m/s and specific enthalpy of 2000 kJ/kg and leaves at 50 m/s with a specific enthalpy of 1500 kJ/kg. If the heat lost to surroundings, as the fluid passes through the turbine, is 40 kJ/kg calculate the power developed.

Steady flow energy equation:

$h_1 + \frac{1}{2}c_1^2 + q = h_2 + \frac{1}{2}c_2^2 + w$

$\therefore\ 2000 \times 10^3 + \frac{1}{2} \times 100^2 - 40 \times 10^3 = 1500 \times 10^3 + \frac{1}{2} \times 50^2 + w$

(*Note*: q is negative as it is heat crossing the boundary out of the system).

$$460 \times 10^3 + \tfrac{1}{2}(100^2 - 50^2) = w$$
$$4{\cdot}6375 \times 10^5 = w$$

$$\therefore\ w = 4{\cdot}64 \times 10^5\ \text{J}$$
$$\text{power} = 4{\cdot}64 \times 10^5\ \text{J/s}$$
$$= 4{\cdot}64\ \text{MW}$$

Note: For perfect gases:

$$\begin{aligned} h_2 - h_1 &= (u_2 + p_2v_2) - (u_1 + p_1v_1) \\ &= c_V(T_2 - T_1) + R\,(T_2 - T_1) \\ &= c_P(T_2 - T_1) \end{aligned}$$

TEST EXAMPLES 5

1. Assuming compression according to the law pV = constant:
(i) Calculate the final volume when 1 m^3 of gas at 120 kN/m^2 is compressed to a pressure of 960 kN/m^2.
(ii) Calculate the initial volume of gas at a pressure of 1·05 bar which will occupy a volume of 5·6 m^3 when it is compressed to a pressure of 42 bar.

2. 0·2 m^3 of gas at a pressure of 1350 kN/m^2 and temperature 177°C is expanded in a cylinder to a volume of 0·9 m^3 and pressure 250 kN/m^2. Calculate the final temperature.

3. A receiver containing 2 m^3 of air at 10 bar, 20°C has a relief valve set to operate at 20 bar. If 8% of the air were to leak out, calculate the temperature at which the relief valve would operate.
Note: for air R = 287 J/kg K

4. An air reservoir contains 20 kg of air at 3200 kN/m^2 gauge and 16°C. Calculate the new pressure and heat energy transfer if the air is heated to 35°C. Neglect any expansion of the reservoir, take R for air = 0·287 kJ/kg K, specific heat at constant volume c_V = 0·718 kJ/kg K, and atmospheric pressure = 100 kN/m^2.

5. The dimensions of a large room are 12 m by 16·5 m by 4 m. The air is completely changed once every 30 minutes and the temperature is maintained at 21°C. If the temperature of the outside atmosphere is 30°C, calculate the quantity of heat required to be extracted from the supply air per hour, and the equivalent power, taking the density of air at atmospheric pressure and 0°C as 1·293 kg/m^3 and the specific heat at constant pressure as 1·005 kJ/kg K.

6. In a steady flow process the working fluid enters and leaves a horizontal system with negligible velocity. The temperature drop from inlet to outlet is 480°C and the heat losses from the system are 10 kJ/kg of fluid. Determine the power output from the system for a fluid flow of 1·7 kg/s.
For fluid c_P = 900 J/kg K.

7. Heat energy is transferred to 1·36 kg of air which causes its temperature to increase from 40°C to 468°C. Calculate, for the two separate cases of heat transfer at (a) constant volume, (b) constant pressure:
(i) the quantity of heat energy transferred,
(ii) the external work done,
(iii) the increase in internal energy

Take c_V and c_P as 0·718 and 1·005 kJ/kg K respectively.

f 8. A closed vessel of 500 cm^3 capacity contains a sample of flue gas at 1·015 bar and 20°C. If the analysis of the gas by volume is 10% carbon dioxide, 8% oxygen, and 82% nitrogen, calculate the partial pressure and mass of each constituent in the sample.

R for CO_2, O_2 and N_2 = 0·189, 0·26 and 0·297 kJ/kg K respectively.

f 9. A gas is discharged from a horizontal convergent nozzle at a steady rate of 1 kg/s. Conditions at inlet are 10 bar and 200°C and at exit 5 bar and 100°C. The change in specific internal energy passing through the nozzle is 80 kJ/kg and heat lost to the surrounds is negligible. If the gas enters the nozzle at 50 m/s determine the exit velocity.

For the gas M is 30 and R_0 may be taken as 8·314 kJ/mol K.

f 10. A vessel of volume 0·4 m^3 contains 0·45 kg of carbon monoxide and 1 kg of air at 15°C. Calculate:

(a) the total pressure in the vessel:
(b) the partial pressure of the nitrogen.

Note: Atomic mass relationships; carbon 12, oxygen 16, nitrogen 14.
Universal Gas Constant = 8·314 kJ/mol K.
Air is 23% oxygen by mass.

CHAPTER 6

EXPANSION AND COMPRESSION OF PERFECT GASES

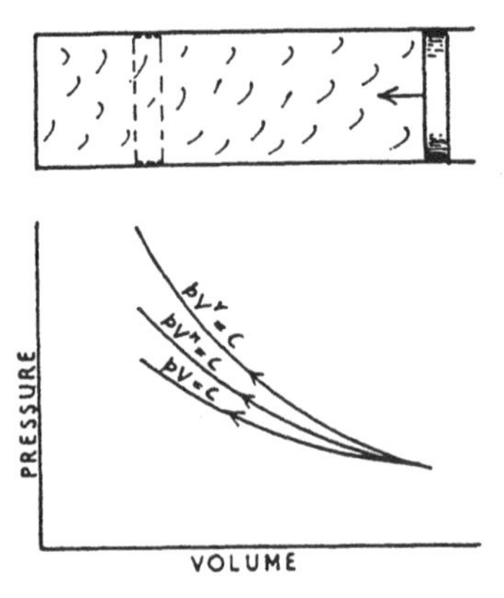

Fig. 13

COMPRESSION OF A GAS IN A CLOSED SYSTEM

When a gas is compressed in a cylinder by the inward movement of a gas-tight piston (Fig. 13), the pressure of the gas increases as the volume decreases. The work done *on* the gas to compress it appears as heat energy in the gas and the temperature tends to rise. This effect can readily be seen with a tyre inflator; in pumping up the tyre the discharge end of the inflator gets hot due to compressing the air.

ISOTHERMAL COMPRESSION. Imagine the piston pushed inward slowly to compress the gas and, at the same time, let heat be taken away via the cylinder walls (by a water-jacket or other means) to avoid any rise in temperature. If the gas could be compressed in this manner, *at constant temperature*, the process would be referred to as *isothermal compression* and the relationship between pressure and volume would follow Boyle's law as stated in the previous chapter:

$$pV = \text{constant} \quad \therefore \; p_1V_1 = p_2V_2$$

ADIABATIC COMPRESSION. Now imagine the piston pushed

inward quickly so that there is insufficient time for any heat energy to be transferred from the gas to the cylinder walls. All the work done in compressing the gas appears as stored up heat energy. The temperature at the end of compression will therefore be high and, for the same ratio of compression as the first case, the pressure will consequently be higher. This form of compression, where no heat energy transfer takes place between the gas and an external source, is known as *adiabatic compression*. The relationship between pressure and volume for adiabatic compression is:

$$pV^{\gamma} = \text{constant} \qquad \therefore\ p_1V_1^{\gamma} = p_2V_2^{\gamma}$$

where γ (gamma) is the ratio of the specific heat of the gas at constant pressure to the specific heat at constant volume, thus,

$$\gamma = \frac{c_p}{c_V}$$

POLYTROPIC COMPRESSION. In practice, neither isothermal nor adiabatic processes can be achieved perfectly. Some heat energy is always lost from the gas through the cylinder walls, more especially if the cylinder is water cooled, but this is never as much as the whole amount of the generated heat of compression. Consequently, the compression curve representing the relationship between pressure and volume lies somewhere between the two theoretical cases of isothermal and adiabatic. Such compression, where a partial amount of heat energy exchange takes place between the gas and an outside source during the process, is termed *polytropic compression* and the compression curve follows the law:

$$pV^{n} = \text{constant} \qquad \therefore\ p_1V_1^{n} = p_2V_2^{n}$$

Thus, the law pV^n = constant may be taken as the general case to cover all forms of compression from isothermal to adiabatic wherein the value of n for isothermal compression is unity, for adiabatic compression $n = \gamma$, and for polytropic compression n generally lies somewhere between 1 and γ.

EXPANSION OF A GAS IN A CLOSED SYSTEM

When a gas is expanded in a cylinder (Fig. 14) the pressure falls and the volume increases as the piston is pushed outward by the energy in the gas.

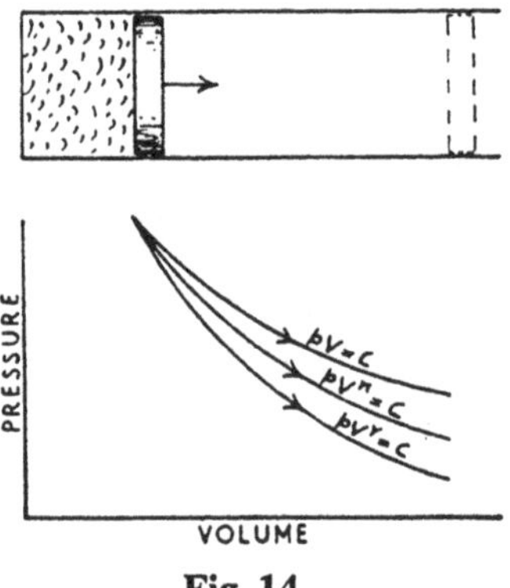

Fig. 14

This is exactly the opposite to compression. Work is done *by* the gas in pushing the piston outward and there is a tendency for the temperature to fall due to the heat energy in the gas being converted into mechanical energy. Therefore to expand the gas *isothermally*, heat energy must be transferred to the gas from an external source during the expansion in order to maintain its temperature constant. The expansion would then follow Boyle's law, pV = constant.

The gas would expand *adiabatically* if no heat energy transfer, to or from the gas, occurs during the expansion, the external work done in pushing the piston forward being entirely at the expense of the stored up heat energy. Therefore the temperature of the gas will fall during the expansion. As for adiabatic compression, the law for adiabatic expansion is pV^γ = constant.

During *polytropic* expansion, a partial amount of heat energy will be transferred to the gas from an outside source but not sufficient to maintain a uniform temperature during the expansion. The law for polytropic expansion is pV^n = constant as it is for polytropic compression.

With reference to Figs. 13 and 14 note that the adiabatic curve is the steepest, the isothermal curve is the least steep, and the polytropic curve lies between the two. Thus, the higher the index of the law of expansion or compression, the steeper will be the curve.

It must also be noted that for any mode of expansion or compression in a closed system, the combination of Boyle's and Charles' laws, and the characteristic gas equation given in the previous chapter, are always true:

$$\frac{pV}{T} = \text{constant} \qquad \therefore \frac{p_1V_1}{T_1} = \frac{p_2V_2}{T_2}$$

$$pV = mRT$$

Example. 0·25 m³ of air at 90 kN/m² and 10°C are compressed in an engine cylinder to a volume of 0·05 m³, the law of compression being $pV^{1.4}$ = constant. Calculate (i) the final pressure, (ii) the final temperature, (iii) the mass of air in the cylinder, taking the characteristic gas constant for air R = 0·287 kJ/kg K.

$$p_1V_1^{1.4} = p_2V_2^{1.4}$$

$$90 \times 0{\cdot}25^{1.4} = p_2 \times 0{\cdot}05^{1.4}$$

$$p_2 = \frac{90 \times 0{\cdot}25^{1{\cdot}4}}{0{\cdot}05^{1{\cdot}4}}$$

$$= 90 \times 5^{1.4} = 856{\cdot}7 \text{ kN/m}^2 \quad \text{Ans. (i)}$$

$$\frac{p_1V_1}{T_1} = \frac{p_2V_2}{T_2}$$

$$\frac{90 \times 0{\cdot}25}{283} = \frac{856{\cdot}7 \times 0{\cdot}05}{T_2}$$

$$T_2 = \frac{283 \times 856{\cdot}7 \times 0{\cdot}05}{90 \times 0{\cdot}25} = 538{\cdot}8 \text{ K}$$

$$= 265{\cdot}8\text{°C} \quad \text{Ans. (ii)}$$

$$p_1V_1 = mRT_1$$

$$m = \frac{90 \times 0{\cdot}25}{0{\cdot}287 \times 283} = 0{\cdot}277 \text{ kg} \quad \text{Ans. (iii)}$$

Example. 0·07 m³ of gas at 4·14 MN/m² is expanded in an engine cylinder and the pressure at the end of expansion is 310 kN/m². If expansion follows the law $pV^{1.35}$ = constant, find the final volume.

$$p_1V_1^{1.35} = p_2V_2^{1.35}$$

$$4140 \times 0{\cdot}07^{1.35} = 310 \times V_2^{1.35}$$

$$V_2 = 0{\cdot}07 \times \sqrt[1{\cdot}35]{\frac{4140}{310}} = 0{\cdot}4774 \text{ m}^3 \quad \text{Ans.}$$

Example. 0·014 m³ of gas at 3·15 MN/m² is expanded in a closed system to a volume of 0·154 m³ and the final pressure is 120 kN/m². If the expansion takes place according to the law pV^n = constant, find the value of n.

$$p_1V_1^n = p_2V_2^n$$

$$3150 \times 0{\cdot}014^n = 120 \times 0{\cdot}154^n$$

$$\frac{3150}{120} = \left\{\frac{0{\cdot}154}{0{\cdot}014}\right\}^n$$

$$26{\cdot}25 = 11^n$$

$$n = 1{\cdot}363 \text{ Ans.}$$

DETERMINATION OF n FROM GRAPH

It will be appreciated that it would be most difficult to obtain two pairs of sufficiently accurate values of pressure and volume from a running engine to enable the law of expansion or compression to be determined as in the previous example. One practical method of finding a fairly close approximation of the law is as follows:

(i) measure a series of connected values of p and V from the curve of an indicator diagram, (ii) reduce the equation $pV^n = C$ to a straight line logarithmic equation, (iii) draw a straight line graph as near as possible through the plotted points of log p and log V to eliminate slight errors of measurement, (iv) determine the law of this graph to obtain the value of n. Thus:

$$p \times V^n = C$$

$$\log p + n \log V = \log C$$

$$\log p = \log C - n \log V$$

This is the same form of equation as,

$$y = a - bx$$

which represents a straight line graph.

The terms log p and log V are the two variables comparable with y and x respectively, and log C and n are constants comparable with a and b respectively. The constant n (like constant b) represents the slope of the straight-line graph and, being a negative value, the line will slope downwards from left to right.

Example. The following related values of the pressure p in kN/m^2 and the volume V in m^3 were measured from the compression curve of an internal combustion engine indicator diagram. Assuming that p and V are connected by the law $pV^n = C$, find the value of n.

p	3450	2350	1725	680	270	130
V	0·0085	0·0113	0·0142	0·0283	0·0566	0·0991

The scales of the graph can be of any convenient choice. In this case both pressure and volume can be expressed in more convenient units, the pressure in bars (1 bar = 10^5 N/m^2 = 10^2 kN/m^2) to

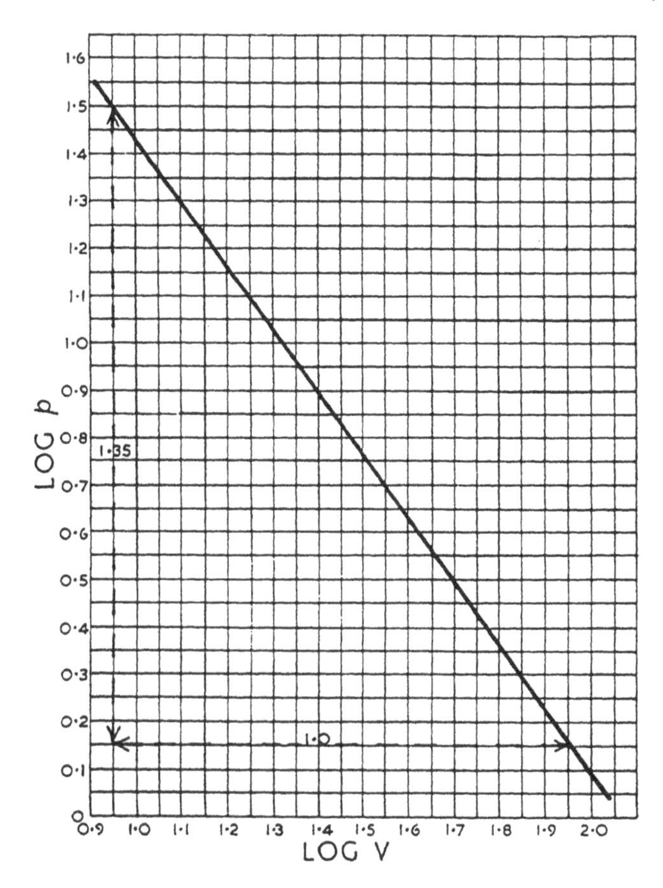

Fig. 15

proportionally reduce the high figures, and the volume in litres (10^3 litres = 1 m^3) to express all volumes above unity, thereby avoiding negative logs and so reducing labour. Hence the following tabulated values of p and V are in bars and litres respectively with their corresponding logarithms to obtain graph plotting points from the respective pairs of log p and log V (base 10 used here).

p[bar]	$\log p$	V [litre]	$\log V$
34·5	1·5378	8·5	0·9294
23·5	1·3711	11·3	1·0531
17·25	1·2368	14·2	1·1523
6·8	0·8325	28·3	1·4518
2·7	0·4314	56·6	1·7528
1·3	0·1139	99·1	1·9961

The graph is then plotted as shown in Fig. 15. Note that it is not necessary to commence at zero origin when only the slope of the line (value of n) is required, a larger graph on the available squared paper can be drawn by starting and finishing to suit the minimum and maximum values to be plotted.

Choosing any two points on the line such as those shown:

$$n = \frac{\text{decrease of } \log p}{\text{decrease of } \log V}$$

$$= \frac{1{\cdot}5 - 0{\cdot}15}{1{\cdot}95 - 0{\cdot}95}$$

$$= \frac{1{\cdot}35}{1} = 1{\cdot}35$$

$$\log p = \log C - 1{\cdot}35 \log V$$

$$p = C \times V^{-1.35}$$

$$pV^{1.35} = C \quad \text{Ans.}$$

RATIOS OF EXPANSION AND COMPRESSION

The ratio of expansion of gas in a cylinder is the ratio of the volume at the end of expansion to the volume at the beginning of expansion. It is usually denoted by r.

$$\text{Ratio of expansion} = r = \frac{\text{final volume}}{\text{initial volume}}$$

The ratio of compression is the ratio of the volume of the gas at the beginning of compression to the volume at the end of compression. This can also be denoted by r.

$$\text{Ratio of compression} = r = \frac{\text{initial volume}}{\text{final volume}}$$

It will be seen that in each of the above ratios, it is the larger volume divided by the smaller, therefore, the ratio of expansion and ratio of compression is always greater than unity.

RELATIONSHIPS BETWEEN TEMPERATURE AND VOLUME, AND TEMPERATURE AND PRESSURE, WHEN $pV^n = C$

As stated previously, the equations,

$$p_1V_1^n = p_2V_2^n \quad \text{and} \quad \frac{p_1V_1}{T_1} = \frac{p_2V_2}{T_2}$$

are always true for any kind of expansion or compression of a perfect gas in a closed system. Some problems arise, however, where neither p_1 nor p_2 are given, and the unknown temperature or volume has to be solved by substituting the value of one of the pressures from one equation into the other. Similarly, where neither volume is given, substitution has to be made for one of the volumes to obtain the unknown temperature or pressure.

Substitution can be made in one of the above equations to eliminate either pressure or volume and so derive relationships for direct solution, as follows.

$$p_1V_1^n = p_2V_2^n \qquad \therefore p_1 = \frac{p_2V_2^n}{V_1^n}$$

Substituting this value of p_1 into the combined law equation:

$$\frac{p_1V_1}{T_1} = \frac{p_2V_2}{T_2}$$

$$\frac{p_2 \times V_2^n \times V_1}{T_1 \times V_1^n} = \frac{P_2V_2}{T_2}$$

$$T_1 \times V_1^n \times p_2 \times V_2 = T_2 \times p_2 \times V_2^n \times V_1$$

$$\frac{T_1}{T_2} = \frac{p_2 \times V_2^n \times V_1}{p_2 \times V_2 \times V_1^n}$$

p_2 cancels,

dividing $V_2{}^n$ by $V_2 = V_2{}^n \div V_2 = V_2{}^{n-1}$

dividing $V_1{}^n$ by $V_1 = V_1{}^n \div V_1 = V_1{}^{n-1}$

$$\frac{T_1}{T_2} = \frac{V_2^{n-1}}{V_1^{n-1}}$$

$$\frac{T_1}{T_2} = \left\{\frac{V_2}{V_1}\right\}^{n-1} \qquad \text{(i)}$$

Again from $p_1V_1^n = p_2V_2^n \qquad V_1^n = \frac{p_2V_2^n}{p_1}$

$$\therefore V_1 = p_2^{1/n} \times \frac{V_2}{P_1^{1/n}}$$

Substituting this value of V_1 into the combined law equation:

$$\frac{p_1V_1}{T_1} = \frac{p_2V_2}{T_2}$$

$$\frac{p_1 \times p_2^{1/n} \times V_2}{T_1 \times p_1^{1/n}} = \frac{p_2 \times V_2}{T_2}$$

$$T_1 \times p_1^{1/n} \times p_2 \times V_2 = T_2 \times p_1 \times P_2^{1/n} \times V_2$$

$$\frac{T_1}{T_2} = \frac{p_1 \times p_2^{1/n} \times V_2}{p_1^{1/n} \times p_2 \times V_2}$$

V_2 cancels,

dividing p_1 by $p_1^{1/n} = p_1 \div p_1^{1/n} = p_1^{1-1/n}$

dividing p_2 by $p_2^{1/n} = p_2 \div p_2^{1/n} = p_2^{1-1/n}$

$$\frac{T_1}{T_2} = \frac{p_1^{1-1/n}}{p_2^{1-1/n}}$$

$$\frac{T_1}{T_2} = \left\{\frac{p_1}{p_2}\right\}^{1-1/n}$$

$$\frac{T_1}{T_2} = \left\{\frac{p_1}{p_2}\right\}^{\frac{n-1}{n}} \qquad \text{(ii)}$$

From (i) and (ii) we have the very useful relationship:

$$\frac{T_1}{T_2} = \left\{\frac{V_2}{V_1}\right\}^{n-1} = \left\{\frac{p_1}{p_2}\right\}^{\frac{n-1}{n}}$$

For an adiabatic process, the adiabatic index γ is substituted for the polytropic index n.

Example. Air is expanded adiabatically from a pressure of 800 kN/m^2 to 128 kN/m^2. If the final temperature is 57°C, calculate the temperature at the beginning of expansion, taking $\gamma = 1{\cdot}4$.

$$\frac{T_1}{T_2} = \left\{\frac{p_1}{p_2}\right\}^{\frac{\gamma-1}{\gamma}}$$

$$\frac{T_1}{330} = \left\{\frac{800}{128}\right\}^{2/7}$$

$$T_1 = 330 \times 6{\cdot}25^{2/7}$$

$$= 557{\cdot}1 \text{ K}$$

$$= 284{\cdot}1^\circ\text{C} \quad \text{Ans.}$$

Example. The ratio of compression in a petrol engine is 9 to 1. Find the temperature of the gas at the end of compression if the temperature at the beginning is 24°C, assuming compression to follow the law pV^n = constant, where $n = 1{\cdot}36$.

$$\frac{T_1}{T_2} = \left\{\frac{V_2}{V_1}\right\}^{n-1} \quad \text{or} \quad \frac{T_2}{T_1} = \left\{\frac{V_1}{V_2}\right\}^{n-1}$$

$$T_2 = 297 \times 9^{0{\cdot}36} = 655{\cdot}1 \text{ K}$$

$$= 382{\cdot}1^\circ\text{C} \quad \text{Ans.}$$

Example. The volume and temperature of a gas at the beginning of expansion are 0·0056 m^3 and 183°C, at the end of expansion the values are 0·0238 m^3 and 22°C respectively. Assuming expansion follows the law $pV^n = C$, find the value of n.

$$\frac{T_1}{T_2} = \left\{\frac{V_2}{V_1}\right\}^{n-1}$$

$$\frac{456}{295} = \left\{\frac{0{\cdot}0238}{0{\cdot}0056}\right\}^{n-1}$$

$$1{\cdot}546 = 4{\cdot}25^{n-1}$$

$$n = 1{\cdot}301 \quad \text{Ans.}$$

WORK TRANSFER

Firstly, consider a gas expanding at *constant pressure* in a cylinder fitted with a gas-tight piston (also refer to Fig. 11 with notes in Chapter 5).

Work done [kJ = kN m] = force [kN] × distance [m]

The total force [kN] on the piston is the product of the pressure p[kN/m^2] of the gas and the area A[m^2] of the piston and, if the piston moves through a distance of S metres as the gas expands, then,

$$\text{Work done} = p \times A \times S$$

The product of the piston area A and the distance it moves S, is the volume swept through by the piston, this is also the increase in volume of the gas in the cylinder. If V_1 is the volume of the gas at the beginning of the expansion, and V_2 is the volume at the end of expansion, the $A \times S$ is equal to $V_2 - V_1$ cubic metres, hence,

$$\text{Work done} = p\,(V_2 - V_1)$$

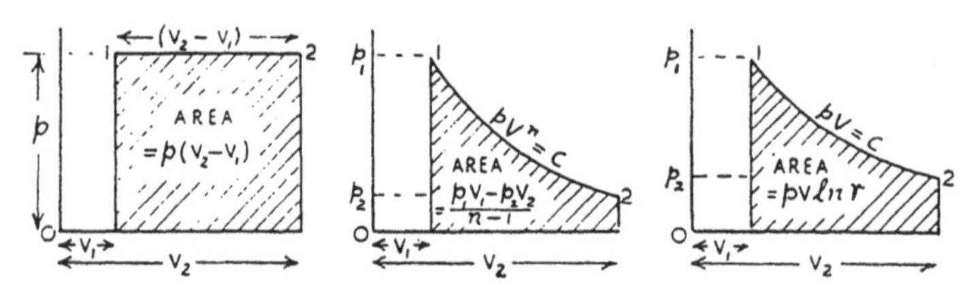

Fig. 16

Fig. 16 shows the pressure-volume diagram representing work done at constant pressure. The graph is a straight horizontal line and the area under it is a rectangle. The area of a rectangle is height × length which, in this case, is $p \times (V_2 - V_1)$. Hence *the area under the pressure-volume line represents work done.*

Now consider cases where the pressure falls during the expansion of the gas. The formula giving the area under the polytropic curve representing the general relationship between pressure and volume, i.e. $pV^n = C$ can only be derived satisfactorily by the use of the calculus. The expression is illustrated in Fig. 16 and derived in the following example:

f Example. A quantity of gas undergoes a non-flow process from an initial pressure of p_1 and volume V_1 to a final pressure p_2 and volume V_2.

(a) Given that the area beneath the pV curve is given by:

$$\int_1^2 p\mathrm{d}V$$

show that the total area beneath the pV curve may be written as:

$$\frac{p_2V_2 - p_1V_1}{1-n}$$

(b) During the process a quantity of heat q per unit mass is transferred. Show that q is given by:

$$q = \left(c_V + \frac{R}{1-n}\right)\left(T_2 - T_1\right)$$

$$pV^n = C$$

$$p = \frac{C}{V^n}$$

$$\text{Area} = \int_1^2 p\mathrm{d}V$$

$$= C\int_{V_1}^{V_2} V^{-n}\,\mathrm{d}V$$

$$= pV^n\left[\frac{V^{-n+1}}{-n+1}\right]_{V_1}^{V_2}$$

$$= \frac{pV^n (V_2^{1-n} - V_1^{1-n})}{1-n}$$

$$\text{Area} = \frac{p_2V_2 - p_1V_1}{1-n} \quad \text{Ans. (a)}$$

Heat transfer = Change of internal energy + External work

$$q = c_V (T_2 - T_1) + \frac{p_2v_2 - p_1v_1}{1-n}$$

$$pv = RT$$

$$q = c_V (T_2 - T_1) + \frac{R (T_2 - T_1)}{1-n}$$

$$q = \left(c_V + \frac{R}{1-n}\right)(T_2 - T_1) \quad \text{Ans. (b)}$$

$$\text{Work done during polytropic expansion} = \frac{p_1V_1 - p_2V_2}{n-1}$$

This is the general expression for work done. For adiabatic expansion, n is replaced by γ. For isothermal expansion however, since the value of n is 1, and as $p_1V_1 = p_2V_2$ then substitution in this expression for work will produce $0 \div 0$ which is indeterminate. A different expression is therefore necessary to obtain work during isothermal expansion (see Fig. 16):

$$pV = C$$

$$p = \frac{C}{V}$$

$$\text{Area} = \int_1^2 p\mathrm{d}V$$

$$= C\int_{V_1}^{V_2} \frac{\mathrm{d}V}{V}$$

$$= C (\ln V_2 - \ln V_1)$$

$$= pV \ln\left(\frac{V_2}{V_1}\right)$$

$$\text{Work done during isothermal expansion} = pV \ln r$$

where r is the ratio of expansion.

The above expressions give the work done *by* the gas during expansion. The same expressions give the work done *on* the gas during compression.

In the case of expansion the initial condition of p_1V_1 will exceed the final condition of p_2V_2 and the expression for work will produce a *positive* result, indicating that work is done *by* the gas in pushing the piston forward. Conversely, for compression, the initial condition of p_1V_1 will be less than the final condition of p_2V_2 and therefore a *negative* result will be obtained, indicating that work is done *on* the gas by the piston.

It is important to bear in mind that in calculating work, the units must be consistent. For example, to express work in kilojoules, the pressure must be in kN/m² and the volume in m³, thus,

$$\text{kN/m}^2 \times \text{m}^3 = \text{kN m} = \text{kJ}$$

Work is transfer of energy, therefore the above can be referred to as *work transfer* from the closed system within the boundary of the cylinder to the external mechanism, or vice-versa. In the first case, when the gas expands, work is being transferred from the energy in the gas to the piston, which, in turn, transmits the work through connecting mechanism to the crank shaft, the work transfer in this case is referred to as being positive. In the second case, when the gas is compressed, work is being transferred from the crank shaft, through the connecting mechanism and piston to the gas, thereby increasing the energy in the gas, and this work transfer is called negative.

Further, since $pV = mRT$, the expressions for work may be stated in terms of mRT instead of pV:

Polytropic expansion

$$\text{Work} = \frac{p_1V_1 - P_2V_2}{n-1} = \frac{mR\,(T_1 - T_2)}{n-1}$$

Isothermal expansion

$$\text{Work} = pV \ln r = mRT \ln r$$

Example. 0·04 m³ of gas at a pressure of 1482 kN/m² is expanded isothermally until the volume is 0·09 m³. Calculate the work done during the expansion.

$$\text{Work done} = pV \ln r$$

$$r = \frac{\text{final volume}}{\text{initial volume}} = \frac{0{\cdot}09}{0{\cdot}04} = 2{\cdot}25$$

$$\text{Work done} = 1482 \times 0{\cdot}04 \times 0{\cdot}81809$$

$$= 48{\cdot}08 \text{ kJ} \quad \text{Ans.}$$

Example. 7·08 litres of air at a pressure of 13·79 bar and temperature 335°C are expanded according to the law $pV^{1{\cdot}32}$ = constant, and the final pressure is 1·206 bar. Calculate (i) the volume at the end of expansion, (ii) the work transfer from the air, (iii) the temperature at the end of expansion, (iv) the mass of air in the system, taking R = 0·287 kJ/kg K.

$$p_1V_1^{1{\cdot}32} = p_2V_2^{1{\cdot}32}$$

$$1379 \times 0{\cdot}00708^{1{\cdot}32} = 120{\cdot}6 \times V_2^{1{\cdot}32}$$

$$V_2 = 0{\cdot}00708 \times \sqrt[1{\cdot}32]{\frac{1379}{120{\cdot}6}}$$

$$= 0{\cdot}04484 \text{ m}^3 \text{ or } 44{\cdot}84 \text{ litres} \quad \text{Ans. (i)}$$

Note that if the units of pressure and volume are of the same kind on each side of the equation, the units cancel each other out and hence any convenient units can be used. The above could therefore be worked in bars of pressure and litres of volume in which the question data is given. However, as pointed out, it is essential to work in fundamental units in such expressions as used in parts (ii) and (iv) of this problem, it is preferential to use fundamental units throughout by expressing the pressure in kN/m^2 and the volume in m^3.

$$\text{Work} = \frac{p_1V_1 - p_2V_2}{n-1}$$

$$= \frac{1379 \times 0{\cdot}00708 - 120{\cdot}6 \times 0{\cdot}04484}{1{\cdot}32 - 1}$$

$$= 13{\cdot}61 \text{ kJ} \quad \text{Ans. (ii)}$$

$$\frac{p_1V_1}{T_1} = \frac{p_2V_2}{T_2}$$

$$\frac{1379 \times 0{\cdot}00708}{608} = \frac{120{\cdot}6 \times 0{\cdot}04484}{T_2}$$

$$T_2 = \frac{608 \times 120{\cdot}6 \times 0{\cdot}04484}{1379 \times 0{\cdot}00708}$$

$$T_2 = 336{\cdot}6 \text{ K}$$

$$= 63{\cdot}6^\circ\text{C} \quad \text{Ans. (iii)}$$

$$p_1V_1 = mRT_1$$

$$m = \frac{1379 \times 0{\cdot}00708}{0{\cdot}287 \times 608}$$

$$= 0{\cdot}05595 \text{ kg} \quad \text{Ans. (iv)}$$

Example. A perfect gas is compressed in a cylinder according to the law $pV^{1\cdot3}$ = constant. The initial condition of the gas is 1·05 bar, 0·34 m^3 and 17°C. If the final pressure is 6·32 bar, calculate (i) the mass of gas in the cylinder, (ii) the final volume, (iii) the final temperature, (iv) the work done to compress the gas, (v) the change in internal energy, (vi) the transfer of heat between the gas and cylinder walls.

Take $c_v = 0{\cdot}7175$ kJ/kg K and $R = 0{\cdot}287$ kJ/kg K.

$$p_1V_1 = mRT_1$$

$$m = \frac{105 \times 0{\cdot}34}{0{\cdot}287 \times 290}$$

$$= 0{\cdot}4289 \text{ kg} \quad \text{Ans. (i)}$$

$$p_1V_1^{1\cdot3} = p_2V_2^{1\cdot3}$$

$$1{\cdot}05 \times 0{\cdot}34^{1\cdot3} = 6{\cdot}32 \times V_2^{1\cdot3}$$

$$V_2 = 0{\cdot}34 \times \sqrt[1\cdot3]{\frac{1{\cdot}05}{6{\cdot}32}}$$

$$= 0{\cdot}08549 \text{ m}^3 \quad \text{Ans. (ii)}$$

$$\frac{p_1V_1}{T_1} = \frac{p_2V_2}{T_2}$$

$$\frac{1{\cdot}05 \times 0{\cdot}34}{290} = \frac{6{\cdot}32 \times 0{\cdot}08549}{T_2}$$

$$T_2 = \frac{290 \times 6{\cdot}32 \times 0{\cdot}08549}{1{\cdot}05 \times 0{\cdot}34}$$

$$T_2 = 438{\cdot}8 \text{ K}$$

$$= 165{\cdot}8^{\circ}\text{C} \quad \text{Ans. (iii)}$$

Alternatively, the final temperature could be obtained from

$$\frac{T_1}{T_2} = \left\{\frac{p_1}{p_2}\right\}^{\frac{n-1}{n}}$$

$$\text{Work done} = \frac{p_1V_1 - p_2V_2}{n-1}$$

$$= \frac{105 \times 0{\cdot}34 - 632 \times 0{\cdot}08549}{1{\cdot}3 - 1}$$

$$= -61{\cdot}1 \text{ kJ}$$

Alternatively, the work done could be obtained from:

$$\frac{mR\,(T_1 - T_2)}{n-1}$$

Note that the minus sign indicates that work is done *on* the gas.

Work to compress gas = 61·1 kJ Ans. (iv)

Increase in internal energy:

$$U_2 - U_1 = mc_V(T_2 - T_1)$$

$$= 0{\cdot}4289 \times 0{\cdot}7175 \times (438{\cdot}8 - 290)$$

$$= 45{\cdot}78 \text{ kJ} \quad \text{Ans. (v)}$$

$$\begin{matrix}\text{Heat supplied} \\ \text{to the gas}\end{matrix} = \begin{matrix}\text{Increase in} \\ \text{internal energy}\end{matrix} + \begin{matrix}\text{Work done} \\ \text{by the gas}\end{matrix}$$

$$= 45{\cdot}78 - 61{\cdot}1$$

$$= -15{\cdot}32 \text{ kJ}$$

The minus sign means that heat is rejected by the gas during compression, that is, this amount of heat energy is transferred *from* the gas *to* the cylinder wall surrounds.

Transfer of heat = 15·32 kJ Ans. (vi)

A RELATIONSHIP BETWEEN HEAT ENERGY SUPPLIED AND WORK DONE

Consider polytropic expansion of a gas in a cylinder from initial conditions represented by state-point 1 to the final conditions of state-point 2, and apply the energy equation:

$$\text{Heat supplied to the gas} = \text{Increase in internal energy} + \text{Work done by the gas}$$

$$= mc_V (T_2 - T_1) + \frac{mR\,(T_1 - T_2)}{n-1}$$

$$\text{Substituting } c_V = \frac{R}{\gamma - 1}$$

$$\text{Heat supplied} = \frac{mR\,(T_2 - T_1)}{\gamma - 1} + \frac{mR\,(T_1 - T_2)}{n-1}$$

$$Q = \frac{mR\,(T_1 - T_2)}{n-1} - \frac{mR\,(T_1 - T_2)}{\gamma - 1}$$

$$= \frac{mR\,(T_1 - T_2)}{n-1}\left\{1 - \frac{n-1}{\gamma - 1}\right\}$$

$$= \frac{mR\,(T_1 - T_2)}{n-1}\left\{\frac{\gamma - 1 - n + 1}{\gamma - 1}\right\}$$

$$= \text{Work} \times \frac{\gamma - n}{\gamma - 1}$$

During expansion of gas, work is done *by* the gas and is a positive quantity, therefore a positive quantity of heat energy is supplied from the cylinder walls to the gas.

In compression of a gas, work is done *on* the gas which is negative work done by the gas, the result of the above expression is negative heat supplied, meaning that heat energy is transferred from the gas to the cylinder walls.

The relations between the properties of a perfect gas in its initial and final states are the same for both reversible (ideal, frictionless) steady-flow (open) and the non-flow (closed) processes as detailed in this chapter.

TEST EXAMPLES 6

1. Gas is expanded in an engine cylinder, following the law pV^n = constant where the value of n is 1·3. The initial pressure is 2550 kN/m^2 and the final pressure is 210 kN/m^2. If the volume at the end of expansion is 0·75 m^3, calculate the volume at the beginning of expansion.

2. The ratio of compression in a petrol engine is 8·6 to 1. At the beginning of compression the pressure of the gas is 98 kN/m^2 and the temperature is 28°C. Find the pressure and temperature at the end of compression, assuming it follows the law $pV^{1 \cdot 36}$ = constant.

3. Gas is expanded in an engine cylinder according to the law $pV^n = C$. At the beginning of expansion the pressure and volume are 1750 kN/m^2 and 0·05 m^3 respectively, and at the end of expansion the respective values are 122·5 kN/m^2 and 0·375 m^3. Calculate the value of n.

4. The temperature and pressure of the air at the beginning of compression in a compressor cylinder are 20°C and 101·3 kN/m^2, and the pressure at the end of compression is 1420 kN/m^2. If the law of compression is $pV^{1 \cdot 35}$ find the temperature at the end of compression.

5. 0·014 m^3 of a gas at 66°C is expanded adiabatically in a closed system and the temperature at the end of expansion is 2°C. Taking the specific heats of the gas at constant pressure and constant volume as 1·005 and 0·718 kJ/kg K respectively, calculate the volume at the end of the expansion.

6. Air is compressed in a diesel engine from 1·17 bar to 36·55 bar. If the temperatures at the beginning and end of compression are 32°C and 500°C respectively, find the law of compression assuming it is polytropic.

7. One kg of air at 20 bar, 200°C is expanded to 10 bar, 125°C by a process which is represented by a straight line on the pV diagram.

   Calculate for the air:
   (a) the work transfer;
   (b) the change in internal energy;
   (c) the heat transfer;
   (d) the change in enthalpy.

Note: for air R = 287 J/kg K and c_p = 1005 J/kg K.

f 8. A closed and insulated vessel contains 1 kg of air at 213°C and 1 bar. A second vessel, which is also insulated, has a volume of 0·2 m^3 and contains air at 6 bar and 412°C. The two vessels are then connected by a pipe of negligible volume.

Calculate:

(a) the final pressure of air in the vessels;

(b) the final temperature of the air in the vessels.

For air: $R = 287\ J/kg\,K$.

f 9. A cylinder fitted with a piston contains 0·1 m^3 of air at 1 bar 15°C. Heat is supplied to the air until the temperature reaches 500°C whilst the piston is fixed. The piston is then released and the air expands according to the law $pV^{1.5} = C$ until the pressure is 1 bar.

Calculate:

(a) the final temperature of the air;

(b) the work done during expansion;

(c) the heat transferred during each process.

For the gas $c_V = 718\ J/kg\,K$, $R = 287\ J/kg\,K$.

f 10. A gas is expanded in a cylinder behind a gas-tight piston. At the beginning of expansion the pressure is 36 bar, volume 0·125 m^3, and temperature 510°C. At the end of expansion the volume is 1·5 m^3 and temperature 40°C. Taking $R = 0{\cdot}284$ kJ/kg K and $c_V = 0{\cdot}71$ kJ/kg K, calculate (i) the pressure at the end of expansion, (ii) the index of expansion, (iii) the mass of gas in the cylinder, (iv) change of internal energy, (v) work done by the gas, (vi) heat transfer during expansion.

CHAPTER 7

I.C. ENGINES – ELEMENTARY PRINCIPLES

Internal combustion engines are so named because combustion of the fuel takes place *inside* the engine. When the fuel burns inside the engine cylinder, it gives out heat which is absorbed by the air previously taken into the cylinder, the temperature of the air is therefore increased with a consequent increase in pressure and/or volume, thus energy is imparted to the piston. The reciprocating motion of the piston is converted into a rotary motion at the crank shaft by connecting rod and crank.

The method of igniting the fuel varies. In diesel engines the air in the cylinder is compressed to a high pressure so that it attains a high temperature, and when oil fuel is injected into this high temperature air the fuel immediately ignites. When the ignition of the fuel is caused solely by the heat of compression, the engine is classed as a *compression-ignition* engine. In petrol and paraffin engines the fuel is usually taken in with the charge of air, compressed and then ignited by an electric spark.

THE FOUR-STROKE DIESEL ENGINE

In this type of engine it takes four strokes of the piston (*i.e.* two revolutions of the crank) to complete one working cycle of operations, hence the name *four-stroke* cycle.

Fig. 17 illustrates each of these four strokes in one cylinder. One the cylinder head is shown the fuel valve (or injector) which lifts to admit oil fuel (under pressure) into the cylinder, the air-induction valve through which air is drawn in, and the exhaust valve through which the exhaust gases are expelled from the cylinder. There are two more valves which are not shown here because they do not operate during the normal working cycle; one is the relief valve which opens against the compression of its spring when the pressure in the cylinder rises too high, the other is the air-starting valve which opens to admit high pressure air into the cylinder to move the piston and start the engine (on large engines).

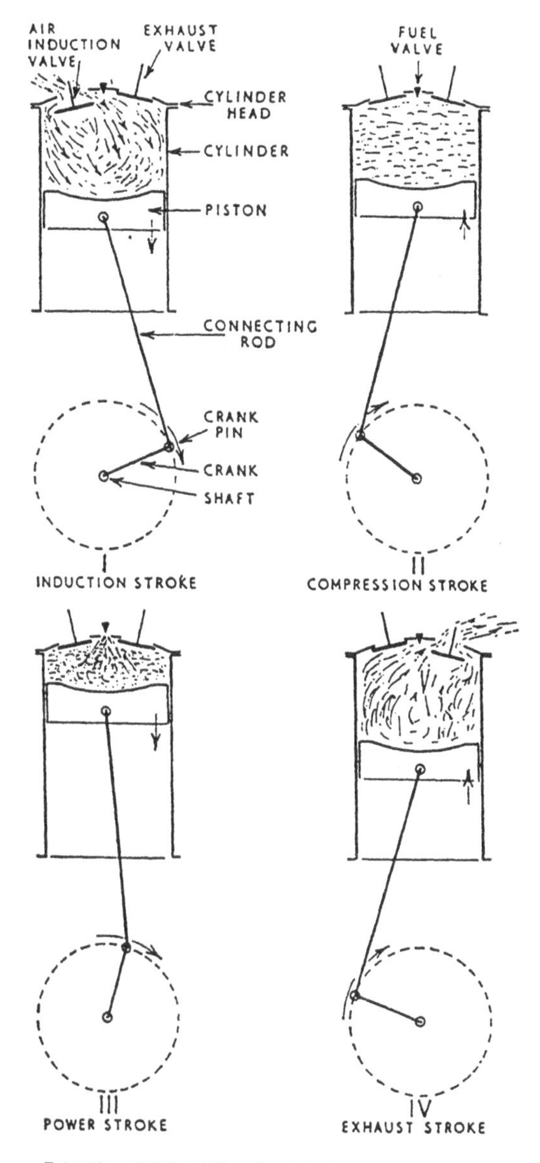

FOUR-STROKE DIESEL ENGINE

Fig. 17

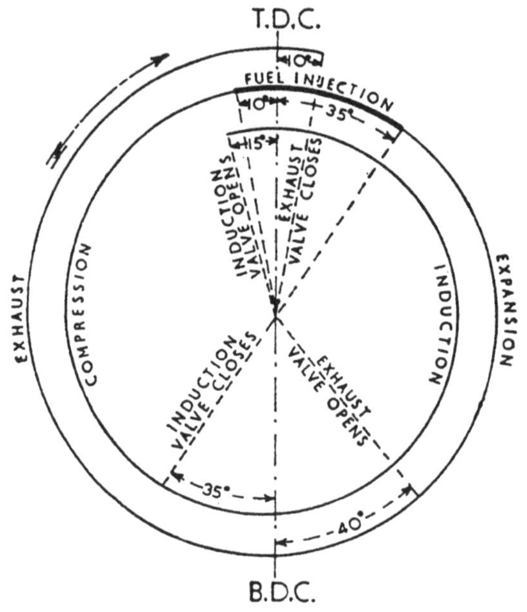

TIMING DIAGRAM FOR A FOUR-STROKE DIESEL ENGINE

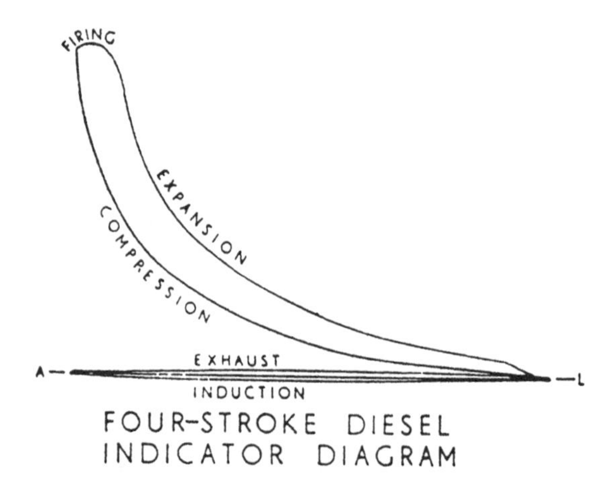

FOUR-STROKE DIESEL INDICATOR DIAGRAM

Fig. 18

THE TWO-STROKE DIESEL ENGINE

The two-stroke diesel engine is so named because it takes two strokes of the piston to complete one working cycle. Every downward stroke of the piston is a power stroke, every upward stroke is a compression stroke, the exhaust of the burned gases from the cylinder and the fresh charge of air is taken in during the late period of the downward stroke and the early part of the upward stroke. The exhaust gases pass through a set of ports in the lower part of the cylinder and the air is admitted through a similar set of ports, the ports are covered and uncovered by the piston itself which must be a long one or have a skirt attached so that the ports are covered when the piston is at the top of its stroke.

As there is no complete stroke to draw the air into the cylinder, the air must be pumped in at a low pressure from a scavenge pump or blower, the air supplied is referred to as scavenge-air and the ports in the cylinder through which the air is admitted are termed scavenge ports. It is the function of this air to sweep around the cylinder and so "scavenge" or clean out the cylinder by pushing the remains of the exhaust gases out, leaving a clean charge of air to be compressed (see Fig. 19).

PETROL ENGINES

Engines which run with petrol (or paraffin) as the fuel are often termed "light oil" engines. The main difference between the majority of petrol engines and the diesel engine is that the petrol engine takes in a charge of air and petrol vapour, this explosive mixture is compressed and ignited by an electric spark; whereas in the diesel engine the cylinder is charged with air only so that only pure air is compressed and the fuel is injected at the moment ignition and burning of the fuel is required, ignition being caused solely by the heat of the compressed air.

When the air is compressed in a diesel engine there is no possibility of firing before the fuel is injected. In a petrol engine, an explosive mixture of petrol and air is compressed and there is danger of the mixture firing spontaneously due to the heat of compression alone and before the electric spark occurs, therefore the ratio of compression must be limited to prevent this. The ratio of compression in diesel engines can be high, such as twelve to one and upwards whereas the ratio of compression in petrol engines is much less, in the region of eight or nine to one.

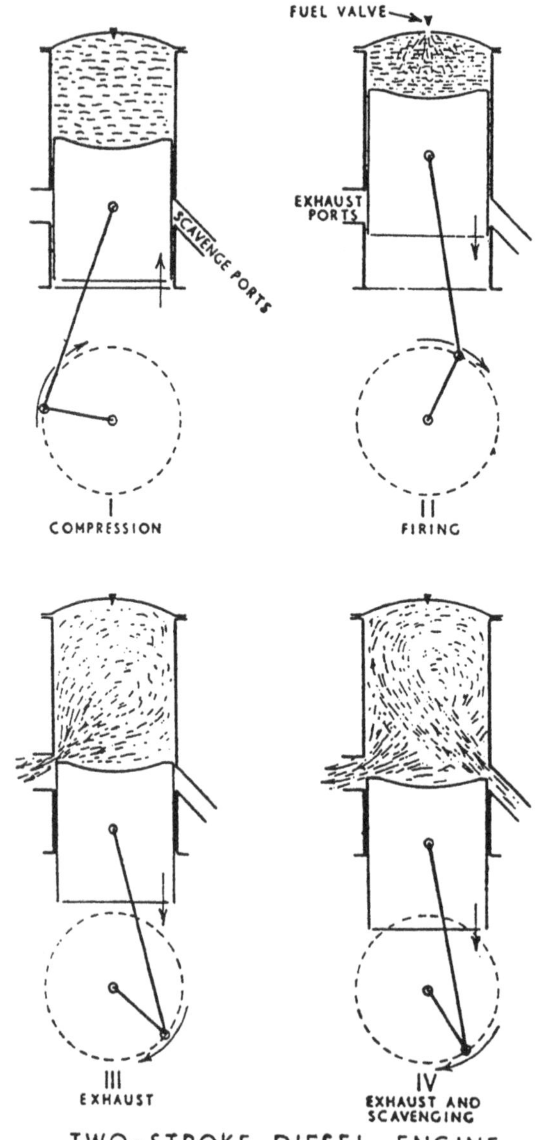

TWO-STROKE DIESEL ENGINE

Fig. 19

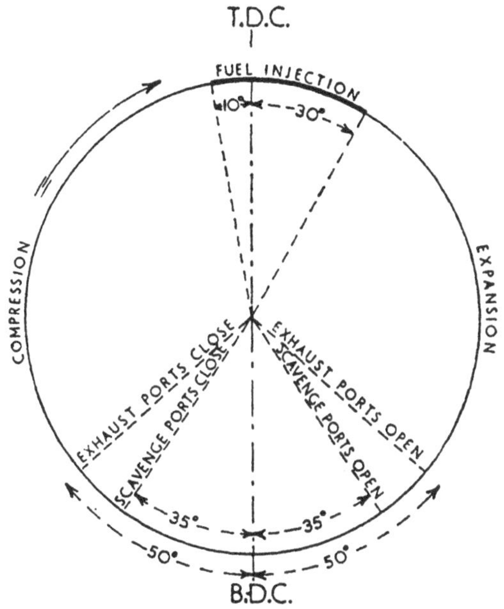

TIMING DIAGRAM FOR A TWO-STROKE DIESEL ENGINE

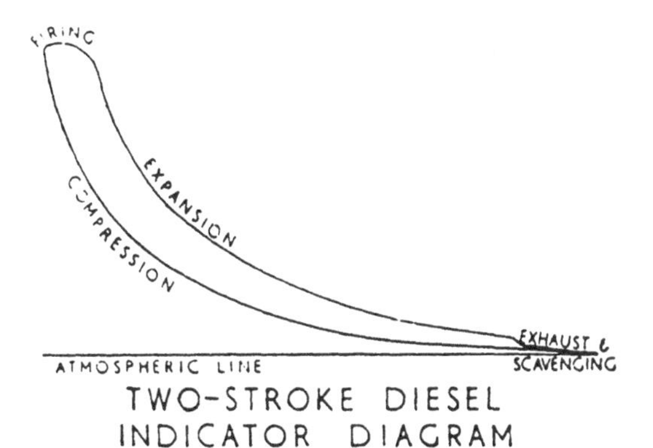

TWO-STROKE DIESEL INDICATOR DIAGRAM

Fig. 20

MEAN EFFECTIVE PRESSURE AND POWER

We have seen in the last chapter that the area of a pV diagram represents work. The indicator diagrams shown in Figs. 18 and 20 are examples of practical pV diagrams taken off engines by means of an engine indicator, the areas of these indicator diagrams represent the work done per cycle.

Fig. 21 shows an indicator which is suitable for taking indicator diagrams off reciprocating engines up to rotational speeds of about 300 rev/min. In this type, the pressure scale spring is anchored at its bottom end to the framework, and the top of the piston spindle bears upwards on the top coil of the spring, the upward motion of the indicator piston thus stretches the spring.

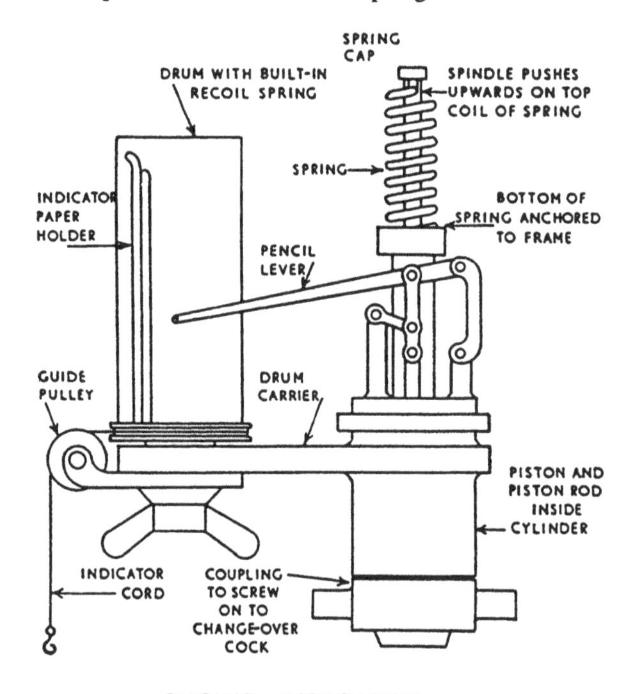

Fig. 21

MEAN EFFECTIVE PRESSURE. Consider the two-stroke diesel engine indicator diagram shown in Fig. 22. The positive work done in one cycle of operations *by* the gas during the burning period and expansion of the gas is shown by the shaded area of Fig. 22b. The work done *on* the air during the compression period, representing negative work done by the engine is shown by the shaded area of Fig. 22c. Hence the net useful work done in one cycle is the difference between positive and negative work and represented by the actual diagram of Fig. 22a. Therefore, if the area of the indicator diagram is divided by its length, the average height is obtained which, to scale, is the average or mean pressure effectively pushing the piston forward and transmitting useful energy to the crank during one cycle. This, expressed in N/m^2 or a suitable multiple of the basic pressure unit, is termed the *indicated mean effective pressure*.

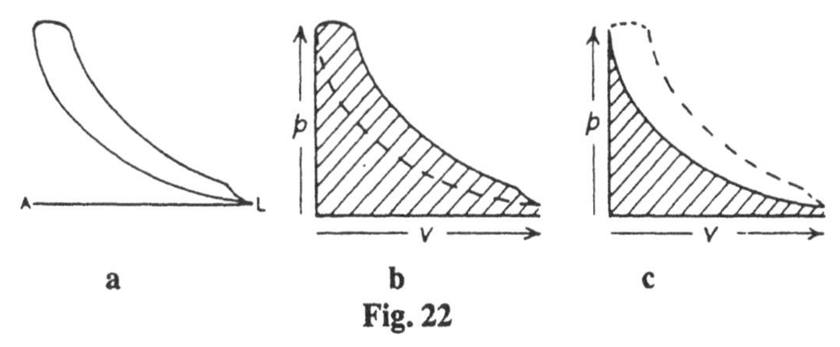

a b c

Fig. 22

The area of the diagram is usually measured by a planimeter. If the area is measured in mm^2 then dividing this by the length in mm gives the mean height in mm. The mean height in mm is now multiplied by the pressure scale of the indicator spring in N/m^2 per mm to obtain the indicated mean effective pressure in N/m^2. The usual convenient multiples of N/m^2 for such pressures are kN/m^2 and bars, and the spring may be graduated in either of these units.

If a planimeter is not to hand, the mean height of the indicator diagram may be obtained by the application of the mid-ordinate rule.

INDICATED POWER. Power is the rate of doing work, that is, the quantity of work done in a given time. The basic unit of power is the *watt* [W] which is equal to the rate of one joule of work being done every second. In symbols:

$$1 \text{ W} = 1 \text{ J/s} = 1 \text{ N m/s}$$

The watt is a small unit and only suitable for expressing the power of small machines. For normal powers in marine engineering, mechanical, electrical or hydraulic, the kilowatt [kW] is usually a more convenient size, and large powers may be expressed in megawatts [MW].

Let p_m = mean effective pressure [N/m^2]

A = area of piston [m^2]

L = length of stroke [m]

n = number of power strokes per second

then,

Average force [N] on piston

$$= p_m \times A \text{ newtons}$$

Work done [J] in one power stroke

$$= p_m \times A \times L \text{ newton-metres} = \text{joules}$$

Work per second [J/s = W]

$$= p_m \times A \times L \times n \text{ watts of power}$$

therefore,

$$\text{Indicated power} = p_m ALn$$

This is the power indicated in one cylinder. The total power of a multi-cylinder engine is that multiplied by the number of cylinders, if the mean effective pressure is the same for all cylinders.

Note that when the mean effective pressure is in N/m^2 the power obtained by the above expression is in watts. If the mean effective pressure in kN/m^2 is inserted, the result will be the power in kW, and this is usually more convenient.

The value of n, the number of power strokes per second, depends upon the working cycle of the engine (two-stroke or four-stroke), its rotational speed, and whether it is a single-acting or double-acting engine.

Referring to single-acting engines, wherein the cycle of operations takes place only on the top side of the piston:

In the four stroke cycle, there is one power stroke in every four strokes, that is, one power stroke in every two revolutions, hence,

$$n = \text{rev/s} \div 2$$

In the two-stroke cycle, there is one power stroke in every two strokes, that is, one power stroke in every revolution, thus,

$$n = \text{rev/s}$$

Example. The area of an indicator diagram taken off one cylinder of a four-cylinder, four-stroke, single-acting internal combustion engine is 378 mm^2, the length is 70 mm, and the indicator spring scale is 1 mm = 2 bar. The diameter of the cylinders is 250 mm, stroke 300 mm, and rotational speed 5 rev/s. Calculate the indicated power of the engine assuming all cylinders develop equal power.

$$\text{Mean height of diagram} = \text{area} \div \text{length}$$
$$= 378 \div 70 = 5{\cdot}4 \text{ mm}$$

$$\text{Indicated } p_m = \text{mean height} \times \text{spring scale}$$
$$= 5{\cdot}4 \times 2 = 10{\cdot}8 \text{ bar}$$
$$10{\cdot}8 \text{ bar} \times 10^2 = 1080 \text{ kN/m}^2$$
$$n = \text{rev/s} \div 2$$
$$= 5 \div 2 = 2{\cdot}5$$
$$\text{Indicated power} = p_m ALn$$
$$= 1080 \times 0{\cdot}7854 \times 0{\cdot}25^2 \times 0{\cdot}3 \times 2{\cdot}5$$
$$= 39{\cdot}74 \text{ kW}$$

Total power for four cylinders

$$= 4 \times 39{\cdot}74 = 158{\cdot}96 \text{ kW Ans.}$$

Example. The diameter of the cylinders of a six-cylinder, single-acting, two-stroke diesel engine, is 635 mm and the stroke is 1010 mm. Indicator diagrams taken off the engine when running at 2·2 rev/s give an average area of 563 mm^2, the length of the diagram being 80 mm and the scale of the indicator spring 1 mm = 160 kN/m^2. Calculate the indicated power.

Indicated mean effective pressure

$$= \text{mean height of diagram} \times \text{spring scale}$$
$$= \frac{\text{area of diagram}}{\text{length of diagram}} \times \text{spring scale}$$
$$= \frac{563}{80} \times 160 = 1126 \text{ kN/m}^2$$

For a single-acting two-stroke

$$n = \text{rev/s} = 2{\cdot}2$$

For a six-cylinder engine,

$$\text{Indicated power} = p_m ALn \times 6$$
$$= 1126 \times 0{\cdot}7854 \times 0{\cdot}635^2 \times 1{\cdot}01 \times 2{\cdot}2 \times 6$$
$$= 4756 \text{ kW} \quad \text{Ans.}$$

BRAKE POWER AND MECHANICAL EFFICIENCY

Power is absorbed in overcoming frictional resistances at the various rubbing surfaces of the engine, such as at the piston rings, crosshead, crank and shaft bearings, therefore only part of the *indicated power* (ip) developed in the cylinders is transmitted as useful power at the engine shaft. The power absorbed in overcoming friction is termed the *friction power* (fp). The power available at the shaft is termed *shaft power* (sp) or, as this is measured by means of a brake it is also called *brake power* (bp).

$$\text{Brake power} = \text{indicated power} - \text{friction power}$$

The *mechanical efficiency* is the ratio of the brake power to the indicated power:

$$\text{Mechanical efficiency} = \frac{\text{brake power}}{\text{indicated power}}$$

Since the brake power is always less than the indicated power, the above expresses the mechanical efficiency as a fraction less than unity. It is common practice to state the efficiency as a percentage, by multiplying the fraction by 100.

Brake power is measured by applying a resisting torque as a brake on the shaft, the heat generated by the friction at the brake being transferred to and carried away by circulating water.

Let F = resisting force of brake, in newtons, applied at a radius of R metres when the rotational speed is in revolutions per second, then:

$$\text{Work absorbed per revolution [Nm = J]}$$
$$= \text{force [N]} \times \text{circumference [m]}$$
$$= F \times 2\pi R$$

$$\text{Work absorbed per second [J/s]} = \text{power absorbed [W]}$$
$$= F \times 2\pi R \times \text{rev/s}$$

$$F \times R = \text{torque in Nm} = T$$

$$\therefore \text{brake power} = T \times 2\pi \times \text{rev/s}$$

$$2\pi \times \text{rev/s} = \text{angular velocity in radians/second}$$
$$= \omega$$
$$\therefore \text{brake power} = T\omega$$

Common types of brakes for measuring brake power are, for small engines, a loaded rope or steel band around a flywheel on the shaft, and, for large engines, a hydraulic dynamometer.

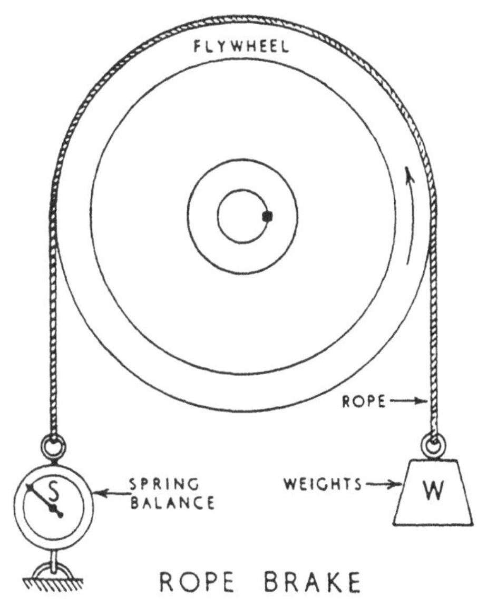

Fig. 23

Fig. 23 illustrates a simple rope brake. The rope passes over the flywheel with one end of the rope anchored to the engine base and the other end hanging freely and loaded against the direction of rotation of the flywheel, the amount of loading depending upon the desired speed.

If W = weight of the load in newtons, and S = reading of spring balance in newtons, then the effective tangential braking force on the flywheel rim is $(W - S)$ newtons. If R = effective radius in metres from centre of shaft to centre of rope, then the braking torque in newton-metres is:

$$T = (W - S) \times R$$

Example. In a single-cylinder four-stroke single-acting gas engine, the cylinder diameter is 180 mm and the stroke 350 mm. When running at 250 rev/min the mean area of the indicator diagrams taken off the engine is 355 mm^2, length of diagram 75 mm, scale of the indicator spring 90 kN/m^2 per mm, and the number of explosions was counted to be 114 per minute. Calculate (i) the indicated power. If the effective radius of the rope brake on the flywheel is 600 mm, load on free end of rope 425 N, and reading of spring balance 72 N, calculate (ii) the brake power, and (iii) the mechanical efficiency.

Indicated mean effective pressure

$$= \text{mean height of diagram} \times \text{spring scale}$$

$$= \frac{\text{area of diagram}}{\text{length of diagram}} \times \text{spring scale}$$

$$= \frac{355}{75} \times 90 = 426 \text{ kN/m}^2$$

$$\text{Indicated power} = p_m ALn$$

$$= 426 \times 0{\cdot}7854 \times 0{\cdot}18^2 \times 0{\cdot}35 \times \frac{114}{60}$$

$$= 7{\cdot}21 \text{ kW} \quad \text{Ans. (i)}$$

$$\text{Braking torque} = (W - S) \times R$$

$$= (425 - 72) \times 0{\cdot}6 = 211{\cdot}8 \text{ Nm}$$

$$\text{Brake power} = T\omega$$

$$= 211{\cdot}8 \times \frac{250 \times 2\pi}{60}$$

$$= 5546 \text{ W} = 5{\cdot}546 \text{ kW} \quad \text{Ans. (ii)}$$

$$\text{Mech. efficiency} = \frac{\text{brake power}}{\text{indicated power}}$$

$$= \frac{5{\cdot}546}{7{\cdot}21}$$

$$= 0{\cdot}7693 \text{ or } 76{\cdot}93\% \quad \text{Ans. (iii)}$$

MORSE TEST

In multi-cylinder internal combustion engines wherein all cylinders are of the same cubic capacity, a reasonable estimate of the indicated power developed in each cylinder can be made by the *morse test*. This is most useful in small high speed engines where indicator diagrams cannot be taken satisfactorily by the standard mechanical indicator.

The test consists of measuring the brake power at the shaft when all cylinders are firing and then measuring the brake power of the remaining cylinders when each one is "cut out" in turn. Cutting out the power of each cylinder is done in petrol engines by shorting the sparking plug, and in diesel engines by by-passing the cylinder fuel supply. The speed of the engine and the petrol throttle or fuel pump setting is kept constant during the test so that friction and pumping losses are approximately constant.

Taking a four-cylinder engine as an example:
With all four cylinders working,

Total bp = total ip – total fp
= sum of ip's of the four cylinders – sum of the fp's of the four cylinders

When the power of one cylinder is cut out,

Total bp = sum of ip's of the three cylinders – sum of the fp's of the four cylinders

Hence, we can see from the above that, when one cylinder is cut out, the loss of *brake* power at the shaft is the loss of the *indicated* power of that cylinder which is not firing.

Example. During a morse test on a four-cylinder four-stroke petrol engine, the throttle was set in a fixed position and the speed maintained constant at 35 rev/s by adjusting the brake, and the following powers in kW were measured at the brake,

With all cylinders working, bp developed = 57
With sparking plug of no. 1 cyl. shorted, bp = 38·5
.. .. .. no. 2 cyl. .. .. bp = 37
.. .. .. no. 3 cyl. .. .. bp = 37·5
.. .. .. no. 4 cyl. .. .. bp = 38

Estimate the indicated power of the engine and the mechanical efficiency.

ip of no. 1 cyl = 57 – 38·5 = 18·5
.. no. 2 cyl = 57 – 37 = 20
.. no. 3 cyl = 57 – 37·5 = 19·5
.. no. 4 cyl = 57 – 38 = 19
total ip = 77·0 kW Ans. (i)

$$\text{Mech. efficiency} = \frac{\text{brake power}}{\text{indicated power}}$$

$$= \frac{57}{77} = 0{\cdot}74 \text{ or } 74\% \quad \text{Ans. (ii)}$$

THERMAL EFFICIENCY

In engine trials it is usually most convenient to base calculations on a running time of one hour. Also, to enable comparisons to be made on the quantity of fuel oil to run the engine under different conditions, or comparisons of one engine with another, the fuel consumption is expressed per unit power developed. The fuel consumed in unit time per unit power developed is termed the *specific fuel consumption* and commonly stated in the units kilogrammes of fuel per kilowatt-hour [kg/kW h].

The thermal efficiency of an engine is the relationship between the quantity of heat energy converted into work and the quantity of heat energy supplied:

$$\text{Thermal efficiency} = \frac{\text{heat energy converted into work}}{\text{heat energy supplied}}$$

In internal combustion engines the heat is supplied directly into the cylinders by the burning of the injected fuel. The heat energy given off during complete combustion of unit mass of the fuel is termed the *calorific value* and may be expressed in kilojoules of heat energy given off during the burning of one kilogramme of fuel [kJ/kg]. However, the calorific value of fuel oil ranges from about 40 000 to 44 000 kJ/kg and is therefore more conveniently expressed in megajoules per kilogramme [MJ/kg], that is, 40 to 44 MJ/kg.

Hence the heat supplied in megajoules is the product of the mass of fuel burned in kilogrammes and its calorific value in megajoules per kilogramme. Therefore, on a basis of one kilowatt-hour:

$$\text{Thermal effic.} = \frac{\text{heat energy equivalent of 1 kW h [MJ/kW h]}}{\text{spec. fuel cons. [kg/kW h]} \times \text{cal. value [MJ/kg]}}$$

The heat energy equivalent of one kilowatt-hour is:

$$\begin{aligned} \text{Energy} &= \text{power} \times \text{time} \\ &= 1000\ [\text{W}] \times 3600\ [\text{s}] \\ &= 3{\cdot}6 \times 10^{6}\ \text{J or } 3{\cdot}6 \times 10^{3}\ \text{kJ or } 3{\cdot}6\ \text{MJ} \end{aligned}$$

Thermal efficiency may be based on the heat energy supplied to develop 1 kW of indicated power in the cylinders, or the heat energy supplied to obtain 1 kW of brake power at the shaft. In the former, the specific fuel consumption (indicated) is expressed as the kilogrammes of fuel per indicated kilowatt-hour [kg/ind. kWh] and the efficiency is the *indicated thermal efficiency*. In the latter, the specific fuel consumption (brake) is expressed as the kilogrammes of fuel per brake kilowatt-hour [kg/brake kWh] and the efficiency is the ***brake thermal efficiency***.

Thus, on the basis of one kilowatt-hour:

$$\text{Indicated thermal effic.} = \frac{3{\cdot}6\ [\text{MJ/kW h}]}{\text{kg fuel/ind. kW h} \times \text{cal. value [MJ/kg]}}$$

$$\text{Brake thermal effic.} = \frac{3{\cdot}6\ [\text{MJ/kW h}]}{\text{kg fuel/brake kW h} \times \text{cal. value [MJ/kg]}}$$

The brake thermal efficiency is also the product of the indicated thermal efficiency and the mechanical efficiency.

The symbol to represent efficiency is η.

Example. The following data were taken during a one-hour trial run on a single-cylinder, single-acting, four-stroke diesel engine of cylinder diameter 175 mm and stroke 225 mm, the speed being constant at 1000 rev/min:

Indicated mean effective pressure = 5·5 bar
Effective diameter of rope brake = 1066 mm
Load on brake = 400 N
Reading of spring balance = 27 N
Fuel consumed = 5·7 kg
Calorific value of fuel = 44·2 MJ/kg

Calculate the indicated power, brake power, specific fuel consumption per indicated kWh and per brake kWh, mechanical efficiency, indicated thermal efficiency and brake thermal efficiency.

$$\text{ip} = p_m ALn$$

$$= 5{\cdot}5 \times 10^2 \times 0{\cdot}7854 \times 0{\cdot}175^2 \times 0{\cdot}225 \times \frac{1000}{60 \times 2}$$

$$= 24{\cdot}8 \text{ kW} \quad \text{Ans. (i)}$$

$$\text{bp} = T\omega$$

$$= (400 - 27) \times \frac{1{\cdot}066}{2} \times \frac{1000 \times 2\pi}{60}$$

$$= 2{\cdot}082 \times 10^4 \text{W} = 20{\cdot}82 \text{ kW} \quad \text{Ans. (ii)}$$

Spec. fuel cons. (indicated)

$$= \frac{5{\cdot}7}{24{\cdot}8} = 0{\cdot}2298 \text{ kg/ind. kW h} \quad \text{Ans. (iii)}$$

Spec. fuel cons. (brake)

$$= \frac{5{\cdot}7}{20{\cdot}82} = 0{\cdot}2738 \text{ kg/brake kWh} \quad \text{Ans. (iv)}$$

$$\text{Mechanical effic.} = \frac{\text{brake power}}{\text{indicated power}}$$

$$= \frac{20{\cdot}82}{24{\cdot}8}$$

$$= 0{\cdot}8395 \text{ or } 83{\cdot}95\% \quad \text{Ans. (v)}$$

$$\text{Ind. thermal effic.} = \frac{3{\cdot}6 \text{ [MJ/kW h]}}{\text{kg fuel/ind. kW h} \times \text{cal. value [MJ/kg]}}$$

$$= \frac{3.6}{0{\cdot}2298 \times 44{\cdot}2}$$

$$= 0{\cdot}3544 \text{ or } 35{\cdot}44\% \quad \text{Ans. (vi)}$$

$$\text{Brake. therm. effic.} = \frac{3{\cdot}6 \text{ [MJ/kW h]}}{\text{kg fuel/brake kW h} \times \text{cal. value [MJ/kg]}}$$

$$= \frac{3.6}{0{\cdot}2738 \times 44{\cdot}2}$$

$$= 0{\cdot}2975 \text{ or } 29{\cdot}75\% \quad \text{Ans. (vii)}$$

Alternatively,

$$\text{Brake therm. effic.} = \text{indicated thermal effic.} \times \text{mech. effic.}$$

$$= 0{\cdot}3544 \times 0{\cdot}8395$$

$$= 0{\cdot}2975 \text{ (as above)}$$

HEAT BALANCE

Of the total heat energy supplied to an engine, only a small proportion is converted into useful work. The heaviest losses are those due to the heat energy transferred to and carried away by the cooling water, and the heat energy remaining in the gases which are released from the cylinders and exhausted up the flue. A clear picture of the distribution of heat is shown by constructing a heat balance chart, based on taking the heat supplied in the fuel as 100%.

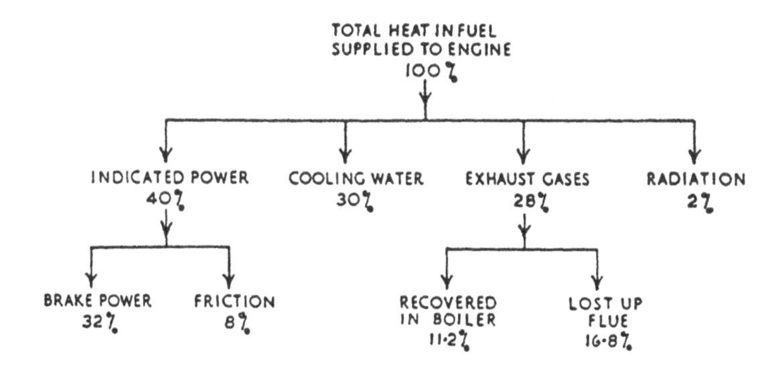

Fig. 24

Fig. 24 illustrates an example of a heat balance of a four-stroke diesel engine. In this particular case the exhaust gases were passed through the tubes of a waste-heat steam boiler before escaping up the funnel, and 40% of the heat energy in the exhaust gases was usefully recovered in generating steam. The radiation loss is usually very small and in most cases it is included with the other losses. Exhaust gas driven turbo-showers further improve efficiencies.

CLEARANCE AND STROKE VOLUME

Mechanical clearance is necessary between the inner face of the cylinder cover and the top of the piston when the piston is at the top of its stroke to avoid contact, and this is measured as the minimum distance between those two parts.

The *clearance volume* is the volume of the enclosed space above the piston when at the top of its stroke, including all cavities up to the valve faces when the valves are closed. In internal combustion engines the clearance volume is the combustion space to accommodate sufficient air for the complete combustion of the fuel and to limit the rise of temperature during burning. It is designed as near a spherical space as practicable, in many cases by concave piston tops and concave cylinder covers.

The *stroke volume*, sometimes termed the *swept volume*, is the volume swept out by the piston as it moves through one complete stroke. It is, therefore, equal to the product of the cross-sectional area of the cylinder and the length of the stroke.

Since the ratio of compression is the ratio of the volume at the beginning of compression to the volume at the end of compression,

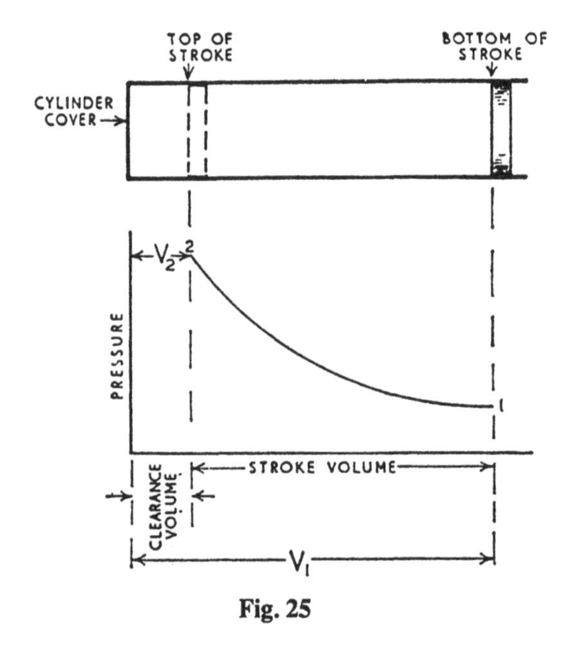

Fig. 25

then, referring to Fig. 25,

$$r = \frac{\text{initial volume}}{\text{final volume}}$$

$$= \frac{V_1}{V_2} = \frac{\text{clearance vol.} + \text{stroke vol.}}{\text{clearance volume}}$$

Therefore, the magnitude of the clearance volume affects the ratio of compression (and expansion) and can be adjusted by shims under the foot of the connecting rod or plates in the clearance space.

Dividing stroke-volume by the cross-sectional area of the cylinder gives the length of the stroke. Similarly, dividing clearance-volume by the cross-sectional area of the cylinder gives the clearance in terms of length. Hence, for convenience, if the stroke is expressed in millimetres, the clearance may be expressed as a length in millimetres. Alternatively, the clearance may be expressed as "a fraction of the stroke" or "a percentage of the stroke", for instance, if the stroke is 200 mm and the clearance length is 20 mm, the clearance could be expressed as "one-tenth of the stroke", or, "10% of the stroke".

Example. The stroke of an engine is 450 mm. The pressure at the beginning of compression is 1·01 bar and at the end of compression it is 11·1 bar. Assuming compression follows the law $pV^{1\cdot36}$ = a constant, calculate the clearance between the piston and cylinder cover at the end of compression in mm of length.

$$\text{Let clearance} = c$$

$$V_1 = \text{stroke} + \text{clearance} = 450 + c$$

$$V_2 = \text{clearance} = c$$

$$p_1 V_1^{1\cdot36} = p_2 V_2^{1\cdot36}$$

$$\left\{\frac{V_1}{V_2}\right\}^{1\cdot36} = \frac{p_2}{p_1}$$

$$\frac{V_1}{V_2} = \sqrt[1\cdot36]{\frac{p_2}{p_1}}$$

$$\frac{450 + c}{c} = \sqrt[1\cdot36]{\frac{11\cdot1}{1\cdot01}}$$

$$\frac{450 + c}{c} = 5\cdot826$$

$$450 + c = 5\cdot826c$$

$$450 = 4\cdot826c$$

$$c = 93\cdot26 \text{ mm} \quad \text{Ans.}$$

TEST EXAMPLES 7

1. The area of an indicator diagram taken off a four-cylinder, single-acting, four-stroke, internal combustion engine when running at 5·5 rev/s is 390 mm^2, the length is 70 mm, and the scale of the indicator spring is 1 mm = 1·6 bar. The diameter of the cylinders is 150 mm and the stroke is 200 mm. Calculate the indicated power of the engine assuming all cylinders develop equal power.

2. The cylinder diameters of an eight-cylinder, single-acting, four-stroke diesel engine are 750 mm and the stroke is 1125 mm. The indicated mean effective pressure in the cylinders is 1172 kN/m^2 when the engine is running at 110 rev/min. Calculate the indicated power and the brake power if the mechanical efficiency is 86%.

3. Calculate the cylinder diameters and stroke of a six-cylinder, single-acting, two-stroke diesel engine to develop a brake power of 2250 kW at a rotational speed of 2 rev/s when the indicated mean effective pressure in each cylinder is 10 bar. Assume a mechanical efficiency of 84% and the length of the stroke 25% greater than the diameter of the cylinders.

4. The flywheel of a rope brake is 1·22 m diameter and the rope is 24 mm diameter. When the engine is running at 250 rev/min the load on the brake is 480 N on one end of the rope and 84 N on the other end. Calculate the brake power. If the rise in temperature of the brake cooling water is 18 K, calculate the quantity of water flowing through the brake in litres per hour assuming that the water carries away 90% of the heat generated at the brake. Take the specific heat of the water as 4·2 kJ/kg K.

5. During a Morse test on a four-cylinder petrol engine, the speed was kept constant at 24·5 rev/s by adjusting the brake and the following readings taken:

With all cylinders firing, torque at brake	= 193·8 Nm
.. no. 1 cyl. cut out	= 130·8 ..
.. no. 2 cyl. cut out	= 130·2 ..
.. no. 3 cyl. cut out	= 129·9 ..
.. no. 4 cyl. cut out	= 131·1 ..

Calculate the bp, ip, and mechanical efficiency.

6. When developing a certain power, the specific fuel consumption of an internal combustion engine is 0·255 kg/kW h (brake) and the mechanical efficiency is 86%. Calculate (i) the indicated ther-

mal efficiency, and (ii) the brake thermal efficiency, taking the calorific value of the fuel as 43·5 MJ/kg. If 35 kg of air are supplied per kg of fuel, the air inlet being at 26°C and exhaust at 393°C, find (iii) the heat energy carried away in the exhaust gases as a percentage of the heat supplied, taking the specific heat of the gases as 1·005 kJ/kg K.

7. A diesel engine uses 27 tonne of fuel per day when developing 4960 kW indicated power and 4060 kW brake power. Of the total heat supplied to the engine, 31·7% is carried away by the cooling water and radiation, and 30·8% in the exhaust gases. Calculate the indicated thermal efficiency, mechanical efficiency, overall efficiency, specific fuel consumption (indicated), and the calorific value of the fuel.

8. The stroke of a petrol engine is 87·5 mm and the clearance is equal to 12·5 mm. A compression plate is now fitted which has the effect of reducing the clearance to 10 mm. Assuming the compression period to be the whole stroke, the pressure at the beginning of compression as 0·97 bar, and the law of compression $pV^{1\cdot35} = C$, calculate the pressure at the end of compression before and after the compression plate is fitted.

f 9. The mean effective pressure measured from the indicator diagram taken off a single-cylinder four-stroke gas engine was 3·93 bar when running at 5 rev/s and developing a brake power of 4·33 kW. The number of explosions per minute was 123 and the gas consumption 3·1 cubic metres per hour. The diameter of the cylinder is 180 mm, stroke 300 mm, and calorific value of the gas 17·6 MJ per cubic metre. Calculate the indicated power, mechanical efficiency, indicated thermal efficiency and brake thermal efficiency.

f 10. A two-stroke cycle compression-ignition engine has a stroke of 1·5 m, mean piston speed 6 m/s and brake mean effective pressure of 7 bar.

A similar engine with the same stroke/bore rate of 2:1 is to develop 370 kW. Both engines have the same power to swept volume ratio and operate at the same speed.

For the engine which is to develop 370 kW, calculate:

(a) the cylinder bore;

(b) brake mean effective pressure.

CHAPTER 8

IDEAL CYCLES

Theoretical cycles are reversible with isentropic (frictionless adiabatic) – sometimes isothermal – processes and can be considered for gas and vapour power cycles. The efficiency of such cycles is called the ideal cycle (thermal) efficiency. By introducing process efficiencies, e.g. utilising polytropics or modifying isentropic to actual compression/expansion utilising isentropic efficiencies, it is possible to estimate the actual cycle efficiencies. The ratio of actual to ideal cycle efficiency is called the efficiency ratio.

Consider now gas power cycles which can be classified into two main groups – IC reciprocating engines (non-flow) and gas turbines (steady-flow). Performance can be assessed using air as the working fluid, i.e. air standard cycles and air standard efficiency.

In any cycle:

Heat converted into work = heat supplied – heat rejected

and,

$$\text{Thermal efficiency} = \frac{\text{heat converted into work}}{\text{heat supplied}}$$

$$= \frac{\text{heat supplied} - \text{heat rejected}}{\text{heat supplied}}$$

$$= 1 - \frac{\text{heat rejected}}{\text{heat supplied}}$$

(Note that the heat converted into work, i.e. work done, is represented by the area of the pV diagram and m.e.p. can be calculated by dividing by the length equivalent – using correct units i.e. kN/m^2 and m^3).

CONSTANT VOLUME CYCLE

This is also known as the Otto cycle and is the basis on which petrol, paraffin, and gas engines usually work.

Designating in sequence, the four cardinal state points of the cycle as 1,2 3 and 4 respectively (Fig. 26), the cycle of operations commence with a volume of air V_1 at pressure p_1 and temperature T_1. The piston moves inward and the air is compressed adiabatically to a volume V_2 and the pressure and temperature rise to p_2 and T_2.

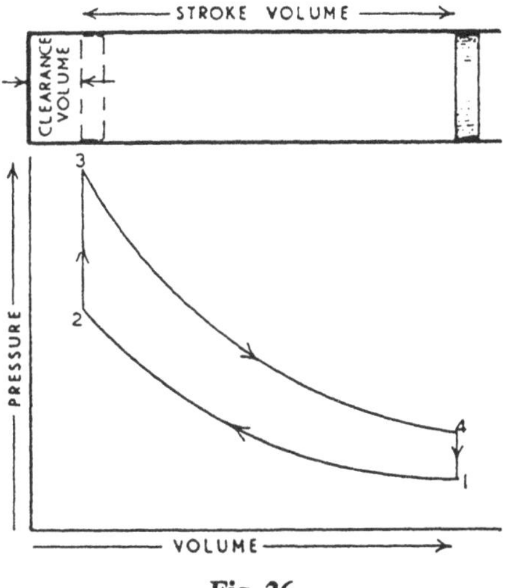

Fig. 26

Heat energy is now given from some outside source and it is assumed that the air receives this heat instantaneously so the there is no time for any change of volume to occur. The pressure and temperature consequently rise to p_3 and T_3 while the volume remains unchanged and, therefore, V_3 is equal to V_2. Adiabatic expansion of the air now takes place while the piston is pushed outward on its power stroke, the volume increasing to V_4 which is the same as the initial volume V_1 and the pressure and temperature during expansion falling to p_4 and T_4. Finally, the cycle is completed by the air rejecting heat (theoretically instantaneously) at constant volume, to an outside source, which causes the pressure and temperature to fall to their initial values of p_1 and T_1.

It should be noted that, since the compression and expansion of the air is adiabatic, then there is no exchange of heat during these operations. This means that all the heat supplied takes place at constant volume between the state points 2 and 3, and all the heat rejected takes place at constant volume between the state points 4 and 1.

Heat supplied or rejected = mass × spec. ht. × temp. change hence,

$$\text{heat supplied} = m \times c_V \times (T_3 - T_2)$$

$$\text{heat rejected} = m \times c_V \times (T_4 - T_1)$$

therefore,

$$\text{Ideal thermal efficiency} = 1 - \frac{\text{heat rejected}}{\text{heat supplied}}$$

$$= 1 - \frac{mc_V(T_4 - T_1)}{mc_V(T_3 - T_2)}$$

$$= 1 - \frac{T_4 - T_1}{T_3 - T_2}$$

In this case the ratios of compression and expansion are the same because:

$$V_1 = V_4 \text{ and } V_2 = V_3$$

$$\text{Also, since } \frac{T_2}{T_1} = \left\{\frac{V_1}{V_2}\right\}^{\gamma-1} = r^{\gamma-1}$$

$$\text{and, } \frac{T_3}{T_4} = \left\{\frac{V_4}{V_3}\right\}^{\gamma-1} = r^{\gamma-1}$$

$$\text{then, } \frac{T_2}{T_1} = \frac{T_3}{T_4} = r^{\gamma-1}$$

$$\text{hence, } T_3 = T_4 r^{\gamma-1} \quad \text{and} \quad T_2 = T_1 r^{\gamma-1}$$

$$\text{therefore, } T_3 - T_2 = r^{\gamma-1}(T_4 - T_1)$$

Substituting this value of $(T_3 - T_2)$ into the general expression for the ideal thermal efficiency, $(T_4 - T_1)$ cancels, leaving:

$$\text{Ideal thermal efficiency} = 1 - \frac{1}{r^{\gamma-1}}$$

If γ is taken as 1·4 (for air) then this is also the Air Standard Efficiency.

$$\text{also, since } r^{\gamma-1} = \frac{T_2}{T_1} = \frac{T_3}{T_4}$$

$$\text{then, Ideal thermal efficiency} = 1 - \frac{T_1}{T_2}$$

$$= 1 - \frac{T_4}{T_3}$$

On examination of the previous expression it will be seen that the greater the value of r, the greater will be the efficiency, hence the trend for higher ratios of compression in modern petrol engines. The majority of petrol engines, however, take in a mixture of petrol vapour and air during the induction stroke and this is compressed during the compression stroke. Being an explosive mixture it will burst into flame without the assistance of an electric spark or other means if it reaches it temperature of spontaneous ignition. Therefore, if the ratio of compression is too high for the grade of petrol used, preignition can take place.

Fig. 27 shows the relationship between the ideal thermal efficiency and the ratio of compression in a constant volume cycle.

From the graph we see that, although the efficiency increases with higher compression ratios, the rate of increase in efficiency becomes less as the compression ratio is increased, and there is no appreciable gain by increasing the compression ratio above about 16 to 1 (practical limitations to about 10 to 1)

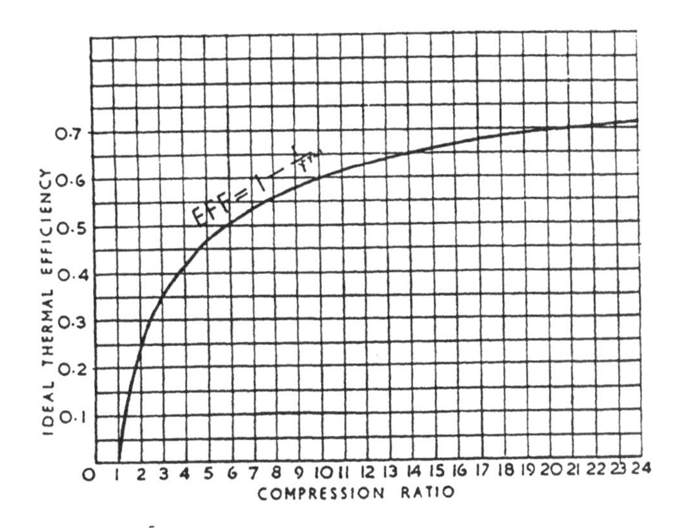

Fig. 27

Example. The compression ratio of an engine working on the constant volume cycle is 9·3 to 1. At the beginning of compression the temperature is 31°C and at the end of combustion the temperature is 1205°C. Taking compression and expansion to be adiabatic and the value of γ as 1·4, calculate (i) the temperature at the end of compression, (ii) the temperature at the end of expansion, (iii) the ideal thermal efficiency.

Referring to Fig. 26

$$V_1 = 9{\cdot}3 \text{ and } V_4 = 9{\cdot}3$$
$$V_2 = 1 \text{ and } V_3 = 1$$
$$T_1 = 304 \text{ K}$$
$$T_3 = 1478 \text{ K}$$

COMPRESSION PERIOD :

$$\frac{T_2}{T_1} = \left\{\frac{V_1}{V_2}\right\}^{\gamma-1}$$

$$\therefore T_2 = 304 \times 9{\cdot}3^{0{\cdot}4} = 741{\cdot}8 \text{ K}$$

$\therefore$ temperature at end of compression

$$= 468{\cdot}8^\circ\text{C} \quad \text{Ans. (i)}$$

EXPANSION PERIOD :

$$\frac{T_3}{T_4} = \left\{\frac{V_4}{V_3}\right\}^{\gamma-1}$$

from which T_4 can be calculated, but, as ratios of compression and expansion are the same, and follow the same law, then an easier method is:

$$\frac{T_2}{T_1} = \frac{T_3}{T_4}$$

$$T_4 = \frac{304 \times 1478}{741{\cdot}8} = 605{\cdot}6 \text{ K}$$

$\therefore$ temperature at end of expansion

$$= 605{\cdot}6 - 273 = 332{\cdot}6^\circ\text{C} \quad \text{Ans. (ii)}$$

The ideal thermal efficiency can now be calculated from any of the expressions given above, thus,

$$\text{Ideal thermal efficiency} = 1 - \frac{T_4 - T_1}{T_3 - T_2} \quad \text{or} \quad 1 - \frac{1}{r^{\gamma-1}}$$

$$\text{or} \quad 1 - \frac{T_1}{T_2} \quad \text{or} \quad 1 - \frac{T_4}{T_3}$$

Taking the last expression,

$$\text{Ideal thermal efficiency} = 1 - \frac{605{\cdot}6}{1478} = 1 - 0{\cdot}4099$$

$$= 0{\cdot}5901 \text{ or } 59{\cdot}01\% \quad \text{Ans. (iii)}$$

f DIESEL CYCLE

The term *constant pressure cycle* refers to one wherein the pressure remains constant during the two periods when heat energy is supplied and rejected. In the diesel cycle however, heat is supplied at constant pressure but rejection of heat takes place at constant volume. Thus, the diesel cycle, that upon which slow-speed diesel engines operate, is usually referred to as the *modified constant pressure cycle.*

Referring to Fig. 28, the cycle of operations commence with a volume of air V_1 at a pressure p_1 and temperature T_1. The air is compressed adiabatically to a volume V_2 and the pressure and temperature rise to p_2 and T_2. The piston is now at the top (inward

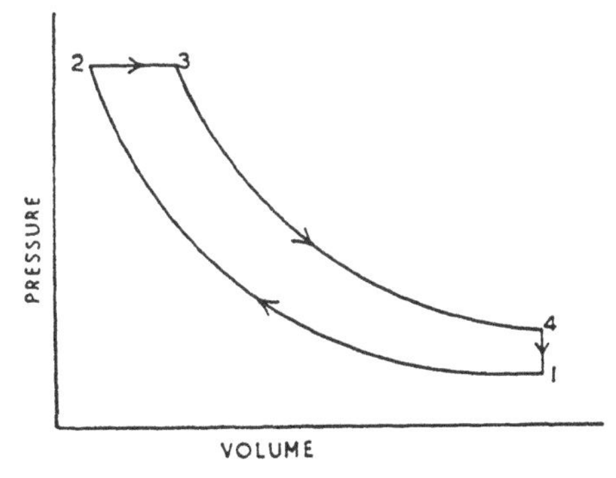

Fig. 28

end) of its stroke and heat is supplied at such a rate to maintain the pressure constant as the piston moves down the cylinder for a fraction of the power stroke. At the end of the heat supply period the volume is V_3, the temperature has been further increased to T_3, and the pressure represented by p_3 is the same as p_2. The air now expands adiabatically for the remainder of the power stroke until the final volume V_4 is the same as the initial volume V_1, the pressure and temperature falling during expansion to p_4 and T_4. Finally, the cycle is completed by the rejection of heat at constant volume to the initial conditions.

$$\text{Ideal thermal efficiency} = 1 - \frac{\text{heat rejected}}{\text{heat supplied}}$$

$$= 1 - \frac{mc_V (T_4 - T_1)}{mc_P (T_3 - T_2)}$$

$$= 1 - \frac{1}{\gamma}\left\{\frac{T_4 - T_1}{T_3 - T_2}\right\}$$

Example. The compression ratio in a diesel engine is 13 to 1 and the ratio of expansion is 6·5 to 1. At the beginning of compression the temperature is 32°C. Assuming adiabatic compression and expansion, calculate the temperatures at the three remaining cardinal points of the cycle, and the ideal thermal efficiency, taking the specific heats at constant pressure and constant volume as 1·005 and 0·718 kJ/kg K respectively.

Referring to Fig. 28,

$$V_1 = 13 \qquad V_4 = 13 \qquad V_2 = 1$$

$$\frac{V_4}{V_3} = \text{ratio of expansion} = 6{\cdot}5$$

$$\therefore\ V_3 = \frac{V_4}{6{\cdot}5} = \frac{13}{6{\cdot}5} = 2$$

$$T_1 = 305\text{ K}$$

$$\gamma = \frac{c_P}{c_V} = \frac{1{\cdot}005}{0{\cdot}718} = 1{\cdot}4$$

FIRST STAGE, ADIABATIC COMPRESSION:

$$\frac{T_2}{T_1} = \left\{\frac{V_1}{V_2}\right\}^{\gamma-1}$$

$$\therefore T_2 = 305 \times 13^{0{\cdot}4} = 850{\cdot}9 \text{ K}$$

$\therefore$ temperature at end of compression

$= 577{\cdot}9°\text{C}$ Ans. (i)

SECOND STAGE, HEATING AT CONSTANT PRESSURE:

$$\frac{T_3}{T_2} = \frac{V_3}{V_2} \quad \text{(Charles' law)}$$

$$T_3 = 850{\cdot}9 \times 2 = 1701{\cdot}8 \text{ K}$$

$\therefore$ temperature at end of combustion

$= 1428{\cdot}8°\text{C}$ Ans. (ii)

THIRD STAGE, ADIABATIC EXPANSION :

$$\frac{T_4}{T_3} = \left\{\frac{V_3}{V_4}\right\}^{\gamma-1}$$

$$T_4 = 1701{\cdot}8 \times \left\{\frac{2}{13}\right\}^{0{\cdot}4}$$

$$= \frac{1701{\cdot}8}{6{\cdot}5^{0{\cdot}4}} = 804{\cdot}8 \text{ K}$$

$\therefore$ temperature at end of expansion

$= 531{\cdot}8°\text{C}$ Ans. (iii)

Ideal thermal efficiency

$$= 1 - \frac{\text{heat rejected}}{\text{heat supplied}}$$

$$= 1 - \frac{1}{\gamma}\left\{\frac{T_4 - T_1}{T_3 - T_2}\right\}$$

$$= 1 - \frac{1}{1{\cdot}4}\left\{\frac{804{\cdot}8 - 305}{1701{\cdot}8 - 850{\cdot}9}\right\}$$

$$= 1 - \frac{1}{1{\cdot}4} \times \frac{499{\cdot}8}{850{\cdot}9}$$

$$= 1 - 0{\cdot}4196$$

$$= 0{\cdot}5804 \text{ or } 58{\cdot}045\% \quad \text{Ans. (iv)}$$

In the ideal diesel cycle, where the compression and expansion are both adiabatic, the ideal thermal efficiency can be expressed in terms of the ratio of compression and a comparison can then be made with the efficiency of a constant volume cycle of the same ratio of compression.

Expressing all temperatures in terms of T_1, substituting and simplifying:

$$\frac{T_2}{T_1} = \left\{\frac{V_1}{V_2}\right\}^{\gamma-1}$$

$$\therefore T_2 = T_1 r^{\gamma-1}$$

$$\frac{T_3}{T_2} = \frac{V_3}{V_2}$$ let this ratio of burning period volumes be represented by ρ, then,

$$\frac{T_3}{T_2} = \rho$$

$$\therefore T_3 = T_2\rho = T_1 r^{\gamma-1}\rho$$

$$\frac{T_4}{T_3} = \left\{\frac{V_3}{V_4}\right\}^{\gamma-1}$$

Since $\frac{V_3}{V_2} = \rho$ and $\frac{V_4}{V_2} = r$, then $\frac{V_3}{V_4} = \frac{\rho}{r}$

$$\therefore \frac{T_4}{T_3} = \left\{\frac{\rho}{r}\right\}^{\gamma-1}$$

$$T_4 = T_3 \times \left\{\frac{\rho}{r}\right\}^{\gamma-1}$$

$$= T_1 r^{\gamma-1}\rho \times \left\{\frac{\rho}{r}\right\}^{\gamma-1}$$

$$= T_1\rho^{\gamma}$$

Ideal thermal efficiency

$$= 1 - \frac{1}{\gamma}\left\{\frac{T_4 - T_1}{T_3 - T_2}\right\}$$

$$= 1 - \frac{1}{\gamma}\left\{\frac{T_1\rho^{\gamma} - T_1}{T_1 r^{\gamma-1}\rho - T_1 r^{\gamma-1}}\right\}$$

$$= 1 - \frac{1}{\gamma} \times \frac{1}{r^{\gamma-1}}\left\{\frac{\rho^{\gamma} - 1}{\rho - 1}\right\}$$

If $\gamma = 1{\cdot}4$, this is also the Air Standard Efficiency.

Comparing this expression with the ideal thermal efficiency of the constant volume cycle in terms of r, it will be seen that, for the same ratio of compression, the constant volume cycle has the higher thermal efficiency. This does not mean, however, that a petrol engine working on the constant volume cycle is more efficient than a diesel engine working on the modified constant pressure cycle, because, in the former, an explosive mixture is compressed and there is a limit to the ratio of compression, whereas air only is compressed in a diesel engine and the ratio of compression can be as high as required.

DUAL COMBUSTION CYCLE

In most high-speed compression-ignition engines, combustion takes place partly at constant volume and partly at constant pressure and therefore the cycle is referred to as *dual-combustion* (or mixed).

Fig. 29 shows the ideal dual-combustion cycle. Commencing with a volume of air, V_1 at pressure p_1 and temperature T_1, the air is compressed adiabatically to a volume V_2 and the pressure and temperature rise to p_2 and T_2. Heat energy is now supplied at constant volume, the pressure and temperature are increased to p_3 and T_3 while the volume remains unchanged so that V_3 is equal to V_2. The supply of heat energy is continued at such a rate as to maintain the pressure constant while the piston moves outward until the volume is V_4, the temperature is further increased to T_4 and the pressure p_4 remains the same as p_3. Now adiabatic expansion takes place until the volume V_5 is the same as the initial volume V_1, the pressure and temperature falling due to expansion to p_5 and T_5. Finally, heat is rejected at constant volume and the pressure and temperature fall to the initial conditions of p_1 and T_1.

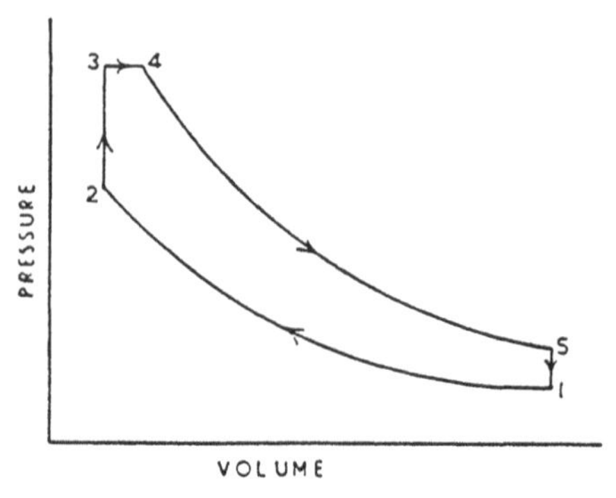

Fig. 29

Ideal thermal efficiency

$$= 1 - \frac{\text{heat rejected}}{\text{heat supplied}}$$

$$= 1 - \frac{mc_V\ (T_5 - T_1)}{mc_V\ (T_3 - T_2) + mc_P\ (T_4 - T_3)}$$

$$= 1 - \frac{(T_5 - T_1)}{(T_3 - T_2) + \gamma (T_4 - T_3)}$$

The above expression can be converted into terms of the ratio of compression in a similar manner as the previous cycles, thus,

If r = ratio of compression = V_1/V_2
γ = ratio of specific heats = c_P/c_V
ρ = ratio of burning period, or "cut-off" ratio = V_4/V_3
α = ratio of pressure increase at constant volume = p_3/p_2

Ideal thermal efficiency

$$= 1 - \frac{1}{r^{\gamma-1}} \left\{ \frac{\alpha\rho^{\gamma} - 1}{(\alpha - 1) + \gamma\alpha(\rho - 1)} \right\}$$

Note (i) if $\alpha = 1$, the above becomes a pure diesel cycle.
(ii) if $\rho = 1$, it becomes a pure constant-volume cycle.

Since compression-ignition oil engines depend upon the temperature of the air at the end of compression to ignite the fuel injected into the cylinder, the compression-ratio must be fairly high, usually not less than about 12 to give the necessary temperature rise during compression.

The higher compression pressures developed in this type of engine limit the use of constant volume combustion, since the maximum pressure in the cycle is limited by the consideration of strength.

As the maximum pressure is limited, increasing the compression ratio reduces the amount of fuel burned at constant volume, so that more must be burned at constant pressure and thus the gain due to increased compression ratio is partly nullified.

f CARNOT CYCLE

This is a purely theoretical cycle devised by the French scientist Sadi Carnot. Although it is not possible from practical considerations for an engine to work on this cycle, it has a higher theoretical thermal efficiency than any other working between the same temperature limits and, therefore, provides a useful standard for comparing the performance of other heat engines.

Referring to the pV diagram, Fig. 30, it is usual to explain this cycle by commencing at state point A. Gas has been previously compressed in the cylinder by the piston moving inward and, at A, the piston is at the "top" of its stroke. The pressure and temperature are high, the value of the latter being represented by T_1. As the piston is pushed outward, doing work, heat is supplied to the gas from an external hot source at such a rate as to maintain its temperature

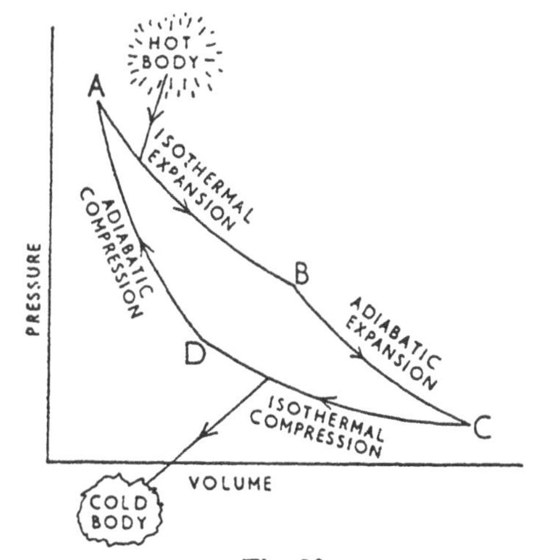

Fig. 30

constant, and during this period the gas therefore expands isothermally until point B is reached. At this point the heat supply is cut off and no heat is given to or rejected from the gas as the piston moves on to the end of the stroke at C. Hence during this period, the gas expands adiabatically as it does work and therefore the temperature falls. The temperature of the gas at point C is represented by T_2. The piston now moves inward to compress the gas from C to D and during this period it is assumed that any generated heat due to compression can flow out of the gas into a cold "sink". That is, the gas rejects heat energy to a cold external source at such a rate to maintain the temperature constant at T_2. This is isothermal compression. At point D, the flow of heat out of the gas is stopped and from D to A the gas is compressed adiabatically while the piston completes its stroke, the temperature of the gas rising to the initial temperature T_1.

Hence, the four stages of the Carnot cycle are briefly as follows:

A to B Isothermal expansion of the gas during which the amount of heat supplied is equal to the work done. Letting r = ratio of isothermal expansion.
Heat supplied $= p_A V_A \ln r = mRT_1 \ln r$

B to C Adiabatic expansion of the gas during which no heat is supplied or rejected.

C to D Isothermal compression. During this period heat is rejected from the gas, the quantity of heat being the equivalent of the work done on the gas and, since the ratio of isothermal compression must be the same as the ratio of isothermal expansion to form a closed cycle, then,
Heat rejected $= p_c V_c \ln r = mRT_2 \ln r$

D to A Adiabatic compression during which no heat is supplied or rejected.

Therefore,

$$\text{Ideal thermal efficiency} = \frac{\text{heat supplied} - \text{heat rejected}}{\text{heat supplied}}$$

$$= 1 - \frac{\text{heat rejected}}{\text{heat supplied}}$$

$$= 1 - \frac{mRT_2 \ln r}{mRT_1 \ln r}$$

$$= 1 - \frac{T_2}{T_1}$$

$$= \frac{T_1 - T_2}{T_1}$$

This expression for the Carnot Efficiency shows that, to obtain the highest efficiency, heat should be taken in at the highest possible temperature (T_1) and rejected at the lowest possible temperature (T_2). This conclusion is applicable in the design of any heat engine.

f REVERSED CARNOT CYCLE

The Carnot cycle is theoretically reversible and if applied in reverse manner would act as a refrigerator by taking heat from a cold region and maintaining it at a low temperature as follows:

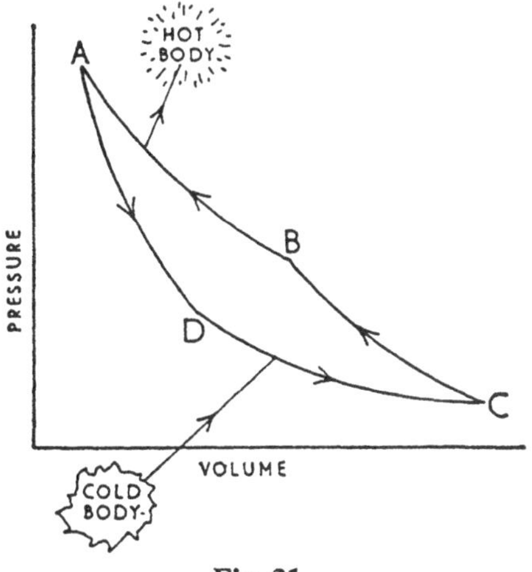

Fig. 31

Referring to Fig. 31 and commencing at state point A, the four stages of the reversed Carnot cycle consist of:

(i) Work done by the gas while it expands adiabatically from A to D and the temperature falls from T_1 to T_2. No heat is given to or taken from the gas during this process.

(ii) Further work done by the gas as it expands isothermally from D to C, a quantity of heat is taken in by the gas (from the cold body) equal to the work done, to maintain the temperature constant at T_2.

(iii) Adiabatic compression of the gas from C to B, no heat being given to or taken from the gas, therefore the temperature increases from T_2 to T_1.

(iv) Isothermal compression from B to A during which heat is rejected from the gas (to the hot body) to maintain the temperature constant at T_1.

Thus an engine working on the reversed Carnot cycle would require to be driven and, as heat would be continually taken from a cold region and sent out to a hotter region, it would therefore act as a refrigerating machine. The measure of the "efficiency" of refrigeration is known as the *coefficient of performance*, and its theoretical value is:

$$\frac{\text{Quantity of heat extracted}}{\text{Heat equivalent of work done to extract the heat}}$$

$$= \frac{mRT_2 \ln r}{mRT_1 \ln r - mRT_2 \ln r}$$

$$= \frac{T_2}{T_1 - T_2}$$

f OTHER IDEAL CYCLES

When considering vapour power cycles, for steam turbines, the **Rankine** cycle is used – see Chapter 12.

With gas power cycles, for gas turbines, the **Joule (Brayton)** cycle is used – also see Chapter 12

Operation on other cycles includes Stirling (constant volumes and isothermals), Ericsson (constant pressures and isothermals), Atkinson (modified constant volume – see Test Example 7) and theoretical variations (see Test Example 8)

f MEAN EFFECTIVE PRESSURE

The ideal cycles can be analysed to calculate the indicated

m.e.p. from the diagram. Consider, as example, the Diesel cycle and refer to Fig. 28.

$$\text{m.e.p.} = \frac{p_2(V_3 - V_2) + \dfrac{p_3V_3 - p_4V_4}{\gamma - 1} - \dfrac{p_2V_2 - p_1V_1}{\gamma - 1}}{V_1 - V_2}$$

(The m.e.p. will be in kN/m^2 if the work done i.e. the area of the diagram is in kJ i.e. $kN/m^2 \times m^3$ and the "length" of the diagram in m^3. Pressures can be in bars, and volumes represented by stroke lengths or ratios giving the m.e.p. in bars – the work done is not then, of course, in consistent units).

f NON-IDEAL CYCLES

The ideal cycle is sometimes used as the model for calculation of properties at cardinal points, evaluation of m.e.p., etc. utilising a diagram and polytropic processes (see Test Examples 9 and 10).

TEST EXAMPLES 8

1. The compression ratio of a petrol engine working on the constant volume cycle is 8·5. The pressure and temperature at the beginning of compression are 1 bar and 40°C and the maximum pressure of the cycle is 31 bar. Taking compression for the air-petrol gas mixture to follow the adiabatic law $pV^{1\cdot35} = C$, calculate (i) the pressure at the end of compression, (ii) temperature at end of compression, (ii) temperature at end of combustion.

2. In an ideal constant volume cycle the temperature at the beginning of compression is 50°C. The volumetric compression ratio is 5:1. If the heat supplied during the cycle is 930 kJ/kg of working fluid, calculate:

   (a) the maximum temperature attained in the cycle,
   (b) work done during the cycle/kg of working fluid, and
   (c) the ideal thermal efficiency of the cycle.
   Take $\gamma = 1{\cdot}4$ and $c_V = 0{\cdot}717$ kJ/kg K.

f 3. In an air-standard Otto cycle the pressure and temperature of the air at the start of compression are 1 bar and 330 K respectively.

The compression ratio is 8:1 and the energy added at constant volume is 1250 kJ/kg.

(a) Calculate:
   (i) the maximum temperature in the cycle;
   (ii) the maximum pressure in the cycle.

(b) Draw the cycle:
   (i) on a pressure-volume diagram;
   (ii) on a temperature-entropy diagram.

For air: $c_P = 1005$ *J/kg K.* $c_V = 718$ *J/kg K.*

f 4. In an air-standard Diesel cycle the pressure and temperature of the air at the start of compression are 1 bar and 330 K respectively. The compression ratio is 16:1 and the energy added at constant pressure is 1250 kJ/kg.

Calculate:
(a) the maximum pressure in the cycle;
(b) the maximum temperature in the cycle;
(c) the cycle efficiency;
(d) the mean effective pressure.

For air: $c_P = 1005$ *J/kg K.* $c_V = 718$ *J/kg K*

f 5. The compression ratio of an engine working on the dual-combustion cycle is 10·7. The pressure and temperature of the air at the beginning of compression is 1 bar and 32°C. The maximum pressure and temperature during the cycle is 41 bar and 1593°C. Assuming adiabatic compression and expansion, calculate the pressures and temperatures at the remaining cardinal points of the cycle and the ideal thermal efficiency. Take the values, $c_V = 0{\cdot}718$, and $c_P = 1{\cdot}005$ kJ/kg K.

f 6. A heat engine is to be operated, using the Carnot cycle, with maximum and minimum temperatures of 1027°C and 27°C respectively.

(a) Calculate the efficiency of the cycle;
(b) The working fluid within the cycle is to be steam;
  (i) state the difficulty associated with the practical operation of the cycle;
  (ii) state the modification required to the cycle to allow practical operation.

f 7. Gas initially at a pressure of 1 bar and temperature 60°C undergoes the following cycle.

(a) Adiabatic compression through a compression ratio of 4·5:1.
(b) Heating at constant volume through a pressure ratio of 1·35:1.
(c) Adiabatic expansion to initial pressure.
(d) Constant pressure cooling to initial volume.

If $c_P = 1000$ J/kg K and $c_V = 678$ J/kg K for the gas determine the thermal efficiency of the cycle.

f 8. One kilogram of air at a pressure and temperature of 1 bar and 15°C initially, undergoes the following processes in a cycle: 1. Isothermal compression to 2 bar. 2. Polytropic compression from 2 bar to 4 bar. 3. Isentropic expansion from 4 bar to initial condition.

Sketch the *p-V* diagram of the cycle and calculate for each process (a) the work transfer (b) the heat transfer.

$R = 287$ J/kg K and $\gamma = 1{\cdot}4$ for air.

f 9. The compression ratio of a diesel engine is 15 to 1. Fuel is admitted for one-tenth of the power stroke and combustion takes place at constant pressure. Exhaust commences when the piston has travelled nine-tenths of the stroke. At the beginning of compression the temperature of the air is 41°C. Assuming compression and expansion to follow the law $pV^n = C$ where $n = 1{\cdot}34$, calculate the temperatures at the end of compression, end of combustion, and beginning of exhaust.

f 10. A two-stroke, single cylinder engine, operating on "diesel cycle", has a volumetric compression ratio of 12:1 and a stroke volume of 0·034 m^3.

The pressure and temperature, at the start of compression are 1 bar and 80°C respectively, while the maximum temperature at the end of heat reception is 1650°C.

If the compression is according to the law $pV^{1 \cdot 36}$ = constant and the expansion follows the law $pV^{1 \cdot 4}$ = constant, calculate:

(a) the indicated mean effective pressure (m.e.p.) for the cycle,

(b) the power developed at 200 cycles/min.

CHAPTER 9

RECIPROCATING AIR COMPRESSORS

Compressed air, at various pressures, is used for many purposes such as scavenging, supercharging, and starting diesel engines, and as the operating fluid for many automatic control systems. Air compressors to produce medium and high pressures are usually of the reciprocating type and may be single or multi-stage. Rotary types are common for large quantities of air at low pressures.

Fig. 32 shows diagrammatically a single-stage, single-acting reciprocating compressor, and its *pV* diagram illustrating the cycle.

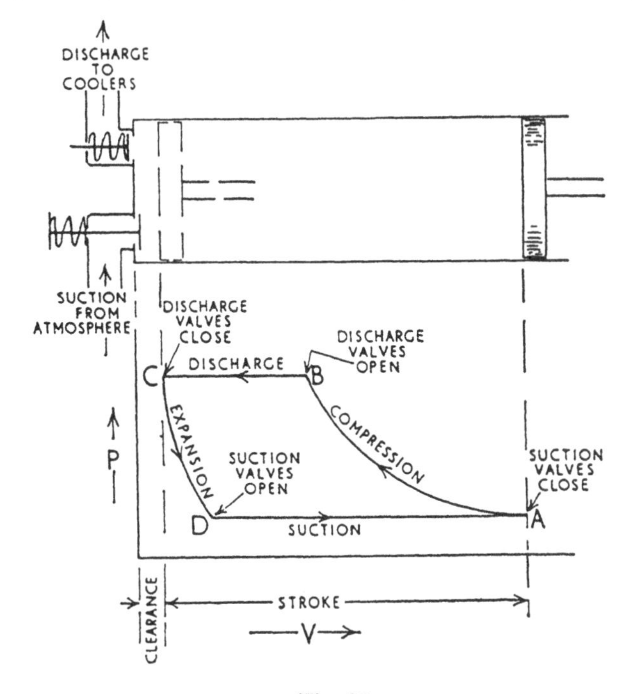

Fig. 32

Commencing at point A, the cycle of operations is as follows: A to B, compression period; with all valves closed the piston moves inward and the air which was previously drawn into the cylinder is compressed. Compression continues until the air pressure is sufficiently high to force the discharge valves open against their pe-set compression springs. Thus, at point B, the discharge valves open, and the compressed air is discharged at constant pressure for the remainder of the inward stroke, *i.e.*, from B to C. At point C the piston has completed its inward stroke and changes direction to move outward. Immediately the piston begins to move back there is a drop in pressure of the compressed air left in the clearance space, the discharge valves close, and from C to D this air expands. At point D the pressure has fallen to less than the atmospheric pressure and the lightly-sprung suction valves are opened by the greater pressure of the atmospheric air. Air is drawn into the cylinder for the remainder of the outward stroke, *i.e.*, from D to A.

Example. The stroke of the piston of an air compressor is 250 mm and the clearance volume is equal to 6% of the stroke volume. The pressure of the air at the beginning of compression is 0·98 bar and it is discharged at 3·8 bar. Assuming compression to follow the law pV^n = constant, where n = 1·25, calculate the distance moved by the piston from the beginning of its pressure stroke before the discharge valves open and express this as a percentage of the stroke.

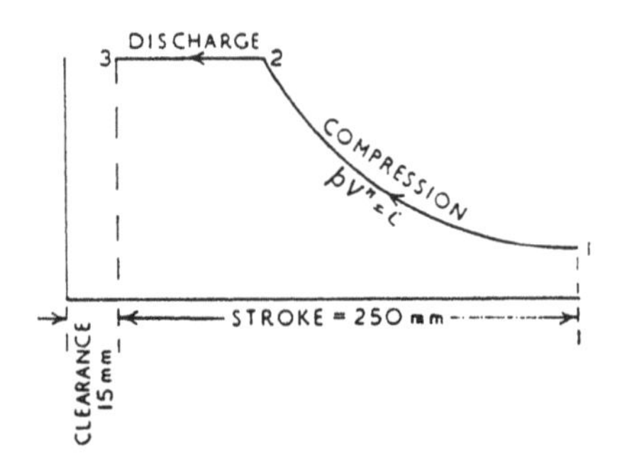

Fig. 33

$$\text{Clearance length} = 6\% \text{ of } 250 = 15$$

$$p_1 = 0{\cdot}98 \quad p_2 = 38 \quad V_1 = 250 + 15 = 265$$

$$p_1 V_1^{1{\cdot}25} = p_2 V_2^{1{\cdot}25}$$

$$0{\cdot}98 \times 265^{1{\cdot}25} = 3{\cdot}8 \times V_2^{1{\cdot}25}$$

$$V_2^{1.25} = \frac{0.98 \times 265^{1.25}}{3.8}$$

$$V_2 = 265 \times \sqrt[1.25]{\frac{0.98}{3.8}} = 89.6$$

Distance moved by piston from beginning of stroke to point where discharge valves open is represented by $V_1 - V_2$

$$V_1 - V_2 = 265 - 89.6 = 175.4 \text{ mm} \quad \text{Ans. (i)}$$

Expressed as a percentage of the stroke of 250 mm

$$= \frac{175.4}{250} \times 100 = 70.16\% \quad \text{Ans. (ii)}$$

Example. The diameter of an air compressor cylinder is 140 mm, the stroke of the piston is 180 mm, and the clearance volume is 77 cm³. The pressure and temperature of the air in the cylinder at the end of the suction stroke and beginning of compression is 0·97 bar and 13°C. The delivery pressure is constant at 4·2 bar. Taking the law of compression as $pV^{1.3}$ = constant, calculate (i) for what length of stroke air is delivered, (ii) the volume of air delivered per stroke, in litres, and (iii) the temperature of the compressed air.

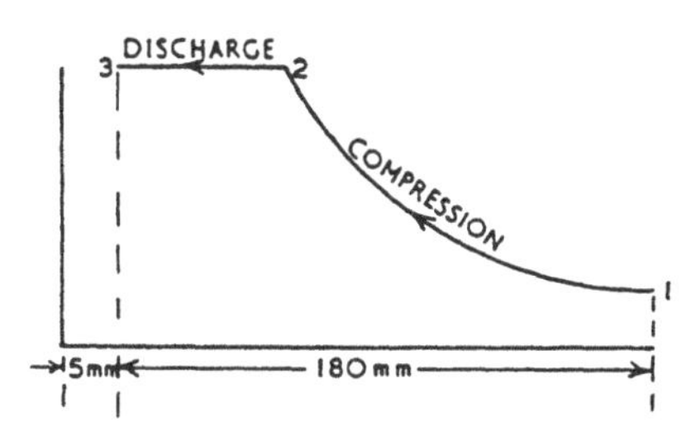

Fig. 34

$$\text{Clearance length [mm]} = \frac{\text{clearance volume [mm}^3\text{]}}{\text{area of cylnder [mm}^2\text{]}}$$

$$= \frac{77 \times 10^3}{0.7854 \times 140^2} = 5 \text{ mm}$$

$$p_1 V_1^{1.3} = p_2 V_2^{1.3}$$

$$0{\cdot}97 \times (180 + 5)^{1{\cdot}3} = 4{\cdot}2 \times V_2^{1{\cdot}3}$$

$$V_2^{1{\cdot}3} = \frac{0{\cdot}97 \times 185^{1{\cdot}3}}{4{\cdot}2}$$

$$V_2 = 185 \times \sqrt[1{\cdot}3]{\frac{0{\cdot}97}{4{\cdot}2}}$$

$$= 59{\cdot}94 \text{ mm}$$

$$\text{Delivery period} = V_2 - V_3 = 59{\cdot}94 - 5$$
$$= 54{\cdot}94 \text{ mm of stroke.} \quad \text{Ans. (i)}$$

$$\text{Volume delivered} = \text{area} \times \text{length}$$
$$= 0{\cdot}7854 \times 140^2 \times 54{\cdot}94$$
$$= 8{\cdot}457 \times 10^5 \text{ mm}^3$$

10^6 mm^3 = 1 litre

$$\text{Volume delivered} = 0{\cdot}8457 \text{ litre} \quad \text{Ans. (ii)}$$

$$\frac{p_1V_1}{T_1} = \frac{p_2V_2}{T_2}$$

$$\frac{0{\cdot}97 \times 185}{286} = \frac{4{\cdot}2 \times 59{\cdot}94}{T_2}$$

$$T_2 = \frac{286 \times 4{\cdot}2 \times 59{\cdot}94}{0{\cdot}97 \times 185} = 401{\cdot}2 \text{ K}$$

$\therefore$ Temperature at end of compression

$$= 128{\cdot}2^\circ\text{C} \quad \text{Ans. (iii)}$$

EFFECT OF CLEARANCE

Clearance is necessary between the piston face and the cylinder head and valves to avoid contact. This should be kept to a minimum because the volume of compressed air left in the clearance space at the end of the inward stroke must be expanded on the outward stroke to below atmospheric pressure before the suction valves can open, thus affecting the volume of air drawn into the cylinder during the suction stroke. This effect is illustrated below.

An ideal compressor would have, theoretically, no clearance as in Fig. 35a. The suction valves would open immediately the piston began to move on its outward stroke and air would be drawn into the cylinder for the whole stroke.

Fig. 35b. shows the effect of a small clearance. The small volume of compressed air left in the clearance space is quickly expanded to sub-atmospheric pressure to allow the suction valves to open, and air is drawn into the cylinder from D to A which is a large proportion of the full stroke E to A.

Fig. 35c shows the effect of excessive clearance. There is a comparatively large volume of air left in the clearance space and it requires a considerable movement of the piston on its outward stroke to expand this air to a pressure below atmospheric. The suction period D to A is an inefficient proportion of the stroke E to A.

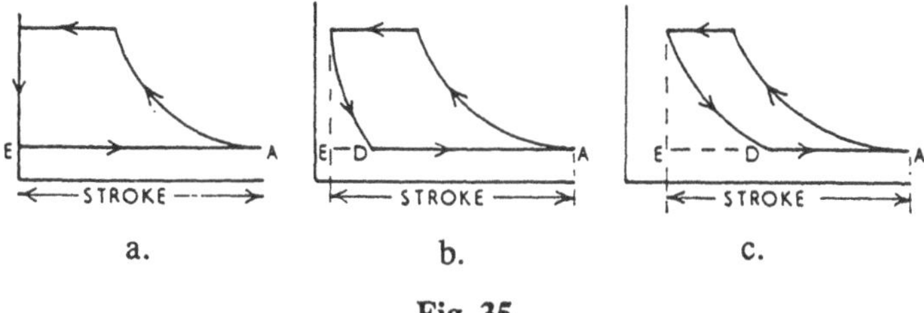

Fig. 35

The ratio between the volume of air drawn into the cylinder during the suction stroke and the full stroke volume swept out by the piston, is the *volumetric efficiency* of the compressor.

$$\text{Volumetric efficiency} = \frac{\text{volume of air drawn in per stroke}}{\text{stroke volume}}$$

and this is represented by the ratio $\frac{DA}{EA}$

WORK DONE PER CYCLE

NEGLECTING CLEARANCE

As previously shown, the area of a pV diagram represents work done, if the pressure is in kN/m^2 and the volume in cubic metres then the area of the pV diagram and the work done per cycle is in kJ.

Referring to Fig. 36 (neglecting clearance):

Net area = Net work done on the air per cycle

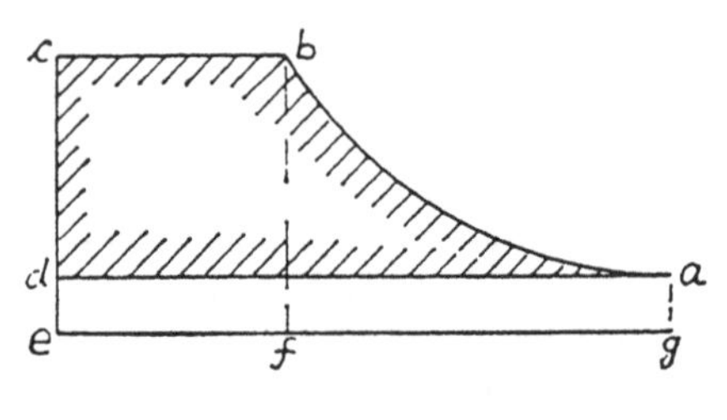

Fig. 36

$$\text{Area } abcd = \text{Area } bcef + \text{Area } abfg - \text{Area } adeg$$

Let p_2 = delivery pressure, V_2 = volume bc
p_1 = suction pressure, V_1 = volume da

$$\text{Area under curve} = \frac{p_2V_2 - p_1V_1}{n-1} \text{ therefore:}$$

$$\text{Work/cycle} = p_2V_2 + \frac{p_2V_2 - p_1V_1}{n-1} - p_1V_1$$

$$= \frac{P_2V_2(n-1) + p_2V_2 - p_1V_1 - p_1V_1(n-1)}{n-1}$$

$$= \frac{np_2V_2 - p_2V_2 + p_2V_2 - p_1V_1 - np_1V_1 + p_1V_1}{n-1}$$

$$= \frac{n}{n-1}(p_2V_2 - p_1V_1)$$

$$p_1V_1 = mRT_1 \qquad p_2V_2 = mRT_2$$

$$\frac{T_2}{T_1} = \left\{\frac{P_2}{p_1}\right\}^{\frac{n-1}{n}} = \left\{\frac{V_1}{V_2}\right\}^{n-1}$$

By substitution, the work per cycle can be expressed in other terms to suit available data, such as:

$$\text{Work/cycle} = \frac{n}{n-1}(p_2V_2 - p_1V_1)$$

$$= \frac{n}{n-1} mR\,(T_2 - T_1)$$

$$= \frac{n}{n-1} mRT_1 \left\{ \frac{T_2}{T_1} - 1 \right\} \text{ or}$$

$$= \frac{n}{n-1} p_1 V_1 \left\{ \frac{T_2}{T_1} - 1 \right\}$$

$$= \frac{n}{n-1} mRT_1 \left[\left\{ \frac{p_2}{p_1} \right\}^{\frac{n-1}{n}} - 1 \right] \text{ or}$$

$$= \frac{n}{n-1} p_1 V_1 \left[\left\{ \frac{p_2}{p_1} \right\}^{\frac{n-1}{n}} - 1 \right]$$

$$= \frac{n}{n-1} mRT_1 \left[\left\{ \frac{V_1}{V_2} \right\}^{n-1} - 1 \right] \text{ or}$$

$$= \frac{n}{n-1} p_1 V_1 \left[\left\{ \frac{V_1}{V_2} \right\}^{n-1} - 1 \right]$$

Example. The cylinder of a single-acting compressor is 225 mm diameter and the stroke of the piston is 300 mm. It takes in air at 0·96 bar and delivers it at 4·8 bar and makes four delivery strokes per second. Assuming that compression follows the law pV^n =

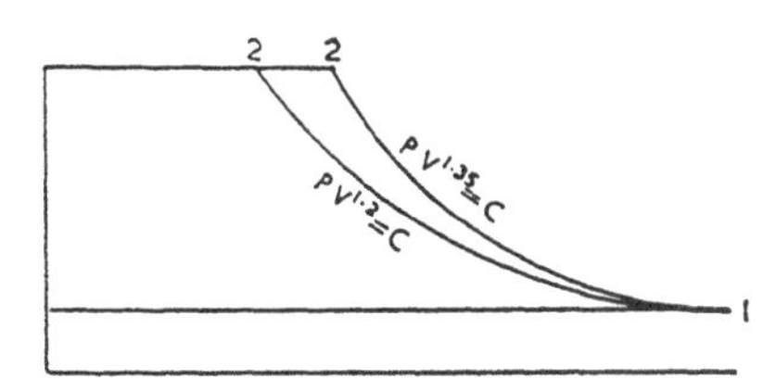

Fig. 37

constant, and neglecting clearance, calculate the theoretical power required to drive the compressor when the value of the index of the law of compression is, (i) 1·2, (ii) 1·35.

$$V_1 = 0{\cdot}7854 \times 0{\cdot}225^2 \times 0{\cdot}3 = 0{\cdot}01193 \text{ m}^3$$

When n = 1·2:

$$p_1 V_1^{1\cdot2} = p_2 V_2^{1\cdot2}$$

$$0{\cdot}96 \times 0{\cdot}01193^{1\cdot2} = 4{\cdot}8 \times V_2^{1\cdot2}$$

$$V_2^{1\cdot2} = \frac{0\cdot96 \times 0\cdot01193^{1\cdot2}}{4\cdot8} = \frac{0\cdot01193^{1\cdot2}}{5}$$

$$V_2 = \frac{0\cdot01193}{\sqrt[1\cdot2]{5}} = 0\cdot00312 \text{ m}^3$$

$$\text{Work done per cycle} = \frac{n}{n-1}(p_2V_2 - p_1V_1)$$

$$= \frac{1\cdot2}{0\cdot2}(4\cdot8 \times 10^2 \times 0\cdot00312 - 0\cdot96 \times 10^2 \times 0\cdot01193)$$

$$= 2\cdot114 \text{ kJ}$$

$$\begin{aligned} \text{Power [kW]} &= \text{work done per second [kJ/s]} \\ &= \text{kJ/cycle} \times \text{cycle/s} \\ &= 2\cdot114 \times 4 = 8\cdot456 \text{ kW.} \quad \text{Ans. (i)} \end{aligned}$$

Note that although any convenient units can be used for pressure and volume when they appear on both sides of an equation, such as pressure in bars, and volume represented by length on a pV diagram (as in the two previous examples), it is again emphasised that the pressure must be in kN/m^2 (1 bar = 10^2 kN/m^2) and the volume in m^3 when kJ of work is required by their product.

When $n = 1\cdot35$

$$p_1V_1^{1\cdot35} = p_2V_2^{1\cdot35}$$

$$0\cdot96 \times 0\cdot01193^{1\cdot35} = 4\cdot8 \times V_2^{1\cdot35}$$

$$V_2 = \frac{0\cdot01193}{\sqrt[1\cdot35]{5}} = 0\cdot003621 \text{ m}^3$$

$$\text{Work done per cycle} = \frac{n}{n-1}(p_2V_2 - p_1V_1)$$

$$= \frac{1\cdot35}{0\cdot35}(4\cdot8 \times 10^2 \times 0\cdot003621 - 0\cdot96 \times 10^2 \times 0\cdot01193)$$

$$= 2\cdot287 \text{ kJ}$$

$$\text{Power} = 2\cdot287 \times 4 = 9\cdot148 \text{ kW.} \quad \text{Ans. (ii)}$$

Note that the *mass* of air delivered per stroke is the same in each case. The greater volume indicated by the value of V_2 when n is $1\cdot35$ is due purely to the air being at a higher temperature at the end of compression than it is when $n = 1\cdot2$.

It can also be seen from the above that the nearer isothermal compression can be approached, the less work will be required to compress and deliver a given mass of air.

WORK DONE PER CYCLE

f INCLUDING CLEARANCE

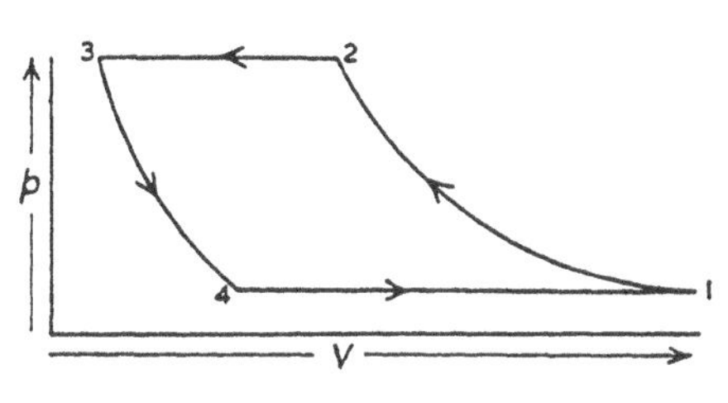

Fig. 38

Referring to Fig. 38 at the end of the compression and delivery stroke, the clearance space of volume V_3 is full of air at the delivery pressure of p_2 and temperature T_2. As the piston moves outward, the air does work on the piston as it expands to volume V_4 and the pressure falls to p_4, theoretically when p_4 is equal to p_1 the suction valves lift to begin the suction period.

The indicated work done by the piston on the air is therefore represented by the net area 1234.

Work/cycle

$$= \frac{n}{n-1}(p_2V_2 - p_1V_1) - \frac{n}{n-1}(p_3V_3 - p_4V_4)$$

$$= \frac{n}{n-1}p_1V_1\left[\left\{\frac{p_2}{p_1}\right\}^{\frac{n-1}{n}} - 1\right] - \frac{n}{n-1}p_4V_4\left[\left\{\frac{p_3}{p_4}\right\}^{\frac{n-1}{n}} - 1\right]$$

Taking the index of expansion to be equal to the index of compression and assuming $p_4 = p_1 \quad p_3 = p_2 \quad T_4 = T_1 \quad T_3 = T_2$ then,

Work/cycle

$$= \frac{n}{n-1}p_1V_1\left[\left\{\frac{p_2}{p_1}\right\}^{\frac{n-1}{n}} - 1\right] - \frac{n}{n-1}p_1V_4\left[\left\{\frac{p_2}{p_1}\right\}^{\frac{n-1}{n}} - 1\right]$$

$$= \frac{n}{n-1} p_1 (V_1 - V_4) \left[\left\{ \frac{p_2}{p_1} \right\}^{\frac{n-1}{n}} - 1 \right]$$

$V_1 - V_4$ is the volume of air taken in per cycle at the pressure p_1 and temperature T_1, letting m = mass of this air:

$$\text{Work/cycle} = \frac{n}{n-1} mRT_1 \left[\left\{ \frac{p_2}{p_1} \right\}^{\frac{n-1}{n}} - 1 \right]$$

Comparing this with the work/cycle when there is no clearance, we see that for a given mass the work is the same in each case, thus, the work done per unit mass of air delivered is not affected by the size of the clearance space.

MULTI-STAGE COMPRESSION

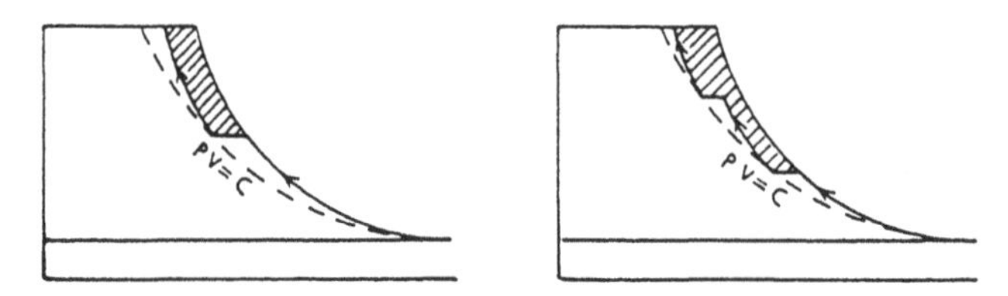

Fig. 39

By compressing the air in more than one stage and intercooling between stages, the practical compression curve can approach the isothermal more closely, hence reducing the work required per kg of air compressed. Fig. 39 shows the pV diagrams neglecting clearance for two-stage and three-stage compression respectively, the shaded areas representing the work saved in each case compared with single-stage compression.

To obtain maximum efficiency from a multi-stage compressor, that is, to do the least work to compress and deliver a given mass of air, (i) the air should be intercooled to as near the initial temperature as possible, and (ii) the pressure ratio in each stage should be the same.

Multi-stage compressors may consist simply of separate compressor cylinders, or may be arranged in tandem. Fig. 40 shows a diagrammatic arrangement of a three-stage tandem air compressor.

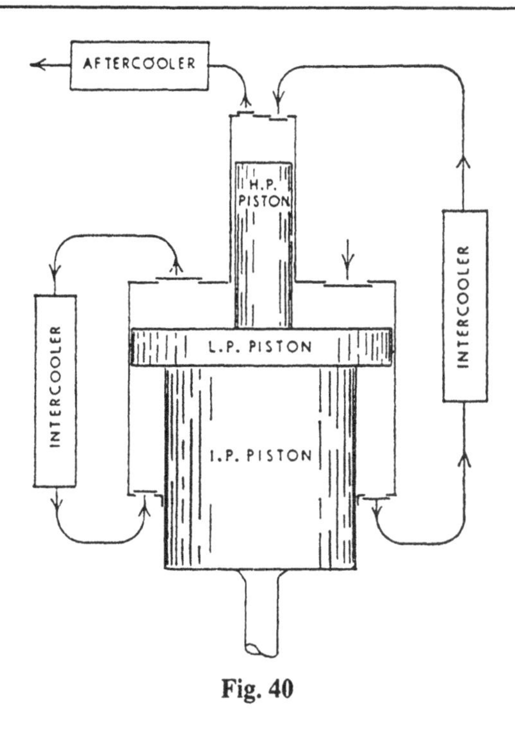

Fig. 40

It is important to note that, when calculating the volume of atmospheric air drawn into the low-pressure cylinder of a tandem compressor, the *effective* area of suction is the annulus between the area of the low pressure and the area of the high pressure. Thus, if D = diameter of low pressure piston, and d = diameter of high pressure, then:

$$\text{Effective area} = 0{\cdot}7854\,(D^2 - d^2)$$

Example. In a single-acting three-stage tandem air compressor, the piston diameters are 70, 335 and 375 mm diameter respectively, the stroke is 380 mm, and it is driven directly from a motor running at 250 rev/min. The suction pressure is atmospheric (1·013 bar) and the discharge is 45 bar gauge pressure. Assuming that the air delivered to the reservoirs is cooled down to the initial suction temperature and taking the volumetric efficiency as 90%, calculate the volume of compressed air delivered to the reservoirs per minute.

$$\begin{aligned}\text{Effective area of L.P.} &= 0{\cdot}7854\,(0{\cdot}375^2 - 0{\cdot}07^2)\\ &= 0{\cdot}7854 \times 0{\cdot}445 \times 0{\cdot}305\ \text{m}^2\end{aligned}$$

Stroke volume = area × stroke

∴ volume [m³] of air drawn in per stroke

$$= 0{\cdot}7854 \times 0{\cdot}445 \times 0{\cdot}305 \times 0{\cdot}38 \times 0{\cdot}9$$
$$= 0{\cdot}03646 \text{ m}^3$$

air drawn in per minute

$$= 0{\cdot}03646 \times 250 = 9{\cdot}115 \text{ m}^3$$

This is the volume of air taken into the compressor per minute at 1·013 bar. The volume delivered per minute at 46·013 bar is at the same temperature, therefore, since pressure × volume = constant, volume varies inversely as the absolute pressure, then:

$$\text{Volume delivered} = 9{\cdot}115 \times \frac{1{\cdot}013}{46{\cdot}013}$$
$$= 0{\cdot}2006 \text{ m}^3/\text{min.} \quad \text{Ans.}$$

Example. The cylinders of a single-acting two-stage tandem air compressor are 55 and 215 mm diameter respectively, and the stroke is 230 mm. It is connected to three air storage bottles of equal size of internal dimensions 300 mm. diameter and 1·5 m long overall with hemispherical ends. Taking atmospheric pressure as 1 bar and assuming a volumetric efficiency of 0·88, calculate the time required to pump up the bottles to a pressure of 31 bar gauge from empty, when running at 125 rev/min.

Length of cylindrical part of bottles

$$= \text{overall length} - \text{diameter}$$
$$= 1{\cdot}5 - 0{\cdot}3 = 1{\cdot}2 \text{ m}$$

$$\text{Volume of three bottles} = 3\left(\frac{\pi}{6} \times 0{\cdot}3^3 + \frac{\pi}{4} \times 0{\cdot}3^2 + 1{\cdot}2\right)$$
$$= 3\pi \times 0{\cdot}3^2\,(0{\cdot}05 + 0{\cdot}3)$$
$$= 0{\cdot}297 \text{ m}^3$$

To produce 0·297 m³ of air at an absolute pressure of 31 + 1 = 32 bar, the volume of atmospheric pressure air (termed "free air") required, at the same temperature, is inversely proportional to the pressure:

Volume of atmospheric air

$$= 0{\cdot}297 \times \frac{32}{1} = 9{\cdot}504 \text{ m}^3$$

However, "empty" bottles contain their own volume of air at atmospheric pressure, therefore, volume of air to be taken into compressor from atmosphere

$$= 9{\cdot}504 - 0{\cdot}207 = 9{\cdot}207\ \text{m}^3$$

Volume of air taken into compressor per minute

$= 0{\cdot}7854\,(0{\cdot}215^2 - 0{\cdot}055^2) \times 0{\cdot}23 \times 0{\cdot}88 \times 125$

$= 0{\cdot}8584\ \text{m}^3/\text{min}$

$\therefore$ Time required $= \dfrac{9{\cdot}207}{0{\cdot}8584} = 10{\cdot}73$ min. Ans.

FREE AIR DELIVERY It is convenient to use free air delivery as a practical comparison between air compressors. Free air delivery is the rate of volume flow measured at inlet pressure and temperature.

At inlet (*i.e.* before entering the compressor) $pV = mRT$

If $\dot{V}$ is the free air delivery in m³/s.

If $\dot{m}$ is the mass flow in kg/s.

$$\therefore\ p\dot{V} = \dot{m}RT$$

$$\text{or}\quad \dot{V} = \frac{\dot{m}RT}{p}$$

or $\dot{m} = \dfrac{p\dot{V}}{RT}$ either may be required when solving problems.

e.g. Power input/cycle $= \dfrac{n}{n-1}\dot{m}R\,(T_2 - T_1)$ [kW]

ISOTHERMAL EFFICIENCY It has been stated that the nearer the isothermal compression can be approached the less will be the work required to compress and deliver the air. Hence another compressor comparison is Isothermal Efficiency.

$$\text{Isothermal Efficiency} = \frac{\text{Power input with isothermal compression}}{\text{Actual power input}} \times 100\%$$

INDICATED MEAN EFFECTIVE PRESSURE Taking the work per cycle as at Fig. 36:

$$\text{m.e.p. (kN/m}^2) = \frac{\text{area of diagram (kJ)}}{\text{length of diagram (m}^3)}$$

MINIMUM WORK For reversible polytropic compression the total work required is a minimum when the work is equally divided between the stages. For a two stage machine with perfect intercooling, by differentiating the total work expression with respect to the intermediate pressure p_2, it can be shown that:

$$p_2 = \sqrt{p_1 p_3}$$

f Example. A two-stage single acting air compressor takes air at 1 bar, 15°C and delivers 0·075 kg/s at 9 bar. The compression and expansion in both stages is according to the law $pV^{1.3}$ = constant. The compressor is designed for minimum work with perfect intercooling.

Calculate:
(a) the power input to the compressor;
(b) the heat rejected in the intercooler.
For air: R = 287 J/kg K, c_p = 1005 J/kg K.

$$p_2 = \sqrt{p_1 p_3} = \sqrt{9} = 3$$

$$\frac{T_2}{T_1} = \left\{\frac{p_2}{p_1}\right\}^{\frac{n-1}{n}}$$

$$T_2 = 288 \times 3^{\frac{0.3}{1.3}} = 370.9 \text{ K}$$

$$\text{Work done per stage} = \frac{n}{n-1}\dot{m}RT\,(T_2 - T_1)$$

$$= \frac{1.3}{0.3} \times 0.075 \times 0.287\,(370.9 - 288)$$

$$= 7.732 \text{ kW}$$

$$\text{Power input to compressor} = 2 \times 7.732 = 15.464 \text{ kW} \quad \text{Ans. (a)}$$

$$\text{Heat rejected in intercooler} = \dot{m}c_p\,(370 - 288) = 0.075 \times 1.005 \times 82.9 = 6.249 \text{ kW} \quad \text{Ans. (b)}$$

TEST EXAMPLES 9

1. In a single-stage air compressor the diameter of the cylinder is 250 mm, the stroke of the piston is 350 mm, and the clearance volume is 900 cm^3. Air is drawn in at a pressure of 0·986 bar and delivered at 4·1 bar. Taking the law of compression to be $pV^{1\cdot25} = C$, calculate the distance travelled by the piston from the beginning of its compression stroke when the delivery valves open.

2. A four cylinder single acting compressor of 152 mm bore and 105 mm stroke runs at 12 rev/s and compresses air from 1 bar to 8 bar. The clearance volume is 8% of the swept volume and both compression and expansion are according to the law $pV^{1\cdot3}$ = constant.
   (a) Sketch the cycle on a pressure volume diagram.
   (b) Calculate:
      (i) the volumetric efficiency;
      (ii) the free air delivery.

3. A single stage air compressor of 180 mm stroke and 140 mm bore compresses air from 1 bar, 15°C to 8·5 bar. The mass of air in the cylinder at the beginning of the compression stroke is 0·0035 kg. The air in the clearance volume expands according to the law $pV^{1\cdot32}$ = constant.
   (a) Sketch the cycle on a pV diagram;
   (b) Calculate:
      (i) the clearance volume;
      (ii) the volumetric efficiency.
   For air $R = 287$ J/kg K.

4. A three cylinder single stage compressor with negligible clearance volume and 100 mm bore, 120 mm stroke produces a free air delivery of 1·2 m^3/min. It compresses air from 1 bar to 6·6 bar and the compression process may be assumed to be polytropic with an index of 1·3. The mechanical efficiency is 90%.
   (a) Sketch the process on a pV diagram.
   (b) Calculate:
      (i) the operating speed;
      (ii) the power input required.

5. A motor driven three-stage single-acting tandem air compressor runs at 170 rev/min. The high pressure and low pressure cylinder diameters are 75 mm and 350 mm diameter respectively, and the stroke is 300 mm. Find the time to pump up air reservoirs of 22 m^3 total capacity from 19 bar gauge to 30 bar gauge, taking the

volumetric efficiency as 0·92 and atmospheric pressure as 1·01 bar.

f 6. In a single-acting air compressor, the clearance volume is 364 cm^3, diameter of cylinder 200 mm, stroke 230 mm, and it runs at 2 rev/s. It receives air at 1 bar and delivers it at 5 bar, the index of compression and expansion being 1·28. Calculate the indicated power, the mean indicated pressure, and the volumetric efficiency.

f 7. The pressure and temperature of the air at the beginning of compression in a single-stage double-acting air compressor are 0·98 bar and 24°C, the pressure ratio is 4·55 and the index of compression and expansion is 1·25. The stroke is 1·2 × cylinder diameter, clearance equal to 5% of the stroke, and the compressor runs at 8 rev/s. If it takes air from the atmosphere at 1·013 bar, 16°C, at the rate of 5 m^3/min, calculate (i) the compressor power, (ii) volumetric efficiency, (iii) the dimensions of the cylinder.

f 8. A single acting two stage air compressor delivers air at 16 bar. Inlet conditions are 1 bar, 33°C and the free air delivery is 17 m^3/min.

If the compressor is designed for minimum work and complete intercooling determine if the compression law is $pV^{1·3}$ = constant:

(a) the power required to drive the first stage.

(b) the heat rejected in the intercooler per minute.

Note: c_p = 1005 J/kg K and R = 287 J/kg K.

f 9. The free air delivery of a single stage, double acting, reciprocating air compressor is 0·6083 m^3/s measured at 1·013 bar and 15°C. Compressor suction conditions are 0·97 bar and 27°C, delivery pressure is 4·85 bar. Clearance volume is 6% of stroke volume, stroke and bore are equal.

If the index of compression and expansion is 1·32 and the machine runs at 5 rev/s determine:

(a) the required cylinder dimensions.

(b) the isothermal efficiency of the compressor.

R = 287 J/kg K.

f 10. A multi-stage air compressor is fitted with perfect intercoolers and is designed for minimum work. It takes in air at 1 bar, 35°C and delivers it at 100 bar. If the maximum temperature at any point during compression is not to exceed 95°C and the compression and expansion in each stage obeys the law $pV^{1·3}$ = constant, determine:

(a) the minimum number of stages required.

(b) the indicated power input required for a mass flow rate of 0·1 kg/s.

Take R = 0·287 kJ/kg K for air.

CHAPTER 10

STEAM

Under normal conditions of a steam engine plant, the engines consume steam at the same rate at which it is generated in the boilers, therefore the steam is generated at constant pressure.

The temperature at which water changes into steam, that is, the boiling point of water depends strictly upon the pressure exerted on it. A few examples are as follows:

PRESSURE [bar]	0·04	1·013	10	20	30
BOILING POINT [°C]	29	100	179·9	212·4	233·8

The temperature of the steam produced at any given pressure is the same as the temperature of the boiling point at the same pressure. Thus, if a boiler is working at a pressure of 30 bar the water begins to boil when its temperature reaches 233·8°C and the steam is generated at the same temperature.

DEFINITIONS. Steam which is in physical contact with the boiling water from which it has been generated is termed *saturated steam*, its temperature is the same as the boiling water and this is referred to as the *saturation temperature* (t_{sat} or t_s). The boiling water is referred to as *saturated liquid* because it is at the saturation temperature corresponding to that particular pressure (p_{sat}) under which evaporation takes place.

If the vapour produced is pure steam, it is called *dry saturated steam*. If the steam contains water (usually very fine particles held in suspension in the form of a mist) it is called *wet saturated steam* which is more often briefly referred to as *wet steam*. The quality of wet steam is expressed by its *dryness fraction* (x) which is the ratio of the mass of pure steam in a given mass of the steam-plus-water mixture.

In order to increase the temperature of steam above its saturation temperature, the steam must be taken away from direct contact with the water from which it was generated and heated externally, usually as it passes through superheater elements heated by high temperature flue gases. If the steam is still wet as it enters the superheaters, the particles of water must first be evaporated to produce dry steam before the temperature is increased. Steam whose temperature is higher than its saturation temperature (which depends upon its pressure) is termed *superheated steam*, and the difference in temperature between the superheated steam and its saturation temperature is referred to as the *degree of superheat.*

FORMATION OF STEAM. Consider generating unit mass of steam in a boiler, at constant pressure p, from the initial stage of pumping in unit mass of the feed water.

Firstly, the water is forced into the boiler at pressure p[kN/m^2] and, representing the volume of unit mass [1 kg] of the water as v_1 [m^3] then the flow energy given to the water to enable it to enter the boiler is pv_1 [kJ].

Heat energy is then transferred to the water and its temperature is raised from feed temperature to evaporation temperature (boiling point). Most of the heat energy received by the water is to increase its temperature and therefore increase its stored up energy, called *internal energy* and represented by u, but some is used to do the work of increasing its volume from v_1 to v_f as it expands against the pressure p, the work done being $p(v_f - v_1)$. If u_1 represents the internal energy initially in the feed water, u_f the internal energy in the water at evaporation temperature, and h_f the heat energy supplied then,

$$\text{heat energy transferred} = \text{increase in internal energy} + \text{external work done}$$

$$h_f = (u_f - u_1) + p(v_f - v_1)$$

On reaching the boiling point, the water evaporates into steam as it absorbs more heat energy, thus changing its physical state from a liquid into a vapour. During the evaporating process the pressure and temperature remain constant, and the steam produced is saturated steam at the same pressure and temperature as the boiling point of water. During the change, considerable expansion takes place, the volume of the steam being many times greater than the volume occupied by the water from which it was generated. Some of the heat energy transferred during this stage is to evaporate the water into steam, freeing the molecules and increasing its internal energy from u_f to u_g. The remainder of the heat energy is used in expanding the volume from v_f to v_g against the pressure p, the work done being $p(v_g - v_f)$. Hence, if h_{fg} represents the heat energy transferred then,

$$h_{fg} = (u_g - u_f) + p(v_g - v_f)$$

ENTHALPY. For a constant pressure process, the heat energy transferred, which is the sum of the internal energy and the work done, is termed *enthalpy*. Thus, enthalpy is an energy function of a constant pressure process and, in quantities per unit mass, is defined by the equation

$$h = u + pv$$

where h = specific enthalpy [kJ/kg]
u = specific internal energy [kJ/kg]
v = specific volume [m^3/kg]
p = absolute pressure [kN/m^2]

For a mass of m kg, the appropriate symbols are written in capitals:

$$H = U + pV$$

Suffixes distinguish between the properties of liquid and vapour, some of these have already been introduced in the previous chapter:

f refers to saturated liquid
g refers to saturated vapour
fg refers to evaporation from liquid to vapour

Thus,

h_f = specific enthalpy of saturated liquid, that is, when the liquid is at its saturation temperature.
h_g = specific enthalpy of dry saturated vapour.
h_{fg} = $h_g - h_f$ = specific enthalpy of evaporation, that is, the change of specific enthalpy to evaporate unit mass of saturated liquid to dry saturated vapour.
h = specific enthalpy of either liquid or vapour at any other state.

Since there is no absolute value for internal energy, there is none for enthalpy, but changes of the property can be measured and the convenient datum of water at 0°C is chosen from which to measure enthalpy and other properties of water and steam. These values are set out in steam tables.

STEAM TABLES

Much experimental work has been done on the properties of steam and the resuts are published in various forms. Those used in this book are:

THERMODYNAMIC AND TRANSPORT PROPERTIES OF FLUIDS, SI UNITS arranged by Y. R. Mayhew and G. F. C. Rogers, published by Basil Blackwell, Oxford.

The student should have a copy by his side for reference while studying the following examples. Note that in these tables, p represents the pressure in bars. 1 bar = 10^5 N/m^2 = 10^2 kN/m^2.

The properties of saturated water and steam are tabulated on pages 2 to 5. The first table, on page 2, is for the convenience of reference to temperature up to 100°C. The remainder are set out with reference to pressure.

The properties of superheated steam at various pressures and temperatures are tabulated on pages 6 to 8.

WATER. The properties of liquids depend almost entirely on the temperature, therefore when looking up the properties of water, we look at its particular temperature and disregard its pressure.

Example. To read from the tables the specific enthalpy of water at,

(a) pressure of 1 bar and temperature 60°C
(b) ... ... 5 bar ... ... 88°C
(c) ... ... 10 bar ... ... 130°C
(d) ... ... 12 bar ... ... 188°C

(a) 60°C: see page 2, h = 251·1 kJ/kg
(b) 88°C: ... ... 3, h = 369 kJ/kg
(c) 130°C: ... ... 4, h = 546 kJ/kg
(d) 188°C: ... ... 4, h_f = 798 kJ/kg

Note that for case (d) we see that the temperature of 188°C is the saturation temperature corresponding to its pressure of 12 bar, therefore we use the suffix f to signify that the water is at its saturation temperature.

STEAM. Example. To find the temperature and specific enthalpy of saturated water and dry saturated steam at pressures of (a) 1·01325 bar, (b) 10 bar, (c) 20 bar, (d) 42 bar.

(a) 1·01325 bar, page 2, t_s = 100°C h_f = 419·1 h_g = 2675·8
(b) 10 bar, ... 4, t_s = 179·9°C, h_f = 763, h_g = 2778
(c) 20 bar, ... 4, t_s = 212·4°C, h_f = 909, h_g = 2779
(d) 42 bar, ... 5, t_s = 253·2°C, h_f = 1102, h_g = 2800

WET STEAM. When steam is "wet" it means that it is a mixture of dry saturated steam and saturated water. Thus, if the dryness fraction is represented by x, then 1 kg of wet steam is composed of

x kg of dry saturated steam, and the remainder, which is $(1 - x)$ kg is water. Hence only x kg out of each 1 kg has been evaporated and the change of enthalpy to do this is $x \times$ specific enthalpy of evaporation, written xh_{fg}.

Therefore, specific enthalpy of wet steam of dryness fraction x is:

$$h = h_f + xh_{fg}$$

Example. Find the enthalpy and volume of 1 kg of wet saturated steam at a pressure of 0·2 bar and dryness fraction 0·85, and find also the additional heat energy required to completely dry the steam.

0·2 bar, page 3,

$$\begin{aligned} h &= h_f + xh_{fg} \\ &= 251 + 0{\cdot}85 \times 2358 \\ &= 251 + 2004{\cdot}3 \\ &= 2255{\cdot}3 \text{ kJ/kg} \quad \text{Ans. (i)} \end{aligned}$$

v_g is the volume occupied by 1 kg of dry saturated steam, therefore the volume occupied by x kg of dry saturated steam is xv_g. The value of v_g is given on page 3 as 7·648 m^3 for a pressure of 0·2 bar.

x_f is the volume occupied by 1 kg of saturated water, therefore the volume occupied by $(1 - x)$ kg of saturated water is $(1 - x)v_f$. The value of v_f is not listed on page 3 but a near figure can be found on page 10 which gives $v_f \times 10^2 = 0{\cdot}1017$ for a pressure of 0·1992 bar.

Hence, volume of 1 kg of wet steam is

$$\begin{aligned} v &= xv_g + (1 - x)v_f \\ &= 0{\cdot}85 \times 7{\cdot}648 + 0{\cdot}15 \times 0{\cdot}001\,017 \\ &= 6{\cdot}5008 + 0{\cdot}000\,152\,55 \end{aligned}$$

We see that the volume of the water is comparatively very small and therefore, for most practical cases can be neglected. Hence, the specific volume of wet steam is usually taken as,

$$v = xv_g$$

and for this example it is:

$$\begin{aligned} v &= 0{\cdot}85 \times 7{\cdot}648 \\ &= 6{\cdot}5008 \text{ m}^3\text{/kg} \quad \text{Ans. (ii)} \end{aligned}$$

To dry the steam, (1 – 0·85) kg of water is to be evaporated,

$$\text{Increase of enthalpy} = 0{\cdot}15 \times h_{fg}$$
$$= 0{\cdot}15 \times 2358$$
$$= 353{\cdot}7 \text{ kJ} \quad \text{Ans. (iii)}$$

Alternatively, it is the difference between the enthalpy of dry saturated steam and wet steam:

$$= h_g - h$$
$$= 2609 - 2255{\cdot}3$$
$$= 353{\cdot}7 \text{ kJ}$$

Example. If the specific enthalpy of wet saturated steam at a pressure of 11 bar is 2681 kJ/kg, find its dryness fraction.

The specific enthalpy of wet steam of dryness x is:

$$h = h_f + xh_{fg}$$

$$\text{hence, } x = \frac{xh_{fg}}{h_{fg}} = \frac{h - h_f}{h_{fg}}$$

$$= \frac{2681 - 781}{2000} \text{ (page 4)}$$

$$= \frac{1900}{2000} = 0{\cdot}95 \quad \text{Ans.}$$

Example. Find the heat transfer required to convert 5 kg of water at a pressure of 20 bar and temperature 21°C into steam of dryness fraction 0·9, at the same pressure.

Water 21°C, page 2, $h = 88$ kJ/kg
Steam 20 bar, page 4,

$$h = h_f + xh_{fg}$$
$$= 909 + 0{\cdot}9 \times 1890 = 2610 \text{ kJ/kg}$$

Change of enthalpy, water to steam

$$= 2610 - 88 = 2522 \text{ kJ/kg}$$

$$\text{Heat transfer} = \text{total change of enthalpy (for 5 kg)}$$
$$= 5 \times 2522 = 12610 \text{ kJ} \quad \text{Ans.}$$

SUPERHEATED STEAM. As previously stated, steam is said to be *superheated* when its temperature is higher than the saturation temperature corresponding to its pressure. In practice, steam is superheated at constant pressure, the saturated steam being taken from the boiler steam space and passed through the superheater tubes where it receives additional heat energy to dry the steam and raise its temperature. The properties of superheated steam are given on pages 6 to 8 of the tables.

Example. Find the specific volume and specific enthalpy of superheated steam (a) at a pressure of 10 bar and temperature 250°C, (b) at a pressure of 100 bar and temperature 400°C.

10 bar 250°C, page 7,

$$\left.\begin{aligned} v &= 0{\cdot}2328\ \text{m}^3/\text{kg} \\ h &= 2944\ \text{kJ/kg} \end{aligned}\right\}\ \text{Ans. (a)}$$

100 bar 400°C, page 8

$$\left.\begin{aligned} v &= 0{\cdot}02639\ \text{m}^3/\text{kg} \\ h &= 3097\ \text{kJ/kg} \end{aligned}\right\}\ \text{Ans. (b)}$$

Note that at the higher pressures of 80 bar and upwards (page 8) the values of the specific volume are given as $v \times 10^2$, that is, the listed values are $10^2 \times$ actual value.

Example. Find the specific volume and specific enthalpy of steam at a pressure of 7 bar having 85°C of superheat, and determine the mean specific heat of the superheated steam over this range.

7 bar, page 7, $t_{sat} = 165°\text{C}$, $h_g = 2764$ kJ/kg

Temperature of superheated steam

$$= 165 + 85 = 250°\text{C}$$

$$v = 0{\cdot}3364\ \text{m}^3/\text{kg} \quad \text{Ans. (i)}$$

$$h = 2955\ \text{kJ/kg} \quad \text{Ans. (ii)}$$

Heat energy [kJ] added to superheat the steam

$$= \text{mass [kg]} \times c_p\text{[kJ/kgK]} \times \theta\text{[K]}$$

$$2955 - 2764 = 1 \times c_p \times 85$$

$$c_p = \frac{191}{85} = 2{\cdot}247\ \text{kJ/kg K} \quad \text{Ans. (iii)}$$

INTERPOLATION. If properties are required at an intermediate pressure or temperature to those given in the tables, an estimate can be made by taking values from the nearest listed pressure or temperature above and below that required and assume a linear variation between the two. The following demonstrate the usual methods.

Example. Find the enthalpy per kg of steam at a pressure of 8 bar and temperature 270°C.

Properties at 8 bar are given on page 7 but 270°C lies between the listed temperatures of 250 and 300°C.

At 8 bar 300°C, $h = 3057$ kJ/kg
At 8 bar 250°C, $h = 2951$

Difference, increase 50°C, $h = 106$ increase
Given temperature of 270°C is 20° higher than 250°C,

$$\text{difference for } 20^\circ\text{C}, = \frac{20}{50} \times 106 = 42{\cdot}4 \text{ kJ/kg}$$

$$\therefore \text{ At 8 bar, } 270^\circ\text{C}, \; h = 2951 + 42{\cdot}4$$
$$= 2993{\cdot}4 \text{ kJ/kg} \quad \text{Ans.}$$

Example. Find the specific enthalpy of steam at a pressure of 12 bar and temperature 350°C.

Values of pressures of 10 bar and 15 bar are given on page 7, temperature of 350°C is listed.

At 10 bar 350°C, $h = 3158$ kJ/kg
At 15 bar 350°C, $h = 3148$
Difference, increase 5 bar $h =$ 10 kJ/kg *decrease*
Given pressure of 12 bar is 2 bar greater than 10 bar

$$\text{Difference for 2 bar increase} = \frac{2}{5} \times 10 = 4 \text{ kJ/kg less}$$

$$\therefore \text{ at 12 bar } 350^\circ\text{C}, \; h = 3158 - 4$$
$$= 3154 \text{ kJ/kg.} \quad \text{Ans.}$$

Example. Find the specific enthalpy of steam at a pressure of 25 bar and temperature 380°C.

Nearest listed pressures and temperatures are 20–30 bar and 350–400°C, on page 7, this requires double interpolation.

20 bar 400°C, $h = 3158$ kJ/kg
30 bar 400°C, $h = 3231$
For 10 bar increase $h = 17$ kJ/kg decrease

$$\text{For 5 bar increase } h = \frac{5}{10} \times 17 = 8{\cdot}5 \text{ decrease}$$

$$\therefore \text{ At 25 bar } 400^\circ\text{C}, \; h = 3248 - 8{\cdot}5$$
$$= 3239{\cdot}5 \text{ kJ/kg} \quad \ldots \ldots \ldots \ldots \quad \text{(i)}$$

20 bar 350°C, $h = 3138$ kJ/kg
30 bar 400°C, $h = 3117$

For 10 bar increase $h = 21$ kJ/kg decrease

For 5 bar increase $h = \frac{5}{10} \times 21 = 10{\cdot}5$ decrease

$\therefore$ At 25 bar 350°C, $h = 3138 - 10{\cdot}5$
$= 3127{\cdot}5$ kJ/kg ... ... (ii)

From (i) 25 bar 400°C, $h = 3239{\cdot}5$
From (ii) 25 bar 350°C, $h = 3127{\cdot}5$
For increase of 50°C, $h = 112$ kJ/kg increase

For increase of 30°C, $h = \frac{30}{50} \times 112 = 67{\cdot}2$ increase

$\therefore$ At 25 bar 380°C, $h = 3127{\cdot}5 + 67{\cdot}2$
$= 3194{\cdot}7$ kJ/kg Ans.

MIXING STEAM AND WATER

In Chapter 2, examples were given involving the mixing of ice and water, and metals in liquids at different temperature, and now further examples will be given to include steam. These have many practical applications as will be seen later.

The same principles apply, namely, that unless otherwise stated it is assumed there is no transfer of heat energy from or to an outside source during the mixing process. That is, the quantity of heat energy absorbed by the colder substance is equal to the quantity of heat energy lost by the hotter substance. In other words, when steam and water are mixed together, the total enthalpy of the water and steam before mixing is equal to the total enthalpy of the resultant mixture.

Example. 3 kg of wet steam at 14 bar and dryness fraction 0·95 are blown into 100 kg of water at 22°C. Find the resultant temperature of the water.

Water 22°C, page 2 of tables, $h = 92{\cdot}2$ kJ/kg
Steam 14 bar, page 4,

$$h = h_f + xh_{fg}$$
$$= 830 + 0{\cdot}95 \times 1960 = 2692 \text{ kJ/kg}$$

Before mixing:
enthalpy of 100 kg water = $100 \times 92{\cdot}2$ = 9220 kJ
enthalpy of 3 kg steam = 3×2692 = 8076
total enthalpy = 17296 kJ

After mixing there are 100 + 3 = 103 kg of water of total

enthalpy 17296 kJ, therefore specific enthalpy of resultant water is:

$$h = \frac{17296}{103} = 167 \cdot 9 \text{ kJ/kg}$$

From page 2 of tables, 167·5 kJ/kg (which is sufficiently close) corresponds to a water temperature of 40°C.

hence, resultant temperature = 40°C. Ans.

An alternative method is to work on the principle of heat energy absorbed by the cold water is equal to the heat energy lost by the steam.

When heat is taken from the steam, it first condenses at its saturation temperature of 195°C (page 4 of tables, at 14 bar) into water at the same temperature, the heat energy lost being the change of enthalpy of condensation which is mass × xh_{fg}. This water condensate falls in temperature from 195°C to the final temperature of the mixture, the further heat energy lost being the product of the mass, specific heat, and fall in temperature.

The heat energy gained by the cold water is the product of its mass, specific heat, and rise in temperature.

Thus, let θ = final temperature of the water mixture, and assuming the mean specific heat of water to be 4·2 kJ/kg K, then:

$$\begin{aligned} \text{Heat energy gained by water} &= \text{Heat energy lost by steam} \\ 100 \times 4 \cdot 2\,(\theta - 22) &= 3\{0 \cdot 95 \times 1960 + 4 \cdot 2(195 - \theta)\} \\ 420\,\theta - 9240 &= 3\{1862 + 819 - 4 \cdot 2\,\theta\} \\ 420\,\theta - 9240 &= 8043 - 12 \cdot 6\,\theta \\ 432 \cdot 6\,\theta &= 17283 \\ \theta &= 39 \cdot 95^\circ\text{C} \end{aligned}$$

Example. Steam is tapped from an intermediate stage of a steam turbine at a pressure of 2·5 bar, its dryness fraction being 0·95, and passed to a contact feed heater. The hot water at 48°C is pumped into the heater where it mixes with the heating steam. Estimate the temperature of the feed water leaving the heater when the amount of steam tapped off is 9% of the steam supplied to the turbine.

Referring to Fig. 41, let the mass of steam supplied to the engine be 1 kg, then 0·09 kg is tapped off to the heater. This leaves (1 − 0·09) = 0·91 kg of steam to continue through the engine, into the condenser, as water pumped into the feed heater. Being a *contact* heater, the 0·09 kg of heating steam makes contact and mixes with the 0·91 kg of condensate, making 1 kg of feed water to be pumped back into the boilers.

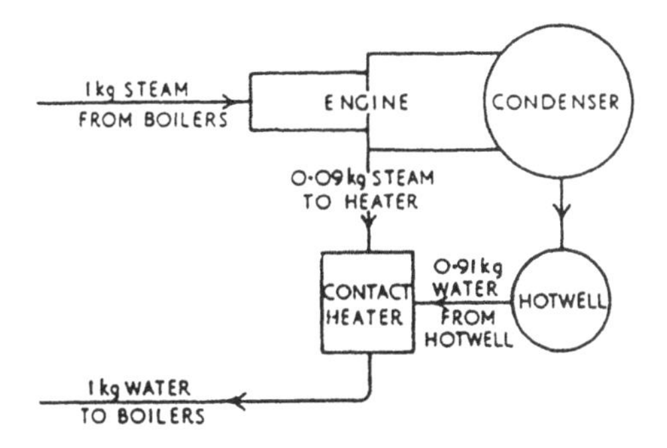

Fig. 41

Tables page 4, steam 2·5 bar, $h_f = 535$, $h_{fg} = 2182$
Tables page 2, water 48°C, $h = 200{\cdot}9$

Enthalpy of steam entering heater:

$$= 0{\cdot}09(h_f + xh_{fg})$$
$$= 0{\cdot}09(535 + 0{\cdot}95 \times 2182) = 234{\cdot}7 \text{ kJ}$$

Enthalpy of water entering heater:

$$= 0{\cdot}91 \times 200{\cdot}9 = 182{\cdot}8 \text{ kJ}$$

Total enthalpy of steam and water entering heater:

$$= 234{\cdot}7 + 182{\cdot}8 = 417{\cdot}5 \text{ kJ}$$

Total enthalpy of water leaving heater is that of 1 kg of feed water, therefore the *specific* enthalpy of the feed water is 417·5 kJ/kg which we now look for in the tables and find the near figure of 417 kJ/kg on page 3 for a temperature of 99·6°C.

∴ Temperature of feed water = 99·6°C Ans.

Alternatively for the latter part, we could assume a mean specific heat of 4·2 kJ/kg K for the feed water and take the enthalpy of water at θ°C as the heat energy to be transferred to it to raise its temperature from 0 to θ°C (mass × spec. heat × temp. rise), then:

$$1 \times 4{\cdot}2 \times \theta = 417{\cdot}5$$

$$\theta = \frac{417{\cdot}5}{4{\cdot}2} = 99{\cdot}4\text{°C}$$

Example. Steam is bled from the main steam pipe line at a pressure of 17 bar and dryness fraction 0·98, to a surface feed heater, and the remainder passes through the engine. The condensate at 42°C from the engine condenser, and the drain from the feed heater, passes to the hotwell. Calculate the percentage of the main steam bled off to the heater if the temperature of the feed water to the

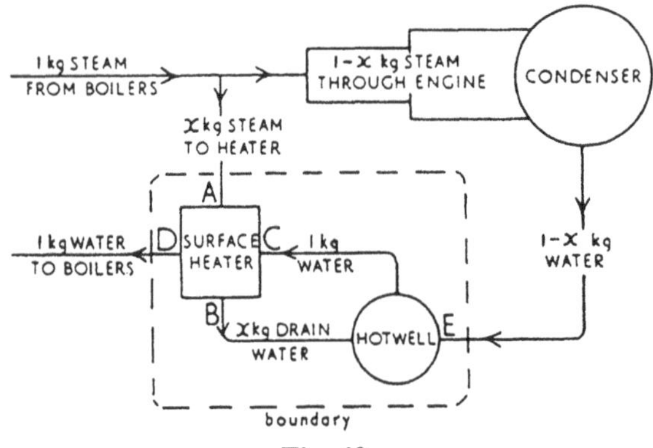

Fig. 42

boilers is 104·8°C.

The arrangement is illustrated in Fig. 42. Being a *surface* heater, there is no mixing of the heating steam and feed water. In this type, the water passes through nests of tubes which are heated on the outside by the steam.

Let H_A = enthalpy of steam entering heater
H_B = ... drain water leaving heater
H_C = ... water entering heater from hotwell
H_D = ... water leaving heater (feed to boilers).
H_E = ... water entering hotwell from condenser.

Enthalpy of drain water and condensate entering hotwell
= Enthalpy of water leaving hotwell

$$H_B + H_E = H_C \quad \text{(i)}$$

Enthalpy of heating steam and hotwell water entering heater
= Enthalpy of drain and feed water leaving heater

$$H_A + H_C = H_B + H_D \quad \text{(ii)}$$

Substituting value of H_C from (i) into (ii):

$$H_A + H_B + H_E = H_B + H_D$$

$$\therefore H_A + H_E = H_D$$

From this we can see that, since there is no heat energy transfer to or from an external source, the heater and hotwell with their connections can be considered as one common system as shown by the dotted boundary line on Fig. 42.

Let the mass of steam supplied from the boilers = 1 kg, then 1 kg of water is fed back to the boilers.

Let x kg of the supply steam be the amount bled off to the heater, then $(1 - x)$ kg is the mass of steam passed on to the engine.

Steam 17 bar, tables page 4, $h_f = 872$ $h_{fg} = 1923$
Water 42°C, ... 2, $h = 175{\cdot}8$
Water 104·8°C ... ... 4, $h = 439$

Enthalpy of x kg of heating steam entering system:

$$H_A = x(872 + 0{\cdot}98 \times 1923) = 2757x \text{ kJ}$$

Enthalpy of $(1 - x)$ kg of condensate entering system:

$$H_E = (1 - x) \times 175{\cdot}8 = 175{\cdot}8 - 175{\cdot}8x \text{ kJ}$$

Total enthalpy entering system:

$$H_A + H_E = 2757x + 175{\cdot}8 - 175{\cdot}8x$$
$$= 2581{\cdot}2x + 175{\cdot}8 \text{ kJ} \quad \ldots \ldots \ldots \ldots \text{(i)}$$

Enthalpy of 1 kg of feed water leaving system:

$$H_D = 439 \text{ kJ} \quad \ldots \ldots \ldots \ldots \ldots \ldots \text{(ii)}$$

$$\text{Total entry } H_A + H_E = \text{Total exit } H_D$$
$$2581{\cdot}2x + 175{\cdot}8 = 439$$
$$2581{\cdot}2x = 263{\cdot}2$$
$$x = 0{\cdot}102$$

As a percentage of the main steam supply,

Bled steam = 10·2% Ans.

Note: Feed heating increases cycle efficiency.

THROTTLING OF STEAM

Throttling is the process of passing the steam through a restricting orifice or a partially opened valve which causes a "wire-drawing" effect and reduces the pressure. There is not sufficient time for any appreciable heat transfer to take place between the steam and its surrounds and, as there are no moving parts, expansion takes place freely and no work is done by the steam. The difference in the velocity of the steam before and after is sufficiently small to be negligible. Consequently, if there is no transfer or conversion of heat energy, then there is no change in the enthalpy of the steam, that is, enthalpy after throttling is equal to enthalpy before throttling.

One important application of throttling takes place in the steam reducing valve. This is a specially designed valve connected to the high pressure steam range, the restricted valve opening is spring controlled and set to maintain a near steady reduced pressure at the outlet, the reduced pressure steam being more suitable for some pumps and other auxiliaries.

Example. Steam is passed through a reducing valve and reduced in pressure from 20 bar to 7 bar. Find the effect of throttling on the temperature and quality of the steam if the high pressure steam was (a) wet, having a dryness fraction of 0·9, (b) dry saturated, (c) superheated to a temperature of 300°C.

Tables pages 4 and 7

7 bar,	$t_s = 165°C$,	$h_f = 697$,	$h_{fg} = 2067$,	$h_g = 2764$
20 bar,	$t_s = 212{\cdot}4°C$,	$h_f = 909$,	$h_{fg} = 1890$,	$h_g = 2799$, $h_{sup} = 3025$

(a) 20 bar, dryness 0·9:

$$h = h_f + xh_{fg}$$
$$= 909 + 0{\cdot}9 \times 1890 = 2610$$

2610 is less than h_g for 7 bar, therefore reduced steam is wet

$\therefore$ 7 bar, dryness x, $h = h_f + xh_{fg}$

enthalpy before = enthalpy after

$$2610 = 697 + x \times 2067$$
$$2067x = 1913$$
$$x = 0{\cdot}9253$$

Effect of throttling:

Temperature is reduced from 212·4°C to 165°C
Dryness is increased from 0·9 to 0·9253 } Ans. (a)

(b) 20 bar dry sat. $h_g = 2799$

7 bar dry sat. $h_g = 2764$

enthalpy before = enthalpy after

hence, enthalpy at 7 bar is 2799, which is 35 kJ/kg higher than its h_g and is therefore superheated.

From page 7 of tables, for 7 bar, $h = 2799$, lies between $h_g = 2764$ at 165°C and $h = 2846$ at 200°C

by interpolation,

7 bar,	$h = 2846$	for 200°C
	$h = 2764$	for 165°C
difference in	$h = 82$ kJ	for 35°C

$$\text{difference for 35 kJ} = \frac{35}{82} \times 35 = 14{\cdot}94° \text{ say } 15°$$

$\therefore$ reduced pressure steam has 15°C of superheat, its temperature being 165 + 15 = 180°C.

Effect of throttling:

Temperature is reduced from 212·4°C to 180°C
Steam has 15° of superheat } Ans. (b)

(c) 20 bar, temperature 300°C, $h = 3025$

(steam has 300 – 212·4 = 87·6°C of superheat)

$h = 3025$ is more than h_g for 7 bar, reduced pressure steam is therefore superheated.

enthalpy before = enthalpy after

From page 7 of tables, for 7 bar, $h = 3025$, lies between $h = 2955$ at 250°C and $h = 3060$ at 300°C.

$$\begin{aligned} h &= 3060 \quad \text{for } 300^\circ\text{C} \\ h &= 2955 \quad \text{for } 250^\circ\text{C} \\ \text{difference in } h &= 105 \text{ kJ for } 50^\circ\text{C} \end{aligned}$$

∴ difference in temperature for (3025 – 2955) = 70 kJ is

$$\frac{70}{105} \times 50 = 33{\cdot}3^\circ\text{C}$$

$$\begin{aligned} \therefore \text{ temperature} = 250 + 33{\cdot}3 &= 283{\cdot}3^\circ\text{C} \\ \text{Degree of superheat} &= 283{\cdot}3 - 165 \\ &= 118{\cdot}3^\circ\text{C} \end{aligned}$$

Effect of throttling:

Temperature is reduced from 300°C to 283·3°C
Degree of superheat is increased from 87·6 to 118·3 } Ans. (c)

Note that in every case the effect of throttling is to (i) reduce the pressure, (ii) reduce the temperature, (iii) increase the quality, that is, either to produce drier steam or to increase the degree of superheat.

With regard to finding the degree of superheat at the lower pressure, if the mean specific heat of superheated steam over this range were given, the degree of superheat could be obtained from,

$$\begin{matrix}\text{Heat energy to} \\ \text{raise temperature}\end{matrix} = \text{mass} \times \text{spec. heat} \times \text{temp. rise}$$

therefore, for 1 kg

$$\begin{matrix}\text{Increase of enthalpy} \\ \text{above } h_g\end{matrix} = \text{mass} \times \text{spec. heat} \times \text{temp. rise}$$

$$\text{or,} \qquad h - h_g = c_P(t - t_{sat})$$

where t = temperature of the steam

Taking the last case (c) as an example, if the mean specific heat was given as 2·21 kJ/kg K:

$$h - h_g = c_P \times \text{degree of superheat}$$

$$\text{degree of superheat} = \frac{3025 - 2764}{2{\cdot}21} = 118{\cdot}1^\circ\text{C}$$

THROTTLING CALORIMETER

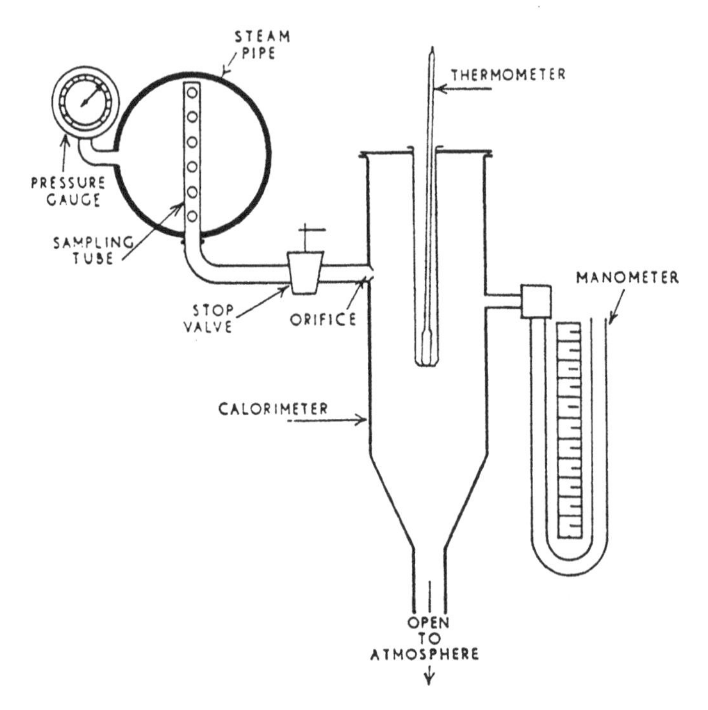

Fig. 43

The throttling calorimeter, an instrument for measuring the dryness fraction of steam, is illustrated diagrammatically in Fig. 43.

The principle of operation depends upon the fact that if steam is reduced in pressure by a throttling process, the enthalpy after throttling is equal to the enthalpy before throttling, as previously explained.

Example. Steam at a pressure of 14 bar from the main steam pipe is passed through a throttling calorimeter. The pressure in the calorimeter is 1·2 bar and the temperature 119°C. Taking the specific heat of the low pressure superheated steam as 2·0 kJ/kgK calculate the dryness fraction of the main steam.

From tables page 4,

1·2 bar, $t_s = 104{\cdot}8$, $h_g = 2683$

14 bar, $h_f = 830$, $h_{fg} = 1960$

$$\text{Enthalpy before throttling} = \text{Enthalpy after}$$
$$830 + x \times 1960 = 2683 + 2(119 - 104{\cdot}8)$$
$$830 + 1960\,x = 2683 + 28{\cdot}4$$
$$1960\,x = 1881{\cdot}4$$
$$x = 0{\cdot}9599 \text{ say } 0{\cdot}96. \quad \text{Ans.}$$

It will be noted that the throttling calorimeter has its limitations. The dryness fraction can only be determined when the temperature of the throttled steam in the calorimeter is higher than its saturation temperature corresponding to its pressure, that is, when it is superheated. If the throttled steam was not superheated, the thermometer would record the saturation temperature whether it was dry or wet and its condition would not be known.

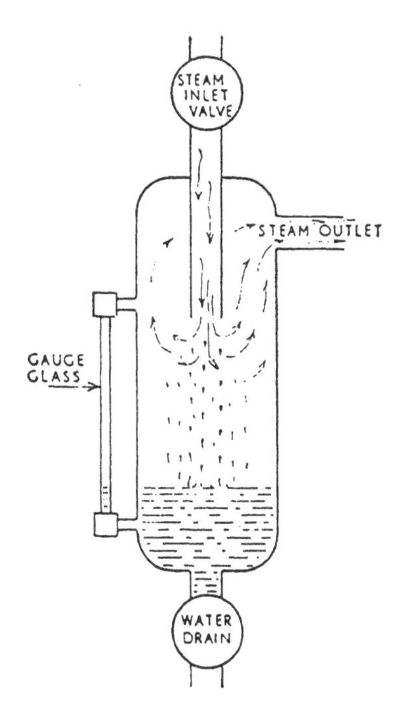

Fig. 44

SEPARATING CALORIMETER

When very wet steam samples are to be tested an *approximate* value for the dryness fraction may be determined by passing the steam through a *separating calorimeter* (Fig. 44).

The separated water collects in the bottom of the separator which is later drained off and measured, let this be m_1. The remaining steam passes out through the outlet and is led to a small condenser where it is condensed into water, collected and measured, let this be m_2:

From the sample of wet steam taken,

m_1 = mass of water separated from the sample
m_2 = mass of steam (assumed dry) remaining

Dryness fraction

$$= \frac{\text{mass of pure steam}}{\text{total mass of wet steam (steam + water mixture)}}$$

$$= \frac{m_2}{m_2 + m_1}$$

However, the dryness fraction determined in this manner is only approximate because, although the steam leaving the separator is assumed to be dry, it will actually be slightly wet since perfect separation is not achieved.

COMBINED SEPARATING AND THROTTLING CALORIMETER

In this arrangement, the outlet from the separator is connected directly to the inlet of the throttling calorimeter, and the condenser is connected to the exit of the throttling calorimeter.

$$x = x_1 \times x_2$$

In words this is:

$$\text{Dryness fraction} = \left\{\begin{array}{l}\text{dryness}\\ \text{fraction as}\\ \text{measured}\\ \text{by separating}\\ \text{calorimeter}\end{array}\right\} \times \left\{\begin{array}{l}\text{dryness}\\ \text{fraction as}\\ \text{measured}\\ \text{by throttling}\\ \text{calorimeter}\end{array}\right\}$$

Example. Steam at a pressure of 8 bar was tested by passing a sample through a combined separating and throttling calorimeter. The mass of water collected in the separator was 0·25 kg and the mass of condensate collected from the condenser after throttling was 2·5 kg. The pressure of the steam in the throttling calorimeter was 1·1 bar and its temperature 106·8°C. Taking the specific heat of the superheated steam after throttling as 2·0 kJ/kgK, find the dryness fraction of the steam sample.

From tables page 4,

8 bar, $h_f = 721$ $\quad h_{fg} = 2048$

1·1 bar, $t_s = 102{\cdot}3°C$, $\quad h_g = 2680$

Dryness fraction by separating calorimeter:

$$x_1 = \frac{m_2}{m_2 + m_1} = \frac{2{\cdot}5}{2{\cdot}5 + 0{\cdot}25} = 0{\cdot}909$$

Dryness fraction by throttling calorimeter:

$$\begin{aligned} \text{Enthalpy before throttling} &= \text{Enthalpy after} \\ 721 + x_2 \times 2048 &= 2680 + 2(106{\cdot}8 - 102{\cdot}3) \\ x_2 \times 2048 &= 1968 \\ x_2 &= 0{\cdot}961 \end{aligned}$$

Dryness fraction of steam sample:

$x = x_1 \times x_2 = 0{\cdot}909 \times 0{\cdot}961 = 0{\cdot}8736$. Ans.

f AIR IN CONDENSERS

It was stated in Chapter 5 under Dalton's law of partial pressures that the pressure exerted in a vessel occupied by a mixture of gases, or a mixture of gases and vapours, is equal to the sum of the pressures that each would exert if it alone occupied the whole volume of the vessel. The pressure exerted by each gas is termed a *partial pressure*.

Applying this to condensers which contain a mixture of leakage air and steam:

$$\text{Total pressure of mixture} = \text{partial pressure due to air} = \text{partial pressure due to steam}$$

This enables an estimate to be made on the amount of air leakage into condensers.

Example. The pressure in a condenser is 0·12 bar and the temperature is 43·8°C. If the internal volume of the condenser is 8·5 m^3, estimate the mass of air in the condenser, taking R for air = 0·287 kJ/kg K. If the dryness fraction of the steam is 0·9, calculate the mass of steam present.

Since there is always some water condensate in the condenser and the steam is in contact with this water, then the steam is at its saturation temperature and the pressure of the steam in the condenser depends upon this temperature. From the tables, page 3, we see that for a saturation temperature of 43·8°C the pressure of the steam is 0·09 bar:

$$\begin{aligned} \text{Partial pressure due to air} &= \text{total pressure} - \text{steam pressure} \\ &= 0{\cdot}12 - 0{\cdot}09 \\ &= 0{\cdot}03 \text{ bar} = 3 \text{ kN/m}^2 \\ \text{From } pV &= mRT, \text{ mass of air present:} \\ m &= \frac{3 \times 8{\cdot}5}{0{\cdot}287 \times (43{\cdot}8 + 273)} \\ &= 0{\cdot}2804 \text{ kg. Ans. (i)} \end{aligned}$$

Also from tables page 3, for saturation temperature of 43·8°C, specific volume of dry saturated steam v_g = 16·2 m^3/kg.

For wet steam

$$v = xv_g = 0{\cdot}9 \times 16{\cdot}2 = 14{\cdot}58 \text{ m}^3\text{/kg}$$

∴ mass of steam in the condenser volume of 8·5 m^3

$$= \frac{8{\cdot}5}{14{\cdot}58} = 0{\cdot}583 \text{ kg} \quad \text{Ans. (ii)}$$

ABSOLUTE HUMIDITY is the ratio of the mass of water vapour to the mass of dry air (sometimes called specific humidity or moisture content).

$$\omega = \frac{m_{WV}}{m_A} = \frac{\rho_{WV}}{\rho_A}$$

$$\omega\% = \frac{v_A}{v_{WV}} \times 100$$

(Relative humidity is sometimes used - ratio of partial pressures of vapour i.e. actual to saturated).

TEST EXAMPLES 10

1. Steam at a pressure of 9 bar is generated in an exhaust gas boiler from feed water at 80°C. If the dryness fraction of the steam is 0·96, determine the heat transfer per kg of steam produced.

2. A turbo-generator is supplied with superheated steam at a pressure of 30 bar and temperature 350°C. The pressure of the exhaust steam from the turbine is 0·06 bar with a dryness fraction of 0·88. (i) Calculate the enthalpy drop per kg of steam through the turbine. (ii) If the tubine uses 0·5 kg of steam per second, calculate the power equivalent of the total enthalpy drop.

3. Steam enters the superheaters of a boiler at a pressure of 20 bar and dryness 0·98, and leaves at the same pressure at a temperature of 350°C. Find (i) the heat energy supplied per kg of steam in the superheaters, and (ii) the percentage increase in volume due to drying and superheating.

4. One kg of wet steam at a pressure of 8 bar and dryness 0·94 is expanded until the pressure is 4 bar. If expansion follows the law pV^n = a constant, where n = 1·2, find the dryness fraction of the steam at the lower pressure.

5. Dry saturated steam at a pressure of 2·4 bar is tapped off the inlet branch of a low pressure turbine to supply heating steam in a contact feed heater. The temperature of the feed water inlet to the heater is 42°C and the outlet is 99·6°C. Find the percentage mass of steam tapped off.

6. Wet saturated steam at 16 bar and dryness 0·98 enters a reducing valve and is throttled to a pressure of 8 bar. Find the dryness fraction of the reduced pressure steam.

7. Steam of mass 0·2 kg trapped within a cylinder is allowed to expand at constant pressure from an initial temperature 165°C and dryness fraction of 0·75 until its volume is doubled.

Calculate:
(a) the final temperature;
(b) the work done;
(c) the heat energy transferred.

8. A combined separating and throttling calorimeter was connected to a main steam pipe carrying steam at 15 bar and the following data recorded:

Mass of water collected in separator = 0·55 kg

Mass of condensate after throttling = 10 kg
Press. of steam in throttling calorimeter = 1·1 bar
Temp. ... ... = 111°C

Taking the specific heat of the throttled superheated steam as 2·0 kJ/kgK, find the dryness fraction of the main steam.

f 9. During the process of raising steam in a boiler, when the pressure was 1·9 bar gauge the temperature inside the boiler was 130°C, and when the pressure was 6·25 bar gauge the temperature was 165°C. If the volume of the steam space is constant at 4·25 m^3 calculate the masses of steam and air present in each case. Take R for air = 0·287 kJ/kgK, atmospheric pressure = 1 bar, and assume the steam is dry in each case.

f 10. A sample of air is saturated with water vapour (i.e. the water vapour in the air is saturated vapour) at 20°C, the total pressure being 1 bar.

Calculate:

(a) the partial pressures of the oxygen, nitrogen and water vapour;

(b) the absolute humidity.

Air is 21% O_2 and 79% N_2 by volume and for air R = 287 J/kgK.

Note: Absolute humidity is the mass of water vapour per unit mass of dry air.

f CHAPTER 11

ENTROPY

Mechanical energy, being the product of two quantities can be represented as an area on rectangular axes. Fig. 45 is a diagram in which the ordinates represent pressure in kN/m^2 and the abscissae represent volume in m^3, the area enclosed represents mechanical work, $kN/m^2 \times m^3 = kNm = kJ$. The availability to do work depends upon the magnitude of the pressure.

In a similar manner, heat energy can be represented as an area. The availability to transfer heat energy depends upon the temperature and therefore the ordinates of a heat energy diagram represent absolute temperature. The abscissae are termed *entropy*. The symbol for entropy is S, change of entropy may be written ΔS and the units are kilojoule per kelvin [kJ/K]. Specific entropy, which is entropy per unit mass, is represented by the symbol s and the units are kilojoule per kilogramme kelvin [kJ/kg K].

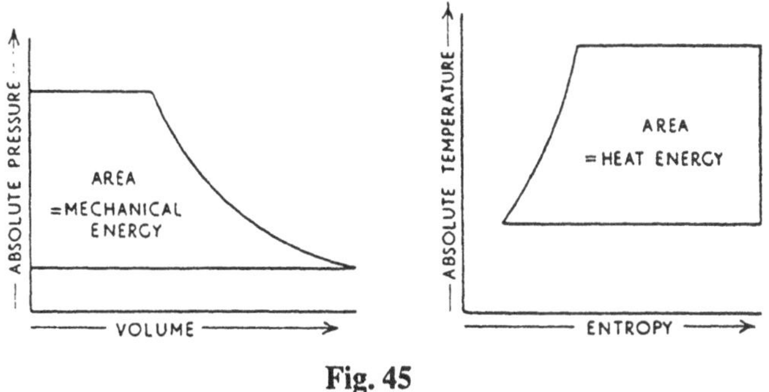

Fig. 45

Thus a diagram whose area represents heat energy, the ordinates represent absolute temperatue (T) and the abscissae represent entropy (S), is referred to as a temperature-entropy (T–S) diagram, as shown in Fig. 45.

ENTROPY OF WATER AND STEAM

Subscripts to distinguish the specific entropy of saturated

liquids and vapours are used as for other properties, explained in the previous chapter, thus:

s_f = specific entropy of saturated liquid
s_g = ... vapour
s_{fg} = $s_g - s_f$ = specific entropy of evaporation from liquid to vapour
s = specific entropy of either liquid or vapour at any other state.

As for internal energy and enthalpy, the values of specific entropy measured from the datum of water at 0°C are listed in the steam tables. (Thermodynamic and Transport Properties of Fluids, SI units, by Y. R. Mayhew and G. F. C. Rogers).

WATER. The properties of liquids depend almost entirely on temperature, therefore when looking up the properties of water we look for those against its particular temperature, disregarding its pressure unless it is saturated.

Example. To read from the tables the specific entropy of water at (a) pressure of 1 bar and temperature 50°C
(b) ... ... 4 bar ... ... 86°C
(c) ... ... 8 bar ... ... 165°C
(d) ... ... 14 bar ... ... 195°C

(a) Tables page 2, water 50°C, $s = 0{\cdot}704$ kJ/kg K
(b) ... ... 3, ... 86°C, $s = 1{\cdot}145$ ...
(c) ... ... 4, ... 165°C, $s = 1{\cdot}992$ ...
(d) ... ... 4, ... 195°C, $s_f = 2{\cdot}284$ ...

Note that for case (d) we see the temperature of 195°C is the saturation temperature corresponding to its pressure of 14 bar, therefore we use the suffix f to signify that the water is at its saturation temperature.

STEAM. Example. To find the specific entropy of saturated water and dry saturated steam at pressures (a) 0·1 bar, (b) 5 bar, (c) 20 bar, (d) 50 bar.

(a) Tables page 2, 0·1 bar, $s_f = 0{\cdot}649$, $s_g = 8{\cdot}149$
(b) ... ... 4, 5 bar, $s_f = 1{\cdot}860$, $s_g = 6{\cdot}822$
(c) ... ... 4, 20 bar, $s_f = 2{\cdot}447$, $s_g = 6{\cdot}340$
(d) ... ... 5, 50 bar, $s_f = 2{\cdot}921$, $s_g = 5{\cdot}973$

Example. To find the specific entropy of wet steam (a) pressure 0·2 bar, dryness 0·8 (b) pressure 6 bar, dryness 0·9 (c) pressure 44 bar, dryness 0·95.

Specific entropy of wet steam of dryness fraction x is obtained in a similar manner as for specific enthalpy, previously explained:

$$s = s_f + xs_{fg}$$

(a) Tables page 3, $s = 0{\cdot}832 + 0{\cdot}8 \times 7{\cdot}075 = 6{\cdot}492$
(b) ... ... 4, $s = 1{\cdot}931 + 0{\cdot}9 \times 4{\cdot}830 = 6{\cdot}278$
(c) ... ... 5, $s = 2{\cdot}849 + 0{\cdot}95 \times 3{\cdot}180 = 5{\cdot}870$

Example. Using the following formula, calculate the entropy per kg of superheated steam at a pressure of 20 bar and temperature 300°C, taking the mean specific heat of water as 4·25 kJ/kg K and the mean specific heat of superheated steam as 2·58 kJ/kg K.

Compare the result with the value of entropy given in the tables.

$$s = c_w \ln\left(\frac{T_{sat}}{273}\right) + \frac{h_{fg}}{T_{sat}} + c_{sup} \ln\left(\frac{T}{T_{sat}}\right)$$

c_w = mean specific heat of water
c_{sup} = mean specific heat of superheated steam
T_{sat} = saturation temperature absolute
T = superheated steam temperature absolute

Superheated steam temp. = 573 K
Tables page 4, 20 bar, h_{fg} = 1890
Saturation temp. = 485·4 K

$$\begin{aligned} s &= c_w \ln\left(\frac{T_{sat}}{273}\right) + \frac{h_{fg}}{T_{sat}} + c_{sup} \ln\left(\frac{T}{T_{sat}}\right) \\ &= 4{\cdot}25 \ln\left(\frac{485{\cdot}4}{273}\right) + \frac{1890}{485{\cdot}4} + 2{\cdot}58 \ln\left(\frac{573}{485{\cdot}4}\right) \\ &= 4{\cdot}25 \ln 1{\cdot}778 + 3{\cdot}894 + 2{\cdot}58 \ln 1{\cdot}18 \\ &= 2{\cdot}446 + 3{\cdot}894 + 0{\cdot}427 \\ &= 6{\cdot}767 \text{ kJ/kg K} \quad \text{Ans. (i)} \end{aligned}$$

From tables, page 7, press. 20 bar, temp. 300°C:

s = 6·768 kJ/kg K. Ans. (ii)

A TEMPERATURE–ENTROPY (T–S) DIAGRAM is a clear method of illustrating the process of generating steam.

Referring to Fig. 46, consider 1 kg of water commencing at a

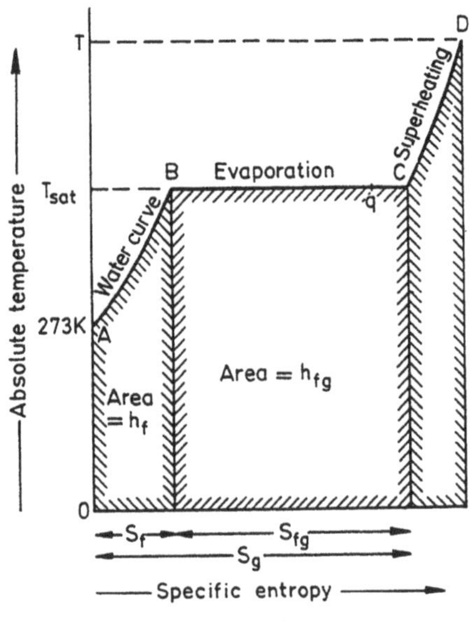

Fig. 46

temperature of 0°C = 273 K, as heat energy is transferred to the water its temperature rises and its entropy is increased, the heating of the water up to its boiling point is represented by the water curve AB. At any point on this water curve, the ordinate is the absolute temperature of the water, and the abscissa is its specific entropy (i.e. for 1 kg here) above that for water at 0°C. The heat energy transferred to the water to raise its temperature is the area of the diagram under the curve AB.

On reaching the saturation temperature corresponding to its pressure, point B, evaporation commences and this continues at constant temperature as the entropy is increased, therefore the evaporation process appears as a horizontal straight line. At point C evaporation is complete and dry saturated steam is produced. The latent heat energy of evaporation is represented by the area of the diagram under the evaporation line BC. Any intermediate point, such as q, along the evaporation line represents incomplete evaporation, that is, the steam is wet, the dryness fraction (x) is represented by the ratio Bq/BC and the heat energy of evaporation is the area under that part of the evaporation line, which is xh_{fg}.

When further heat energy is transferred to dry saturated steam at constant pressure, it becomes superheated and the temperature rises as its entropy increases. This is shown by the superheat curve CD, the area of the diagram under this curve is the heat energy added to superheat the steam.

TEMPERATURE-ENTROPY CHART FOR STEAM

A temperature-entropy (T–S) chart is a complete diagram drawn up within the range of temperatures normally required. A simplified chart, for steam, is illustrated in Fig. 47. The temperature scale here is in °C but this is not usual. Entropy here is specific.

The WATER LINE gives the relation between the temperature of the water and its specific entropy. Curve AB on Fig. 46 is part of this line.

The DRY SATURATED STEAM LINE is produced by drawing a curve through points scaled off to the right from the water line, representing the increase in entropy for complete evaporation of one kg of water into steam at the various temperatures. Any point on this line represents dry saturated steam conditions at the given level of temperature. Point C on Fig. 46 lies on this line. Inside this line the steam is wet, outside this line the steam is superheated.

SUPERHEAT LINES slope steeply upwards from points on the dry saturated steam line. These give the relation between the temperature of the steam and its specific entropy when the steam is superheated at constant pressure. CD on Fig. 46 is one of these lines.

LINES OF CONSTANT DRYNESS FRACTIONS are drawn within the wet steam region, that is, between the two main boundary lines of the water and dry saturated steam. These are plotted by scaling off the values of xs_{fg} from the water line. For instance, if 0·9 of the values of specific entropy of complete evaporation for a series of temperatures were scaled off from the water line, the line drawn through these points is the dryness fraction line of 0·9. Lines for dryness fractions of 0·8, 0·7, etc., down to 0·1 are obtained in a similar manner.

CONSTANT VOLUME LINES are also included. Each constant volume line represents one particular volume occupied by one kg of saturated steam under varying conditions of quality from dry to very wet.

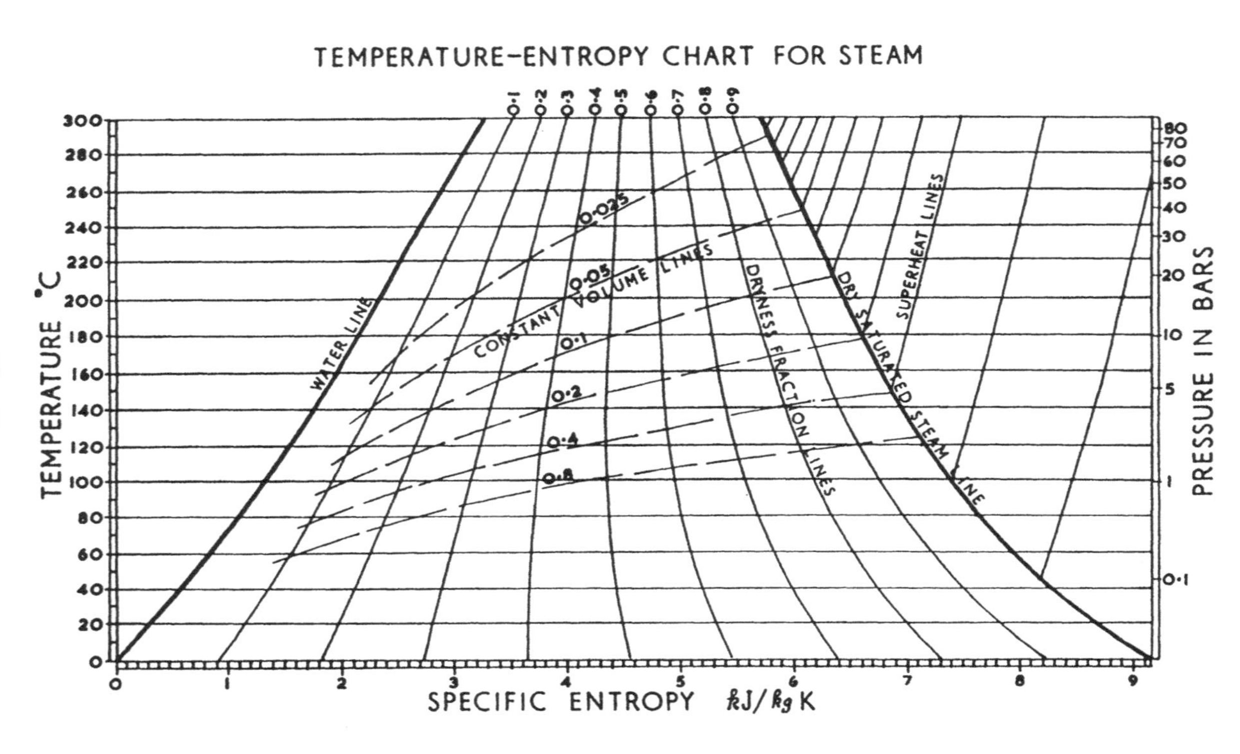
TEMPERATURE-ENTROPY CHART FOR STEAM
0·1
0·2
0·3
0·4
0·5
0·6
0·7
0·8
0·9
TEMPERATURE °C
300
280
260
240
220
200
180
160
140
120
100
80
60
40
20
0
PRESSURE IN BARS
80
70
60
50
40
30
20
10
5
1
0·1
WATER LINE
CONSTANT VOLUME LINES
0·025
0·05
0·1
0·2
0·4
0·8
DRYNESS FRACTION LINES
DRY SATURATED STEAM LINE
SUPERHEAT LINES
SPECIFIC ENTROPY kJ/kg K
0
1
2
3
4
5
6
7
8
9

Fig. 47

Taking a volume of 0·1 m^3 as an example, we read from the tables that 0·1 m^3 is the volume occupied by 1 kg of dry saturated steam when its temperatue is 212·2°C. The curve for a volume of 0·1 m^3 therefore begins at this point on the dry saturated steam line at the level of 212·2°C of temperature.

Now taking some other temperature, say 195°C, we note that the specific volume of dry saturated steam is given as 0·1408 m^3/kg hence, for 1 kg of steam of 195°C to occupy a volume of 0·1 m^3 it must be wet, and its dryness fraction will be 0·1 ÷ 0·1408 = 0·7103. At another temperature, say 165°C, v_g = 0·2728 m^3/kg for 1 kg of steam at 165°C to occupy a volume of 0·1 m^3 the dryness fraction will be 0·1 ÷ 0·2728 = 0·3665. Plotting these three corresponding points of temperature and dryness fractions and drawing a curve through them, the constant volume line for 0·1 m^3 is produced.

The above neglects the volume occupied by the water in the wet steam as being negligible, and obviously more than three plotted points are needed to obtain a true curve. Constant volume lines for various volumes are given covering such a range as likely to be met in practice.

ISOTHERMAL AND ISENTROPIC PROCESSES

During an isothermal process the temperature remains constant, therefore an isothermal operation is represented on a temperature-entropy diagram by a straight horizontal line.

During an adiabatic process, no heat transfer takes place to or from the surroundings, a reversible adiabatic process is *isentropic,* that is, it takes place without change of entropy. Therefore, the entropy after the process is equal to the entropy before, hence, an isentropic process appears on the temperature-entropy diagram as a straight vertical line. The Carnot cycle for a vapour, which is the same for a gas (see Fig. 48), shows these processes.

EFFECT OF ISENTROPIC EXPANSION ON QUALITY. By drawing vertical lines down from the initial steam temperature on the temperature-entropy chart to represent isentropic expansion, it will be seen that dry saturated steam becomes wet during isentropic expansion, steam of normal wetness becomes wetter, and steam which is very wet initially becomes slightly dryer. The T–s diagram of the Rankine cycle (Fig. 61 of Chapter 12) illustrates this.

Example. Steam at a pressure of 14 bar expands isentropically to a pressure of 2·7 bar, find the dryness fraction at the end of expansion if the dryness of the steam at the beginning of expansion is (a) 1·0, (b) 0·8, (c) 0·6, (d) 0·4.

Tables page 4,

14 bar, $s_f = 2{\cdot}284$ $s_{fg} = 4{\cdot}185$ $s_g = 6{\cdot}469$

2·7 bar, $s_f = 1{\cdot}634$ $s_{fg} = 5{\cdot}393$

Entropy after expansion = Entropy before

$$\text{(a) } 1{\cdot}634 + x \times 5{\cdot}393 = 6{\cdot}469$$
$$x \times 5{\cdot}393 = 4{\cdot}835$$
$$x = 0{\cdot}8964 \quad \text{Ans. (a)}$$

$$\text{(b) } 1{\cdot}634 + x \times 5{\cdot}393 = 2{\cdot}284 + 0{\cdot}8 \times 4{\cdot}185$$
$$x \times 5{\cdot}393 = 3{\cdot}998$$
$$x = 0{\cdot}7415 \quad \text{Ans. (b)}$$

$$\text{(c) } 1{\cdot}634 + x \times 5{\cdot}393 = 2{\cdot}284 + 0{\cdot}6 \times 4{\cdot}185$$
$$x \times 5{\cdot}393 = 3{\cdot}161$$
$$x = 0{\cdot}5861 \quad \text{Ans. (c)}$$

$$\text{(d) } 1{\cdot}634 + x \times 5{\cdot}393 = 2{\cdot}284 + 0{\cdot}4 \times 4{\cdot}185$$
$$x \times 5{\cdot}393 = 2{\cdot}324$$
$$x = 0{\cdot}4309 \quad \text{Ans. (d)}$$

The above results can be checked graphically on the temperature-entropy chart, by drawing a vertical line from the temperature level corresponding to 14 bar, which is 195°C, and the appropriate dryness, down to the temperature level corresponding to 2·7 bar, which is 130°C, and reading off the dryness fraction at the lower temperature.

Example. Superheated steam at a pressure of 20 bar and temperature 300°C is expanded isentropically. (i) At what pressure will the steam be just dry and saturated? (ii) What will be the dryness fraction if it is expanded to a pressure of 0·04 bar?

Tables page 7, 20 bar 300°C, $s = 6{\cdot}768$

Tables page 3, 0·04 bar, $s_f = 0{\cdot}422$ $s_{fg} = 8{\cdot}051$

(i) Since entropy at 20 bar 300°C is 6·768, dry saturated steam at a lower pressure is to have the same entropy.

Tables page 4, gives $s_g = 6{\cdot}761$ for 6 bar, therefore, pressure when steam is dry sat. = practically 6 bar. Ans. (i).

(ii) Final entropy = Initial entropy

$$0{\cdot}422 + x \times 8{\cdot}051 = 6{\cdot}768$$
$$x \times 8{\cdot}051 = 6{\cdot}346$$
$$x = 0{\cdot}7881 \quad \text{Ans. (ii)}$$

T–s DIAGRAM FOR GASES

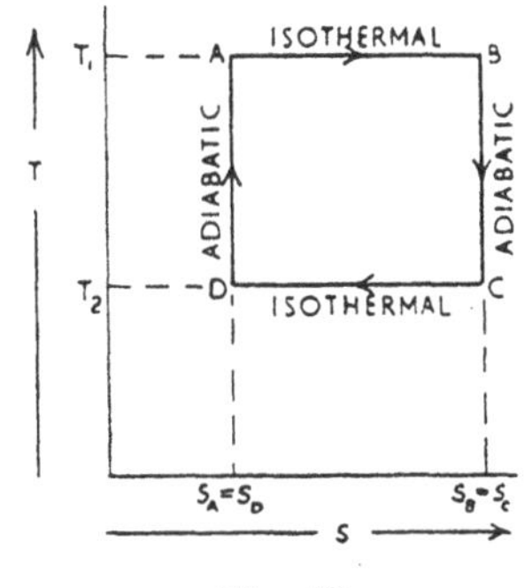

Fig. 48

This provides a good illustration of isothermal and isentropic processes.

As an example, Fig. 30 in Chapter 8 shows the pressure-volume diagram of the Carnot cycle for a gas, this on a temperature-entropy diagram appears as shown in Fig. 48, and, referring to this:

$$\text{Efficiency} = \frac{\text{heat supplied} - \text{heat rejected}}{\text{heat supplied}}$$

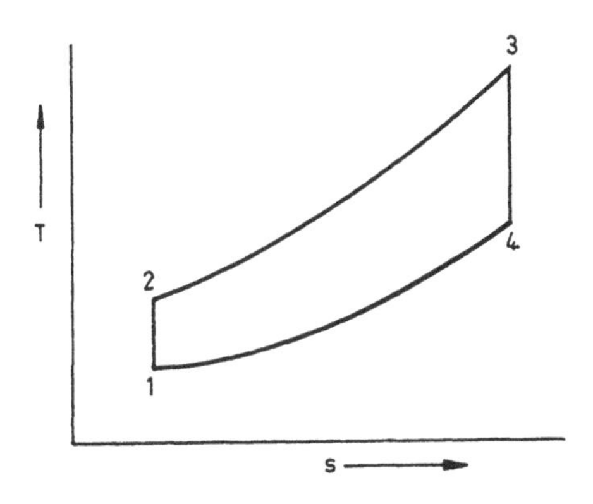

Fig. 49

$$\text{Efficiency} = 1 - \frac{\text{heat rejected}}{\text{heat supplied}}$$

$$= 1 - \frac{T_2(s_C - s_D)}{T_1(s_B - s_A)}$$

$$= 1 - \frac{T_2}{T_1} \quad \text{(as given in Chapter 8)}$$

Other ideal cycles can be similarly illustrated - for example, Fig. 49 is for the Otto cycle (see the *p-V* diagram, Fig. 26 of Chapter 8 (also see for the Joule cycle the *T–s* diagram, Fig. 67 of Chapter 12; there the lines of constant pressure are less steep than the lines of constant volume here).

ENTHALPY–ENTROPY CHART FOR STEAM

A diagram drawn with specific entropy of steam as the base, and the vertical ordinates representing specific enthalpy, is a most useful chart for solving problems or checking results of calculations. Fig. 50 is a simplified chart. Complete full-size enthalpy-entropy (*h–s*) charts to scale are used.

That part under the saturation line is referred to as the wet steam region, and above it represents superheated conditions. Lines of constant dryness are plotted and drawn within the wet steam region, and constant temperature lines are drawn in the superheat region. Lines of constant pressure are drawn throughout the complete diagram.

Note the simplicity of reading off the enthalpy drop and final condition when steam is expanded isentropically, isentropic expansion being represented by drawing a straight vertical line from the initial pressure and condition of steam to the final pressure line. Throttling, being a constant enthalpy process, is represented by a straight horizontal line from the initial pressure and condition to the final pressure line.

Similar charts can be used for other vapours (e.g. refrigerants) and gases. The theoretical basis is also the same.

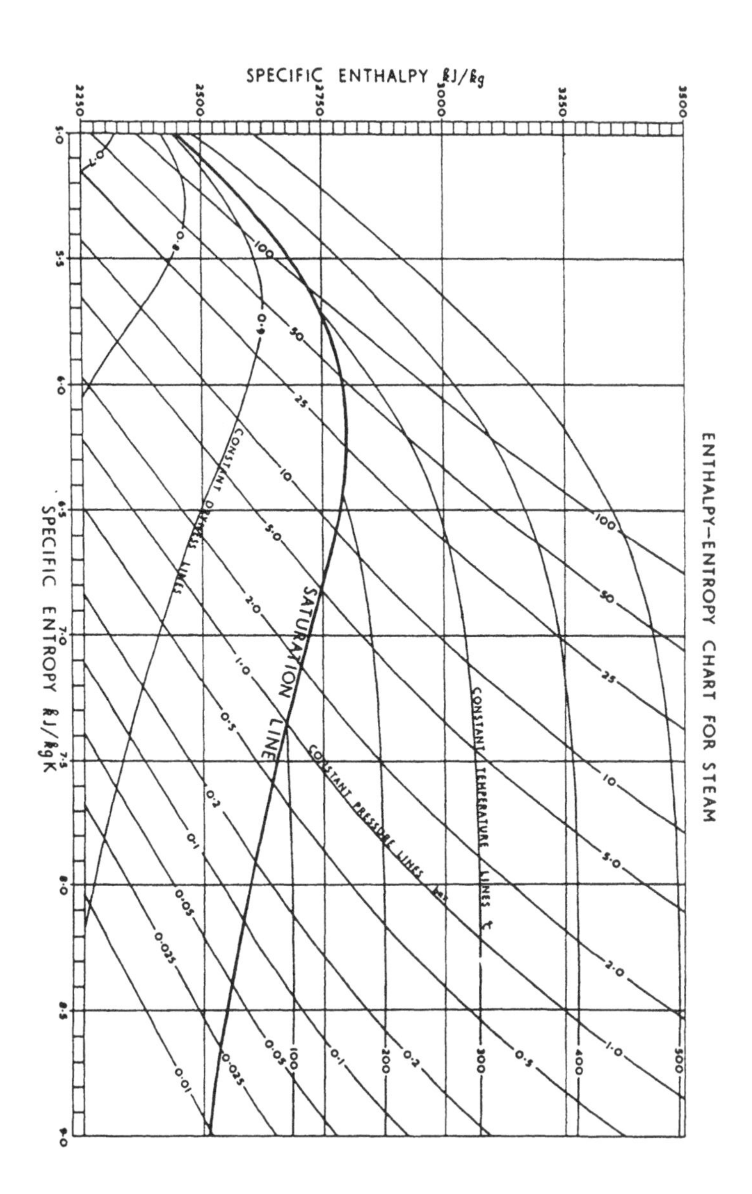
ENTHALPY–ENTROPY CHART FOR STEAM
SPECIFIC ENTHALPY kJ/kg
2250
2500
2750
3000
3250
3500
SPECIFIC ENTROPY kJ/kg K
5·0
5·5
6·0
6·5
7·0
7·5
8·0
8·5
9·0
SATURATION LINE
CONSTANT DRYNESS LINES
CONSTANT PRESSURE LINES bar
CONSTANT TEMPERATURE LINES °C

Fig. 50

f TEST EXAMPLES 11

1. Find the entropy of one kilogramme of wet steam at a pressure of 17 bar if the dryness fraction is 0·95.

2. Find the specific entropy of wet steam of temperature 195°C and dryness 0·9.

3. Dry saturated steam at 5·5 bar is expanded isentropically to 0·2 bar, find the dryness fraction of the steam at the end of expansion.

4. Superheated steam at 17 bar and 350°C is expanded in an engine and the final pressure is 1·7 bar. If the expansion is isentropic, find the dryness fraction of the expanded steam.

5. One kilogramme of dry saturated steam at 22 bar is throttled to a pressure of 7 bar, then expanded at constant entropy to a pressure of 1·4 bar. Calculate (i) the degree of superheat at 7 bar, (ii) the increase in entropy, (iii) the dryness fraction at 1·4 bar.

CHAPTER 12

TURBINES

A turbine is a machine for converting the heat energy in the working fluid (gas or steam) into mechanical energy at the shaft. This series of events is reversed in the rotary compressor. Input drive gives increased velocity to the fluid in the blades which is converted to pressure rise in a fixed diffuser ring. In the axial-flow turbine the rotor coupled to the shaft receives its rotary motion direct from the action of a high velocity jet impinging on blades fitted into grooves around the periphery of the rotor.

There are two types of turbine, the impulse and the "reaction". In both cases the fluid is allowed to expand from a high pressure to a lower pressure so that the fluid acquires a high velocity at the expense of pressure, and this high velocity fluid is directed on to curved section blades which absorb some of its velocity; the difference is in the methods of expanding the fluid.

In impulse turbines the fluid is expanded in nozzles in which the high velocity of the fluid is attained before it enters the blades on the turbine rotor, the pressure drop and consequent increase in velocity therefore takes place in these nozzles. As the fluid passes over the rotor blades it loses velocity but there is no fall in pressure.

In "reaction" turbines, expansion of the fluid takes place as it passes through the moving blades on the rotor as well as through the guide blades fixed to the casing. For a 50% reaction design (usual) the pressure (and enthalpy) ratios across fixed and moving blades are approximately equal.

THE IMPULSE TURBINE

As stated above, in the impulse turbine the high pressure fluid passes into nozzles wherein it expands from a high pressure to a lower pressure and thus the heat energy is converted into velocity energy (kinetic energy). The high velocity jet is directed on to blades fitted around the turbine wheel, the blades being of curved section so that the direction of the steam is changed thereby imparting a force to the blades to push the wheel around. The simplest form of impulse is the single-stage De-Laval. A number of such stages in series is a Rateau turbine; velocity compounding is a Curtis stage.

The best efficiency is obtained when the linear speed of the blades is half of the velocity of the fluid entering the blades, thus, when one set of nozzles is used to expand the fluid from its high supply pressure right down to the final low pressure, the resultant velocity of the fluid leaving the nozzles is very high, say about 1200 m/s. To obtain a high efficiency it means therefore that the wheel should run at a very high speed so that the linear velocity of the blades approaches 600 m/s, for example, in the case of a turbine wheel diameter of one metre, the speed would be about 191 rev/s. Lower speeds, which are more suitable, can be obtained by pressure-compounding, or velocity-compounding, or a combination of these termed pressure-velocity-compounding.

In the pressure-compounded impulse turbine, the drop in pressure is carried out in stages, each stage consisting of one set of nozzles and one bladed turbine wheel, the series of wheels being keyed to the one shaft with nozzle plates fixed to the casing between the wheels.

In the velocity-compounded impulse turbine, the complete drop in pressure takes place in one set of nozzles but the drop in velocity is carried out in stages, by absorbing only a part of the velocity in each row of blades on separate wheels and having guide blades fixed to the casing at each stage between the wheels to guide the fluid in the proper direction on to the moving blades.

The pressure-velocity-compounded turbine is a combination of the two.

CONSTRUCTION. In marine engines where the shaft is required to run in the astern direction as well as ahead, a separate astern turbine is necessary. Fig. 51 shows the principal parts of a pressure-velocity-compounded impulse steam turbine in which there are four pressure stages consisting of four sets of nozzles and four wheels in the ahead turbine, and two similar pressure stages in the astern turbine. Each wheel carries two rows of blades and there is one row of guide blades fixed to the casing protruding radially inwards between each row of moving blades to drop the velocity in two steps from each set of nozzles.

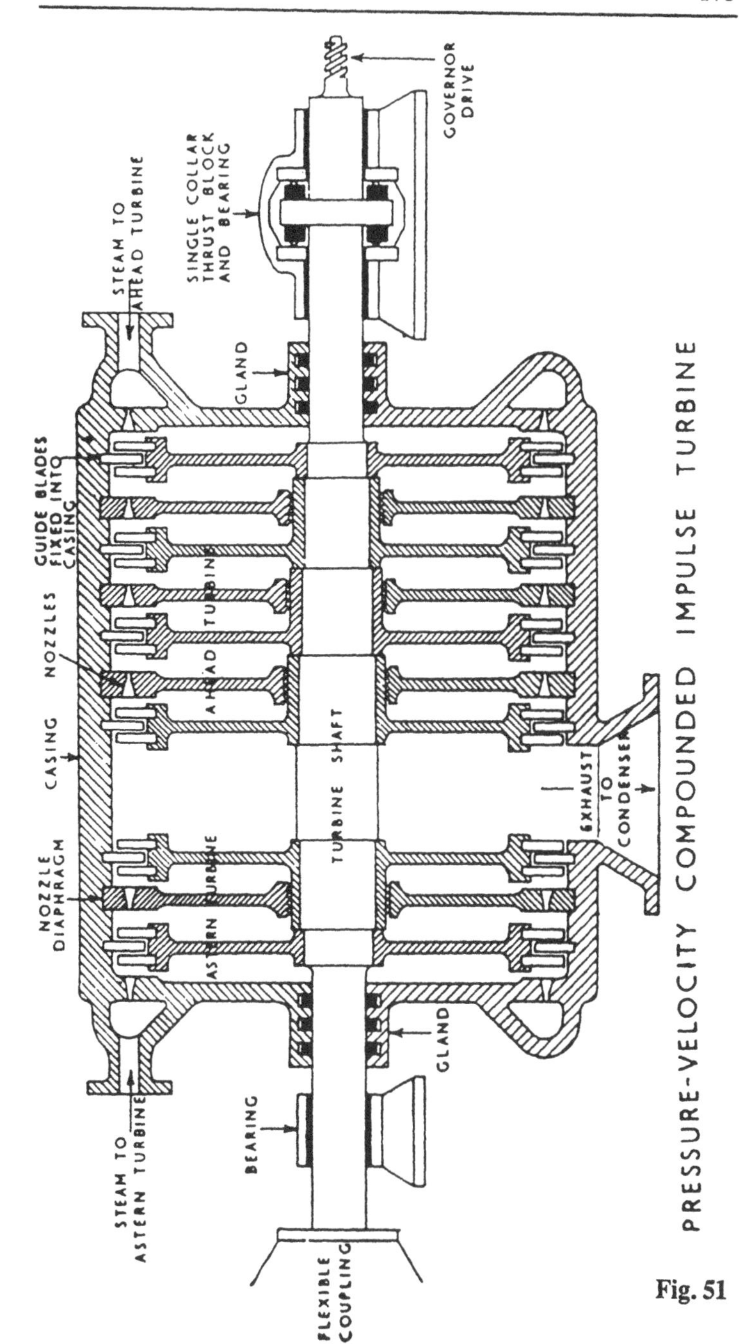

PRESSURE-VELOCITY COMPOUNDED IMPULSE TURBINE

Fig. 51

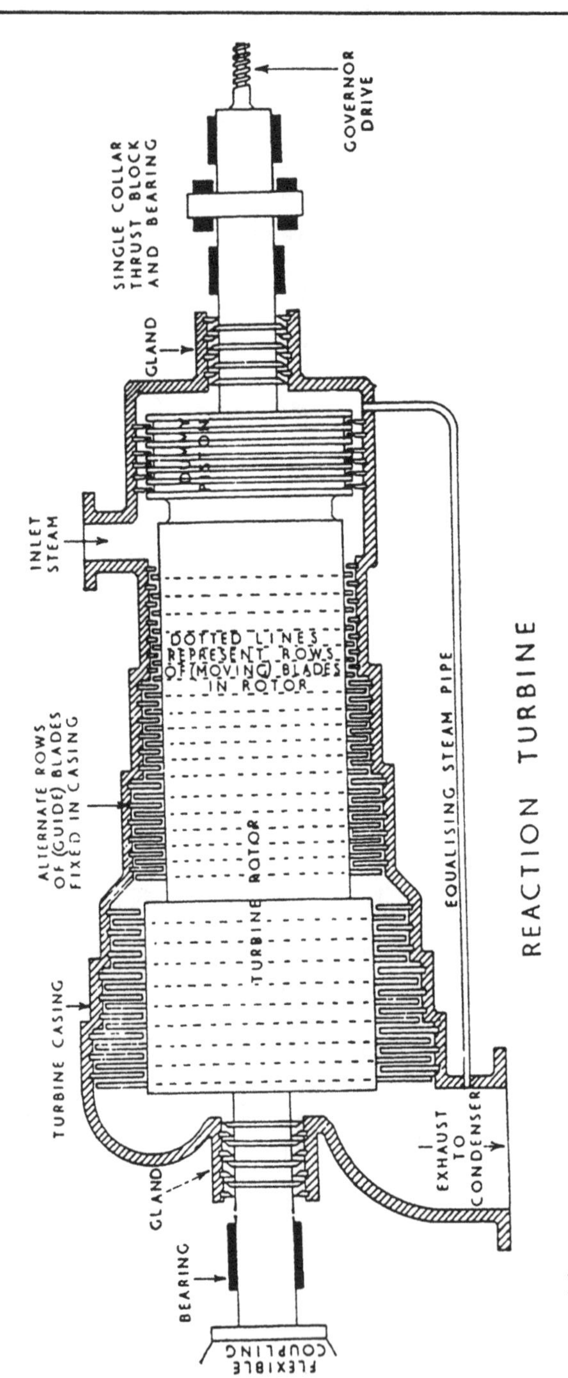
GOVERNOR DRIVE
SINGLE COLLAR THRUST BLOCK AND BEARING
GLAND
INLET STEAM
DOTTED LINES REPRESENT ROWS OF (MOVING) BLADES IN ROTOR
ALTERNATE ROWS OF (GUIDE) BLADES FIXED IN CASING
TURBINE ROTOR
EQUALISING STEAM PIPE
REACTION TURBINE
TURBINE CASING
GLAND
EXHAUST TO CONDENSER
BEARING
FLEXIBLE COUPLING

Fig. 52

THE REACTION TURBINE

This, for a steam turbine, is shown in Fig. 52. The steam is expanded continuously through guide blades fixed to the casing and also as it passes through the moving blades on the rotor, on its way from the inlet end to the exhaust end of the turbine. There are no nozzles as in the impulse turbine. When the high pressure steam enters the reaction turbine, it is first passed through a row of guide blades in the casing through which the steam is expanded slightly, causing a little drop in pressure with a resulting increase in velocity, the steam being guided on to the blades in the first row of the rotor gives an impulse effect to these blades. As the steam passes through the rotor blades it is allowed to expand further so that the steam issues from them at a high relative velocity in a direction approximately opposite to the movement of the blades, thus exerting a further force due to reaction. This operation is repeated through the next pair of rows of guide blades and moving blades, then through the next and so on throughout a number of rows of guide and moving blades until the pressure has fallen to exhaust pressure. As explained, the action of the steam on the blades is partly impulse and partly reaction, and a more correct name for this type of turbine might be "impulse-reaction" but it is generally known as "reaction" to distinguish it from the pure impulse type.

NOZZLES

The function of a nozzle is to produce a jet of high velocity which can be directed on to the blades of a turbine. The velocity is produced by allowing the working fluid to expand from a high pressure at the inlet to a lower pressure at the open exit, as no external work is done the decrease in enthalpy is converted into an increase in kinetic energy. The expansion is adiabatic because practically no heat energy is transferred in the very short time it takes the steam to pass along the nozzle. The ideal expansion would produce the maximum velocity at exit, this would be obtained if there were no friction and the expansion was isentropic.

Let conditions at entrance be:

Kinetic energy $= \frac{1}{2} m v_1^2$

Enthalpy $= H_1$

and conditions at exit:

Kinetic energy $= \frac{1}{2} m v_2^2$

Enthalpy $= H_2$

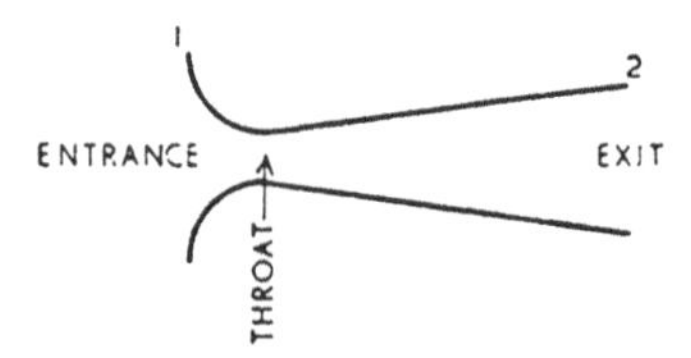

Fig. 53

then, working in joules of energy,

$$\text{Gain in kinetic energy} = \text{Enthalpy drop}$$
$$\tfrac{1}{2} m (v_2^2 - v_1^2) = H_1 - H_2$$

for unit mass,

$$\tfrac{1}{2} (v_2^2 - v_1^2) = h_1 - h_2$$
where h = specific enthalpy.

The inlet velocity is usually small compared with the exit velocity, therefore v_1 is negligible,

$$\tfrac{1}{2} v_2^2 = h_1 - h_2$$
$$v_2 = \sqrt{2(h_1 - h_2)}$$
$$\therefore v\,[\text{m/s}] = \sqrt{2 \times \text{spec. enthalpy drop [J/kg]}}$$

Since $\sqrt{2 \times 10^3} = 44{\cdot}72$, the above may be written

$$v = 44{\cdot}72 \sqrt{\text{spec. enthalpy drop [kJ/kg]}}$$

MASS AND VOLUME FLOW. At any point along the nozzle:

$$\text{Volume flow [m}^3\text{/s]} = \text{area [m}^2\text{]} \times \text{velocity [m/s]}$$

$$\text{mass flow [kg/s]} = \frac{\text{volume flow [m}^3\text{/s]}}{\text{specific volume [m}^3\text{/kg]}}$$

$\therefore$ area [m^2] $\times$ velocity [m/s]
= mass flow [kg/s] $\times$ spec. vol. [m^3/kg].

Example. Dry saturated steam enters a nozzle at 7 bar and leaves at 4 bar 0·98 dry. Find the velocity of the steam at exit. If the exit area of the nozzle is 300 mm^2, calculate the mass flow rate.

Tables page 4, 7 bar, $h_g = 2764$
4 bar, $h_f = 605$ $h_{fg} = 2134$

$$\text{Spec. enthalpy drop} = 2764 - (605 + 0{\cdot}98 \times 2134)$$
$$= 2764 - 2696$$
$$= 68 \text{ kJ/kg} = 68 \times 10^3 \text{J/kg}$$

$$v = \sqrt{2 \times 68 \times 10^3} = 368{\cdot}7 \text{ m/s} \quad \text{Ans.}$$

$$\begin{aligned} \text{Volume flow [m}^3\text{s]} &= \text{area [m}^2] \times \text{velocity [m/s]} \\ &= 300 \times 10^{-6} \times 368{\cdot}7 \\ &= 0{\cdot}1106 \text{ m}^3\text{/s} \end{aligned}$$

Tables page 4, 4 bar $v_g = 0{\cdot}4623$ m³/kg
Spec. volume at 0·98 dry

$$= 0{\cdot}98 \times 0{\cdot}4623 = 0{\cdot}453 \text{ m}^3\text{/kg}$$

$$\text{mass flow} = \frac{0{\cdot}1106}{0{\cdot}453} = 0{\cdot}2442 \text{ kg/s} \quad \text{Ans.}$$

Care must be taken to avoid confusion with the symbols. v usually represents velocity, and the same symbol is used to represent specific volume. In such cases where both velocity and specific volume would appear in the one expression or equation it is therefore advisable to write out the words fully or use an understandable abbreviation.

f Example. Dry saturated steam enters a convergent-divergent nozzle at a pressure of 3·5 bar. The specific enthalpy drop between entrance and throat is 97 kJ/kg and the pressure there is 2·0 bar. The pressure at exit from the nozzle is 0·1 bar, the total possible specific enthalpy drop from entrance to exit is 534 kJ/kg and this is reduced by 12% due to the effect of friction in the divergent part of the nozzle. Calculate the nozzle area at the throat and the mouth to pass 0·113 kg of steam per second.

Tables page 4, 3·5 bar, $h_g = 2732$ $v_g = 0{\cdot}5241$
2 bar, $h_f = 505$ $h_{fg} = 2202$ $v_g = 0{\cdot}8856$
page 3, 0·1 bar, $h_f = 192$ $h_{fg} = 2392$ $v_g = 14{\cdot}67$

$$\begin{aligned} h \text{ of steam at 2 bar} &= h_g \text{ at 3·5 bar} - 97 \\ 505 + x \times 202 &= 2732 - 97 \\ x \times 2202 &= 2130 \\ \text{dryness at throat } x &= 0{\cdot}9674 \end{aligned}$$

Spec. vol. of steam at throat

$$= 0{\cdot}9674 \times 0{\cdot}8856 = 0{\cdot}8567 \text{ m}^3\text{/kg}$$

Velocity through throat [m/s]

$$= \sqrt{2 \times \text{spec. enthalpy drop [J/kg]}}$$

$$= \sqrt{2 \times 10^3 \times 97} = 440{\cdot}4 \text{ m/s}$$

$$\text{area [m}^2] \times \text{velocity [m/s]} = \text{mass flow [kg/s]} \times \text{spec. vol. [m}^3\text{/kg]}$$

$$\therefore \text{ Area [mm}^2] = \frac{0{\cdot}113 \times 0{\cdot}8567}{440{\cdot}4} \times 10^6$$

$$= 219{\cdot}9 \text{ mm}^2 \quad \text{Ans. (i)}$$

Spec. enthalpy drop between entrance and exit

$$= 0{\cdot}88 \times 534 = 469{\cdot}9 \text{ kJ/kg}$$

$$h \text{ of steam at } 0{\cdot}1 \text{ bar} = h \text{ at } 3{\cdot}5 \text{ bar} - 469{\cdot}9$$

$$192 + x \times 2392 = 2732 - 469{\cdot}9$$

$$x \times 2392 = 2070$$

$$\text{dryness at exit } x = 0{\cdot}8654$$

Spec. vol. of steam at exit

$$= 0{\cdot}8654 \times 14{\cdot}67 = 12{\cdot}7 \text{ m}^3/\text{kg}$$

$$\text{velocity at exit} = \sqrt{2 \times 10^3 \times 469{\cdot}9}$$

$$\text{area [m}^2] \times \text{velocity [m/s]} = \text{mass flow [kg/s]} \times \text{spec. vol. [m}^3/\text{kg]}$$

$$\therefore \text{ Area [mm}^2] = \frac{0{\cdot}113 \times 12{\cdot}7}{969{\cdot}4} \times 10^6$$

$$= 1480 \text{ mm}^2 \quad \text{Ans. (ii)}$$

CRITICAL PRESSURE RATIO. Fig. 53 shows a convergent (entrance to throat) divergent nozzle. In the convergent part the fluid (steam, air or gas) is expanding according to the law pV^n = constant gaining in velocity as the area of the nozzle is reducing. Eventually a point is reached when any further reduction in nozzle area does not increase the velocity. The pressure ratio p_2/p_1, where p_2 (or p_T) is the pressure at the throat and p_1 is the inlet pressure, at which the minimum area of nozzle is reached is called the *critical pressure ratio*. At this point mass flow per unit area of nozzle is a maximum.

$$\text{Critical pressure ratio } \frac{p_T}{p_1} = \left(\frac{2}{n+1}\right)^{\frac{n}{n-1}}$$

This may vary from about 0·57 to 0·48 depending upon the value of n. If $n = \gamma = 1{\cdot}4$ for air then

$$\frac{p_T}{p_1} = 0{\cdot}528$$

ISENTROPIC EFFICIENCY

$$\text{Isentropic efficiency} = \frac{\text{actual KE at nozzle exit}}{\text{isentropic KE at nozzle exit}}$$

$$\text{also, isentropic efficiency} = \frac{\text{actual enthalpy change}}{\text{isentropic enthalpy change}}$$

$$\text{and for a perfect gas } h_1 - h_2' = c_P (T_1 - T_2')$$

$$h_1 - h_2 = c_P (T_1 - T_2)$$

$$\text{hence isentropic efficiency} = \frac{c_P (T_1 - T_2')}{c_P (T_1 - T_2)}$$

f Example. Steam at 5·5 bar, 250°C expands isentropically to 1 bar through a convergent-divergent nozzle. The flow is in equilibrium throughout and the inlet velocity is negligible. Determine, if the critical pressure ratio is 0·546, the throat area of the nozzle for a mass flow rate of 0·15 kg/s and the condition of the steam at the exit.

$$p_T = 5{\cdot}5 \times 0{\cdot}546 = 3 \text{ bar}$$

from an h–s chart

$$h_1 = 2960 \text{ kJ/kg}$$
$$h_2 = 2830 \text{ kJ/kg}$$

To determine h_1 and h_2 using tables; h_1 can be read directly, interpolation of entropy values is used for h_2 i.e.:
Isentropic expansion to throat $\therefore$ $s_1 = s_2 = 7{\cdot}227$ kJ/kg K
If t = degrees of superheat.

$$\text{then } s_1 = s_{150} + \frac{t}{50}(s_{200} - s_{150}) \text{ at 3bar}$$

$$\text{is } 7{\cdot}227 = 7{\cdot}078 + \frac{t}{50}(7{\cdot}312 - 7{\cdot}078)$$

$$\therefore t = 50\frac{(7{\cdot}227 - 7{\cdot}078)}{(7{\cdot}312 - 7{\cdot}078)} = 31{\cdot}84\text{°C}$$

$$h_2 = h_{150} + \frac{t}{50}(h_{200} - h_{150})$$

$$= 2762 + \frac{31{\cdot}84}{50}(2866 - 2762)$$

$$= 2828{\cdot}2 \text{ kJ/kg}$$

$$\text{Velocity at throat} = \sqrt{2(h_1 - h_2)}$$

$$= \sqrt{2\,(2962 - 2828) \times 10^3}$$

$$= 513{\cdot}8 \text{ m/s}$$

Specific volume at throat by interpolation

$$= v_{150} + \frac{t}{50}(v_{200} - v_{150})$$

$$= 0{\cdot}6342 + \frac{31{\cdot}84}{50}(0{\cdot}7166 - 0{\cdot}6342)$$

$$= 0{\cdot}6867 \text{ m}^3/\text{kg}$$

$\therefore$ Nozzle area $\times$ 513·8 = 0·6867 $\times$ 0·15
Nozzle area = 0·0002 m² or 2 cm². Ans.

Isentropic expansion to inlet $\therefore s_1 = s_2 = s_3$
At 1 bar $s_1 = s_f + xs_{fg}$
$7{\cdot}227 = 1{\cdot}303 + x \times 6{\cdot}056$
$x = 0{\cdot}98$ dry. Ans.

VELOCITY DIAGRAMS FOR IMPULSE TURBINES

Fig. 54 illustrates the vector diagram of velocities at the entrance side of the rotor blades ("moving" blades). v_1 represents the absolute velocity of the fluid (illustrated here as steam) directed towards the blades at an angle α_1 (as close as practicable) to the direction of their movement. The linear velocity of the blades is represented by u. Drawing the vectors of the steam velocity and blade velocity towards a common point, the vector joining these to form a closed figure is the velocity of the steam relative to the moving blades, represented by v_{r1}.

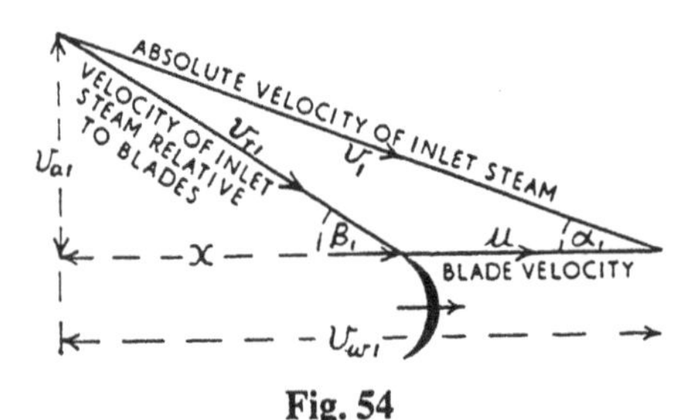

Fig. 54

The relative direction of the steam to the blades is β_1 and in order that the steam should glide on to the blades without shock, the entrance edge of the blades must be in this direction. Therefore β_1 is the *entrance angle of the blades.*

$v_{w1} = v_1 \cos \alpha_1$ is the component of the velocity of the steam jet in the direction of the blade movement, and is referred to as the *velocity of whirl at entrance.*

$v_{\alpha 1} = v_1 \sin \alpha_1$ this is the axial component of the steam jet, that is, the component in the direction of the axis of the turbine.

Fig. 55 is the vector diagram of velocities at the exit side of the moving blades. β_2 is the *exit angle of the blades* and the relative velocity of the exit steam v_{r2} is in this direction. In impulse turbines, since there is no fall in steam pressure as it passes over the rotor blades, if friction is neglected, the relative velocity at exit is the same magnitude as the relative velocity at entrance, that is, $v_{r2} = v_{r1}$. Friction between the steam and the blade surface reduces the velocity and, to take friction into account, $v_{r2} = kv_{r1}$ where the velocity

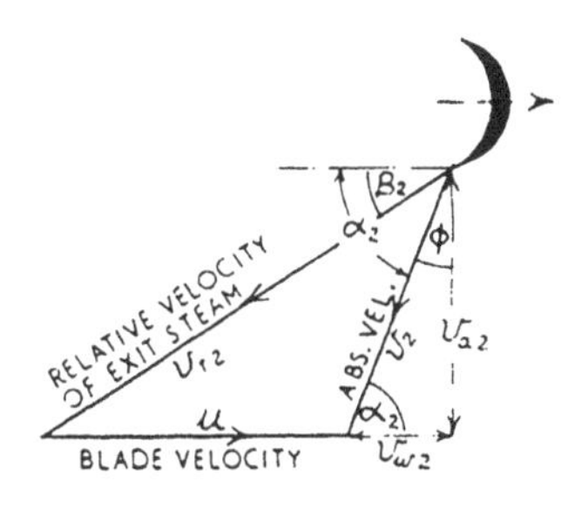

Fig. 55

coefficient k is in the region of 0·8 to 0·95. Loss of kinetic energy in the steam due to friction over the blade surfaces is converted into heat energy.

In the simple impulse turbine the blades are often symmetrical, whence $\beta_1 = \beta_2$.

u is the vector of the blade velocity, v_2 is the vector of the absolute velocity of the exit steam. The direction of the exit steam is at α_2 to the direction of the blade movement, or at angle ϕ to the axis of the turbine.

$v_{w2} = v_2 \cos \alpha_2$ this is the component of the exit steam velocity in the direction of blade movement and is referred to as the *velocity of whirl at exit.*

The axial component of the exit steam is $v_{a2} = v_2 \sin \alpha_2$

Gas, for steam, would be treated in the same way.

Example. Gas at a velocity of 600 m/s from a nozzle is directed on to the blades at 20° to the direction of blade movement. Calculate

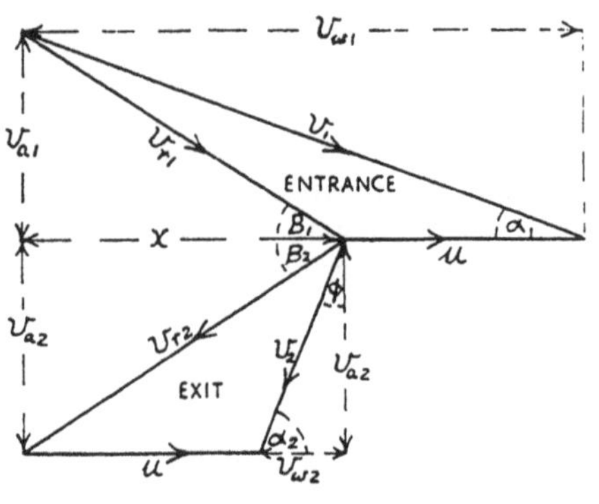

Fig. 56

the inlet angle of the blades so that the gas will enter without shock when the linear velocity of the blades is 240 m/s. If the exit angle of the blades is the same as the inlet angle, find, neglecting blade friction, the magnitude and direction of the gas leaving the blades.

Referring to Fig. 56:

$$v_{a1} = v_1 \sin \alpha_1 = 600 \times \sin 20° = 205{\cdot}2 \text{ m/s}$$
$$v_{w1} = v_1 \cos \alpha_1 = 600 \times \sin 20° = 563{\cdot}9 \text{ m/s}$$
$$x = v_{w1} - u = 563{\cdot}9 - 240 = 323{\cdot}9 \text{ m/s}$$

$$\tan \beta_1 = \frac{v_{a1}}{x} = \frac{205{\cdot}2}{323{\cdot}9}$$

Inlet angle of blades $\beta_1 = 32°22'$. Ans. (i)

Neglecting friction across the blades, $v_{r2} = v_{r1}$ and since $\beta_2 = \beta_1$ then x is common to both entrance and exit diagrams, and $v_{a2} = v_{a1}$.

$$v_{w2} = x - u$$
$$= 323{\cdot}9 - 240 = 83{\cdot}9 \text{ m/s}$$

$$\tan \phi = \frac{v_{w2}}{v_{a2}} = \frac{83{\cdot}9}{205{\cdot}2} = 0{\cdot}4089$$

$\phi = 22°\ 14'$, and $\alpha_2 = 90° - 22°\ 14' = 67°\ 46'$

$$v_2 = \frac{v_{w2}}{\sin \phi} = \frac{83{\cdot}9}{0{\cdot}3783} = 221{\cdot}8 \text{ m/s}$$

Abs. velocity of gas at exit = 221·8 m/s at 67° 46′ to blade movement or 22° 14′ to turbine axis } Ans. (ii)

Both entrance and exit velocity diagrams contain the vector of the blade velocity u therefore they can be combined together by using the blade velocity as a common base, as shown in Fig. 57. This is a convenient diagram to solve either graphically or by calculation and will be used for all future problems of this type. A diagram is essential for reference in the calculations, students are advised to draw the diagram to scale rather than just a rough sketch, it will then serve as a check on the calculated results.

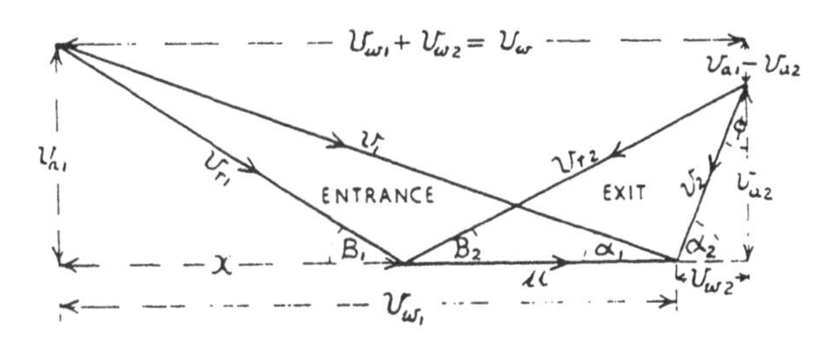

Fig. 57

f VELOCITY COMPOUNDING. Reference was made earlier to velocity compounding, it is a system of alternate moving and fixed blades which enables a large pressure and enthalpy drop to take place in the nozzles. This reduces pressure on the turbine casing and simplifies sealing at the glands, especially for H.P. steam turbines.

The fluid leaves the nozzles with a high velocity and enters a moving row of blades whose velocity is low compared to that of the fluid.

Upon leaving the moving blades at an angle of α_2 to the plane of rotation, the fluid enters a row of stationary blades whose entrance and exit angles are α_2. *i.e.* symmetrical blades. Ideally, as the fluid passes through the fixed blades, there will be no alteration in the pressure, velocity or enthalpy and the fluid will be guided without shock into the next row of moving blades.

f Example. A velocity compounded impulse wheel has two rows of moving and one row of fixed blades, all of which are symmetrical.

Steam leaves the nozzles at 16° to the plane of rotation of the wheel with a velocity of 600 m/s. If the mean blade velocity is 125 m/s and the steam loses 10% of its inlet relative velocity in each of the moving blade passages determine: (a) the blade angles in each of the fixed and moving blades and (b) the direction of the absolute velocity of the final exit steam.

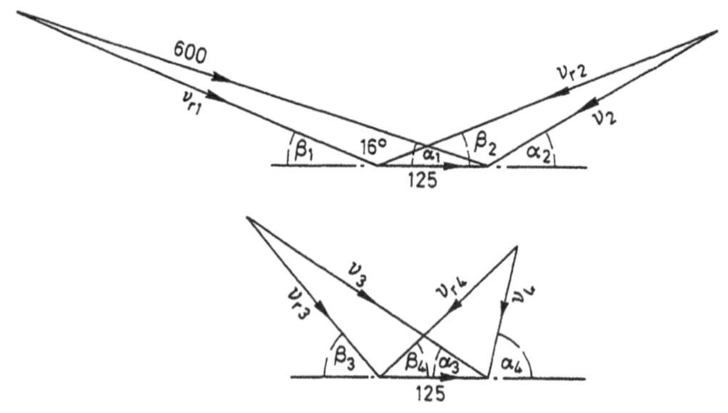

Fig. 58

$$\tan \beta_1 = \frac{600 \sin 16°}{600 \cos 16° - 125}$$

$$\beta_1 = 20{\cdot}1° \quad \beta_1 = \beta_2. \quad \text{Ans. (a) (i)}$$

$$v_{r1} = \frac{600 \sin 16°}{\sin \beta_1} = 481{\cdot}2 \text{ m/s}$$

$$v_{r2} = 0{\cdot}9 \times 481{\cdot}2 = 433{\cdot}1 \text{m/s}$$

$$\tan \alpha_2 = \frac{433{\cdot}1 \sin 20{\cdot}1°}{433{\cdot}1 \cos 20{\cdot}1° - 125}$$

$$\alpha_2 = 27{\cdot}85° \quad \alpha_2 = \alpha_3 \quad \text{Ans. (a) (ii)}$$

$$v_2 = \frac{433{\cdot}1 \sin 20{\cdot}1°}{\sin 27{\cdot}85°} = 318{\cdot}6 \text{ m/s}$$

$v_2 = v_3$ as no friction in fixed blade (assumed)

$$\tan \beta_3 = \frac{318{\cdot}6 \sin 27{\cdot}85°}{318{\cdot}6 \cos 27{\cdot}85° - 125}$$

$$\beta_3 = 43{\cdot}52° \quad \beta_3 = \beta_4 \quad \text{Ans. (a) (iii)}$$

$$v_{r3} = \frac{318{\cdot}6 \sin 27{\cdot}85°}{\sin 43{\cdot}52°} = 216 \text{ m/s}$$

$$v_{r4} = 0{\cdot}9\, v_{r3} = 194{\cdot}5 \text{ m/s}$$

$$\tan \alpha_4 \quad \frac{194{\cdot}5 \sin 43{\cdot}52^\circ}{194{\cdot}5 \cos 43{\cdot}52^\circ - 125} \quad \alpha_4 = 83^\circ \quad \text{Ans. (b)}$$

FORCE ON BLADES

The effective velocity of the fluid causing motion of the rotor blades is the component in the direction of movement of the blades. This is the velocity of whirl. The velocity of whirl at entrance is $v_{w1} = v_1 \cos \alpha_1$. The velocity of whirl at exit is $v_{w2} = v_2 \cos \alpha_2$. The effective change of velocity of the fluid is the algebraic difference between v_{w1} and v_{w2}, let this be represented by v_w. In the previous example, and in most cases, v_{w2} is in the opposite direction to v_{w1} then

$$\text{effective change of velocity} = v_{w1} - (-v_{w2})$$
$$\therefore v_w = v_{w1} + v_{w2}$$

If v_{w2} had been in the same direction as v_{w1} then the effective change of velocity would be $v_{w1} - v_{w2}$.

Since the motion of the fluid is changed as it passes over the blades, the the *blades* must exert a force *on the fluid* to cause this change. From Newton's third law of motion which states that action and re-action are equal in magnitude and opposite in direction, this force is also the magnitude of the force exerted by the fluid on the *blades*.

$$\begin{aligned} \text{Force [N]} &= \text{mass [kg]} \times \text{acceleration [m/s}^2\text{]} \\ &= \text{mass} \times \text{change of velocity per second} \\ &= \text{change of momentum per second} \end{aligned}$$

Therefore, if the mass rate of flow [kg/s] of the fluid is represented by $\dot{m}$, then:

$$\begin{aligned} \text{Force on blades [N]} &= \dot{m}\text{[kg/s]} \times \text{effective change of velocity} \\ &= \dot{m} v_w \end{aligned}$$

$$\text{Work done} = \text{force} \times \text{distance}$$

Since distance through which the force acts on the blades is the linear distance moved by the blades, we have:

Work done per second [N m/s] = force on blades [N] × blade velocity [m/s].

Also,

Work done per second [N m/s] = Power [J/s = W]

therefore,

$$\text{Power [W]} = \dot{m} v_w u$$

The work supplied to the blades is the kinetic energy of the fluid jet,

Work supplied per second = $\frac{1}{2}\dot{m}v_1^2$

hence,

$$\text{Diagram efficiency} = \frac{\text{work done on blades [J/s]}}{\text{work supplied [J/s]}}$$

$$= \frac{\dot{m}v_w u}{\frac{1}{2}\dot{m}v_1^2} = \frac{2uv_w}{v_1^2}$$

This is sometimes called the blade efficiency.

The *axial* force on the blades is due to the difference between the axial components of the fluid velocities at entrance and exit, thus,

$$\text{axial thrust} = \dot{m}\,(v_{a1} - v_{a2})$$

Example. Steam leaves the nozzles of a single stage impulse turbine at a velocity of 670 m/s at 19° to the plane of the wheel, and the steam consumption is 0·34 kg/s. The mean diameter of the blade ring is 1070 mm. Find (i) the inlet angle of the blades to suit a rotor speed of 83·3 rev/s. If the velocity coefficient of the steam across the blades is 0·9 and the blade exit angle is 32°, find (ii) the force on the blades, (iii) the power given to the wheel, and (iv) the diagram efficiency.

Referring to Fig. 57:

$$\text{Linear velocity of blades} = \text{mean circumference} \times \text{rev/s}$$

$$u = \pi \times 1{\cdot}07 \times 83{\cdot}3 = 280 \text{ m/s}$$

$$v_{a1} = v_1 \sin\alpha_1 = 670 \times \sin 19° = 218{\cdot}1 \text{ m/s}$$

$$v_{w1} = v_1 \cos\alpha_1 = 670 \times \cos 19° = 633{\cdot}6 \text{ m/s}$$

$$x = v_{w1} - u = 633{\cdot}6 - 280 = 353{\cdot}6 \text{ m/s}$$

$$\tan\beta_1 = \frac{v_{a1}}{x} = \frac{218{\cdot}1}{353{\cdot}6} = 0{\cdot}6169$$

$$\text{Entrance angle} = 31°40' \quad \text{Ans. (i)}$$

$$v_{r1} = \frac{v_{a1}}{\sin\beta_1} = \frac{218{\cdot}1}{\sin 31°40'} = 415{\cdot}6 \text{ m/s}$$

$$v_{r2} = 0{\cdot}9v_{r1} = 0{\cdot}9 \times 415{\cdot}6 = 374 \text{ m/s}$$

$$v_{w2} = v_{r2}\cos\beta_2 - u = 374\cos 32° - 280$$

$$= 317{\cdot}1 - 280 = 37{\cdot}1 \text{ m/s}$$

$$\text{Effective change of velocity} = v_w = v_{w1} + v_{w2}$$

$$\text{Force on blades [N]} = \dot{m}\text{[kg/s]} \times v_w\text{[m/s]}$$

$$= 0{\cdot}34 \times 670{\cdot}7$$
$$= 228{\cdot}1 \text{ N} \quad \text{Ans. (ii)}$$

$$\text{Power [W = J/s = N m/s]} = \text{force [N]} \times \text{linear velocity [m/s]}$$
$$= 228{\cdot}1 \times 280$$
$$= 63870 \text{ W or } 63{\cdot}87 \text{ kW} \quad \text{Ans. (iii)}$$

$$\text{Diagram efficiency} = \frac{\text{work done on blades [J/s]}}{\text{work supplied [J/s]}}$$

$$= \frac{\dot{m}v_w u}{\frac{1}{2}\dot{m}v_1^2} = \frac{2uv_w}{v_1^2}$$

$$= \frac{2 \times 280 \times 670{\cdot}7}{670^2}$$

$$= 0{\cdot}8368 \text{ or } 83{\cdot}68\% \quad \text{Ans. (iv)}$$

f Example. An impulse turbine has a row of nozzles set at angle α_1, to the plane of motion of the moving blades and issues steam at a velocity of C. The blades have a tangential velocity of u and are symmetrical. Neglecting the effects of friction show that the diagram efficiency is given by:

$$\eta = \frac{4u\,(C\cos\alpha_1 - u)}{C^2}$$

$$v_w = v_{r1}\cos\beta_1 + v_{r2}\cos\beta_2$$
$$v_{r1} = v_{r2} \text{ as no blade friction}$$
$$\beta_1 = \beta_2 \text{ as blades symmetrical}$$
$$v_{r1}\cos\beta_1 = C\cos\alpha_1 - u$$
$$P = \dot{m}v_w u$$
$$= \dot{m} \times 2(C\cos\alpha_1 - u) \times u$$
$$\text{Work supplied} = \tfrac{1}{2}\dot{m}C^2$$

$$\eta = \frac{2\dot{m}u\,(C\cos\alpha_1 - u)}{0{\cdot}5\,\dot{m}C^2}$$

$$= \frac{4u\,(C\cos\alpha_1 - u)}{C^2} \quad \text{Ans.}$$

VELOCITY DIAGRAMS FOR REACTION TURBINES

It has been seen that, in the impulse turbine, the fluid expands whilst passing through fixed nozzles and there is no expansion on

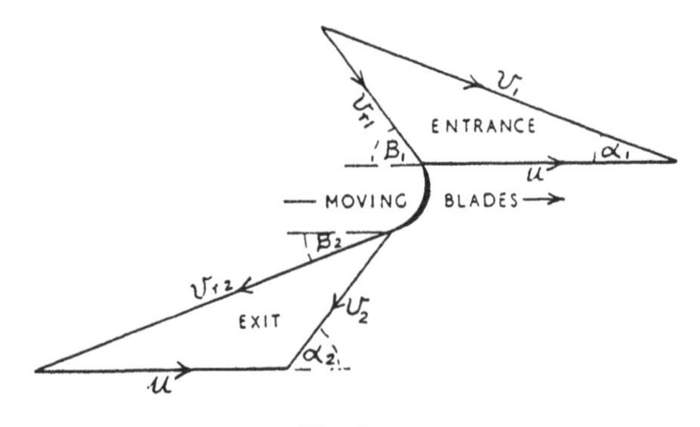

Fig. 59

its passage through the channels between the moving blades. All generation of velocity takes place in the nozzles.

In the reaction turbine, expansion of the fluid takes place during its passage through the fixed (guide) blades which take the place of nozzles, and it also expands as it passes through the moving blades. Therefore the velocity of the fluid is increased as it passes through the fixed blades, and the relative velocity of the fluid to the moving blades is increased as it passes through the moving blades, so that v_{r2} is greater than v_{r1}.

In the reaction turbine, the fixed and moving blades are often of the same section and reversed in direction. In such cases, the entrance and exit angles of the fixed blades are the same as those of the moving blades, and the velocity vector diagram at entrance is identical with the velocity vector diagram at exit, therefore the combined diagram is symmetrical. See Figs. 59 and 60. Thus the relative velocity of the fluid at exit from the moving blades is equal to the absolute velocity at entrance, $v_{r2} = v_1$ and the absolute velocity of the fluid at exit from the moving blades is equal to the relative velocity at entrance, $v_2 = v_{r1}$. Hence, $\beta_2 = \alpha_1$ and $\alpha_2 = \beta_1$.

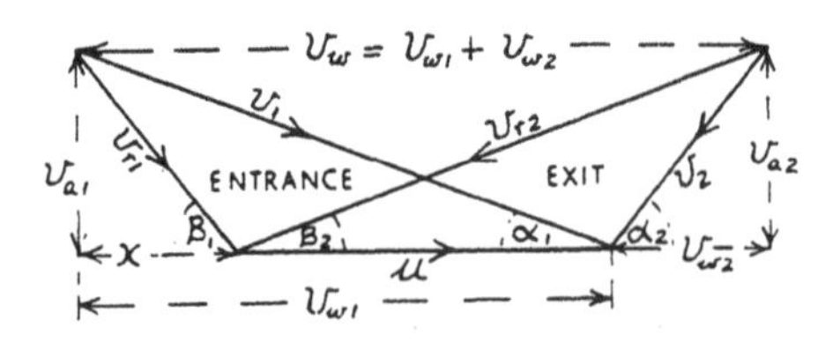

Fig. 60

Example. At one stage of a reaction turbine the velocity of the steam leaving the fixed blades is 90 m/s and the exit angle is 20°. The linear velocity of the moving blades is 60 m/s and the steam consumption is 1·1 kg/s. Assuming the fixed and moving blades to be of identical section, calculate (i) the entrance angle of the blades, (ii) the force on the blades, and (iii) the stage power.

$v_{a1} = v_1 \sin\alpha_1 = 90 \times \sin 20° = 30{\cdot}78$ m/s
$v_{w1} = v_1 \cos\alpha_1 = 90 \times \sin 20° = 84{\cdot}57$ m/s
$x = v_{w1} - u = 84{\cdot}57 - 60 = 24{\cdot}57$ m/s

$$\tan\beta_1 = \frac{30{\cdot}78}{24{\cdot}57} = 1{\cdot}253$$

Entrance angle = 51°24′ Ans. (i)

Effective change of velocity = $v_w = v_{w1} + v_{w2}$
(note that $v_{w2} = x$)
$v_w = 84{\cdot}57 + 24{\cdot}57$ = 109·14 m/s
Force on blades [N] = $\dot{m}$[kg/s] × v_w[m/s]
= 1·1 × 109·19
= 120 N Ans. (ii)

Power [W = J/s = N m/s] = force [N] × velocity [m/s]
= 120 × 60
= 7200 W or 7·2 kW Ans. (iii)

f IDEAL CYCLES

The Carnot cycle has been described for a gas and a vapour - see Fig. 30 (pressure-volume diagram) of Chapter 8 and Fig. 48 (temperature-entropy) of Chapter 11. The wider the range of temperature the more efficient the cycle. The lowest practical condensing temperature is governed by the coolant temperature and the highest practical temperature is governed by metallurgical limits of pressure and temperature. There are two major reasons why the Carnot cycle is not used as the theoretical basis for vapour (steam) cycles. Firstly, it has a low work ratio and secondly it is difficult to compress a wet vapour efficiently; it is easier to fully condense the vapour and compress the liquid to boiler pressure using a feed pump. The resulting cycle is known as the Rankine cycle.

The Rankine cycle is illustrated in Fig. 61 on a pV diagram and on a T-s diagram and consists of:

(i) a to b, feed water pumped into the boiler and receiving heat

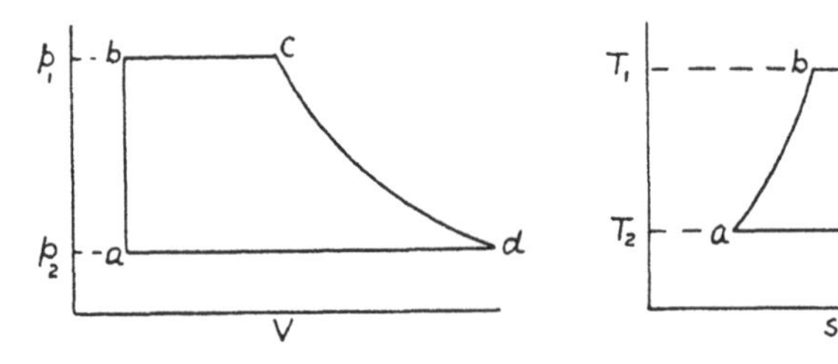

Fig. 61

energy as it is increased in pressure from p_2 to p_1 and in temperature from T_2 to T_1.

(ii) b to c. The water is completely evaporated into steam in the boiler at constant pressure p_1 and constant temperature T_1 and the steam is supplied to the engine as it is generated.

(iii) c to d. The steam supply to the engine expands isentropically from the highest to the lowest limits of pressure and temperature.

(iv) d to a. The steam is exhausted from the engine into the condenser and condensed into water at constant pressure p_2 and constant temperature T_2.

In the ideal engine there would be no energy losses due to friction or heat transfer, no leakage of steam, and no undercooling of the condensate in the condenser, *i.e.* the condensate would be at the saturation temperature corresponding to the condenser steam pressure.

Therefore the work which would be done in the ideal engine is equal to the heat energy given up by the steam on its passage through the engine. This is, per unit mass flow, the enthalpy drop, *i.e.* the difference between enthalpy of supply steam (h_1) and the enthalpy of exhaust steam (h_2).

The heat energy supplied per unit mass flow to the feed water in the boiler to produce the steam is the difference between the enthalpy of the supply steam (h_1) and the enthalpy of the feed water (h_{f2}), the feed water temperature being equal to the condensate temperature.

Thus, neglecting the work done by the feed pump in compressing the water as being comparatively small:

$$\text{Rankine efficiency} = \frac{h_1 - h_2}{h_1 - h_{f2}}$$

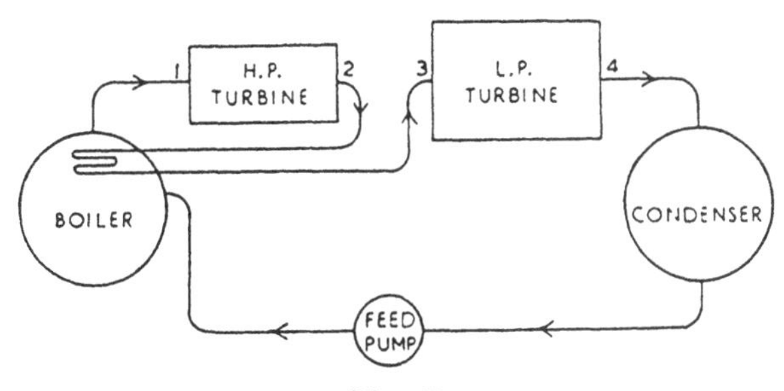

Fig. 62

A cycle operating with steam in the wet region will be limited to a maximum at the critical state (22·1 bar, 374·15°C) which limits efficiency. Superheating raises the temperature (above saturation) without raising the boiler pressure and this increases the ideal cycle efficiency.

It is also common practice to re-heat steam between stages which reduces wetness at low pressures and increases power output for a given size of units. Fig. 62 and the example following illustrate both applications and a resulting Rankine cycle. Examples are often solved more quickly using an h–s chart.

Rankine efficiency

$$= \frac{\text{Heat energy given up by steam through engine}}{\text{Heat energy supplied by boiler and re-heater}}$$

$$= \frac{(h_1 - h_2) + (h_3 - h_4)}{(h_1 - h_{f4}) + (h_3 - h_2)}$$

f Example. In an ideal steam reheat cycle the steam is expanded in the first stage of a turbine from 40 bar 400°C to 3·0 bar. At this pressure the steam is passed through a reheater and its temperature is raised to 300°C at constant pressure. It then passes through the remainder of the turbine and expanded to 0·75 bar. Calculate the Rankine efficiency.

h_1 at 40 bar 400°C = 3214

Isentropic expansion from 40 bar 400°C to 3 bar:

$$\text{Entropy after expansion} = \text{Entropy before}$$

$$1{\cdot}672 + x \times 5{\cdot}321 = 6{\cdot}769$$

$$x \times 5{\cdot}321 = 5{\cdot}097$$

$$x = 0{\cdot}9579$$

h_2 at 3 bar, 0·9579 dry

$$= 561 + 0{\cdot}579 \times 2164 = 2634$$

$$h_3 \text{ at 3 bar } 300^\circ\text{C} = 3070$$

Isentropic expansion from 3 bar 300°C to 0·075 bar:

$$\text{Entropy after expansion} = \text{Entropy before}$$

$$0{\cdot}576 + x \times 7{\cdot}674 = 7{\cdot}702$$

$$x \times 7{\cdot}674 = 7{\cdot}126$$

$$x = 0{\cdot}9287$$

h_4 at 0·075 bar, 0·9287 dry

$$= 169 + 0{\cdot}9287 \times 2405 = 2403$$

h_{f4} at 0·075 bar = 169

$$\text{Rankine efficiency} = \frac{(h_1 - h_2) + (h_3 - h_4)}{(h_1 - h_{f4}) + (h_3 - h_2)}$$

$$= \frac{(3214 - 2634) + (3070 - 2403)}{(3214 - 169) + (3070 - 2634)}$$

$$= \frac{580 + 667}{3045 + 436} = \frac{1247}{3481}$$

$$= 0{\cdot}3583 \text{ or } 35{\cdot}83\% \quad \text{Ans.}$$

ACTUAL STEAM CYCLES

An estimate of the actual vapour (steam) cycle efficiency can be arrived at by amending the isentropic process of expansion (and compression) utilising an efficiency factor. Net work is the turbine (expansion) actual work minus the actual compressor (compression) or pump work.

ISENTROPIC EFFICIENCY. When an actual process is compared with an isentropic process the resulting efficiency is called "isentropic efficiency".

If we consider steam passing through, for example, a turbine. The ratio of the actual enthalpy drop to the isentropic enthalpy drop is the isentropic efficiency. This can be easily shown on a h–s chart.

h = Specific Enthalpy kJ/kg. s = Specific Entropy kJ/kg K.

$$\text{Isentropic efficiency} = \frac{\text{Actual Enthalpy drop}}{\text{Isentropic Enthalpy drop}} \times 100\%$$

$$= \frac{h_a}{h_s} \times 100\%$$

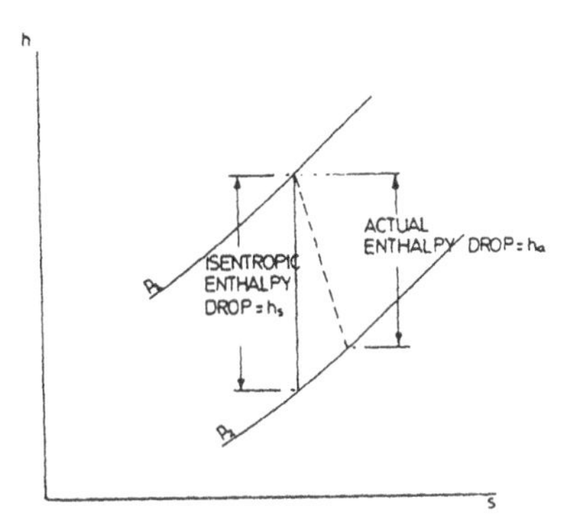

Fig. 63

f Example. Steam at 50 bar and 400°C is expanded in a turbine to 2 bar and 0·96 dry. It is then reheated at constant pressure to 250°C and finally expanded to 0·06 bar with isentropic efficiency of 0·8. Using the h-s chart, find (a) the isentropic efficiency of the first expansion (b) the total power developed for a mass flow of 3 kg/s.

If the reheater was by-passed determine, if its isentropic efficiency was unaltered, the loss of power in the second stage.

Specific enthalpy values read from the chart (Fig. 64) as follows:

$h_1 = 3195$ kJ/kg $h_3 = 2970$ kJ/kg $h_5 = 2120$ kJ/kg
$h_2' = 2615$ kJ/kg $h_4' = ?$
$h_2 = 2512$ kJ/kg $h_4 = 2370$ kJ/kg

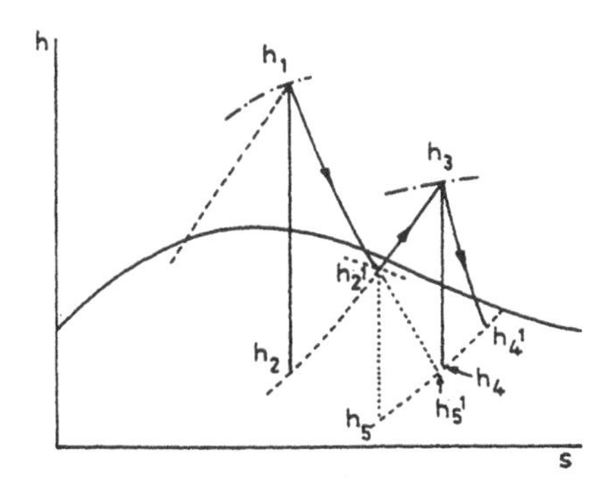

Fig. 64

$$\text{Isentropic efficiency of first stage} = \frac{h_1 - h_2'}{h_1 - h_2}$$

$$= \frac{3195 - 2615}{3195 - 2515} \times 100\%$$

$$= 85\%$$

Actual enthalpy drop in second stage = Isentropic enthalpy drop × Isentropic efficiency

$$= (h_3 - h_4) \times 0{\cdot}8$$
$$= (2970 - 2370) \times 0{\cdot}8$$
$$= 480 \text{ kJ/kg}$$

$$\text{Total power developed} = \dot{m}[(h_1 - h_2') + (h_3 - h_4')]$$
$$= 3[580 + 480]/10^3$$
$$= 3{\cdot}18 \text{ MW.}$$

With re-heater by-passed

$$\text{Actual enthalpy drop in second stage} = (h_2' - h_5) \times 0{\cdot}8$$
$$= (2615 - 2120) \times 0{\cdot}8$$
$$= 396 \text{ kJ/kg}$$

Power loss in second stage = Original power – New power

$$= 3[480 - 396]$$
$$= 252 \text{ kW.}$$

THERMAL EFFICIENCY

The thermal efficiency of an engine is the ratio of the heat energy converted into work in the engine to the heat energy supplied. For a steam engine the heat energy is that transferred to the steam in the boilers from feed water at condenser condensate temperature, thus,

$$\text{energy [kJ] supplied} = \text{steam consumption [kg]} \times (h_1 - h_{f2}) \text{ [kJ/kg]}$$

where, h_1 = enthalpy per kg of supply steam

h_{f2} = enthalpy per kg of water at saturation temperature corresponding to condenser pressure.

hence, on a time basis of one second:

$$\text{Thermal efficiency} = \frac{\text{power [kW = kJ/s]}}{\text{steam consumption [kg/s]} \times (h_1 - h_{f2})}$$

or, on a time basis of one hour (3600 seconds):

$$\text{Thermal efficiency} = \frac{\text{power [kW]} \times 3600}{\text{steam consumption [kg/h]} \times (h_1 - h_{f2})}$$

or, on a basis of one kW h:

$$\text{Thermal efficiency} = \frac{3600}{\text{specific steam cons. [kg/kW h]} \times (h_1 - h_{f2})}$$

f GAS TURBINE CYCLES

Marine type gas turbines work on the ideal constant pressure (Joule) cycle. Fig. 65 is a diagrammatic sketch of a simple open cycle gas turbine plant which consists of three essential parts – air compressor, combustion chamber, and turbine. Referring to Fig. 65, air is drawn in from the atmosphere and compressed from $p_1V_1T_1$ to the higher pressure, smaller volume and higher temperature $p_2V_2T_2$. The compressed air is delivered to the combustion chamber. Some of this air is used for burning the fuel which is admitted through the burner into the combustion chamber, the remainder of the air passes through the jacket surrounding the burner housing, mixes with the products of combustion, and is heated at constant pressure while the volume and temperature increases, the conditions now being $p_3V_3T_3$. The mixture of hot air and gases now passes through the

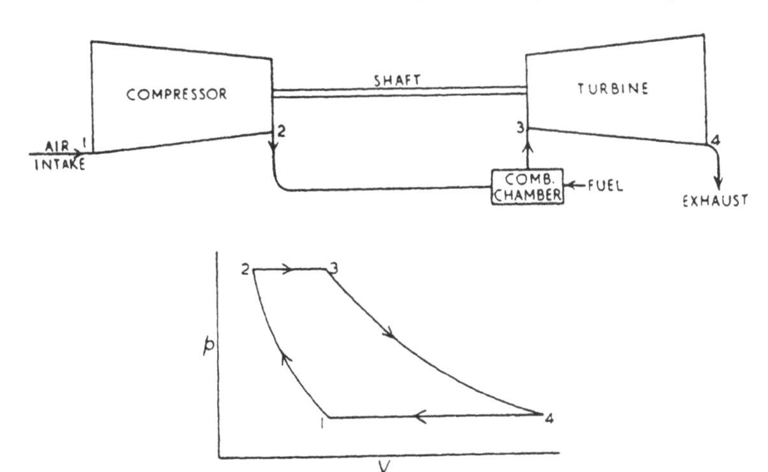

Fig. 65

turbine where it expands to $p_4V_4T_4$ as it does work in driving the rotor. Finally, the gases exhaust at constant pressure.

Of the power developed in the turbine, some is absorbed in driving the compressor, the remainder being available for external use such as for propulsion, or driving an electric generator. A starting motor is fitted at the opposite end of the shaft to that for the external drive.

In the ideal cycle, compression and expansion are isentropic between the same pressures p_2 and p_1 following the law pV^γ = a constant. Referring to Fig. 65:

$$\text{Ideal thermal efficiency} = \frac{\text{heat energy converted into work}}{\text{heat energy supplied}}$$

$$= \frac{\text{heat supplied} - \text{heat rejected}}{\text{heat supplied}}$$

$$= 1 - \frac{\text{heat rejected}}{\text{heat supplied}}$$

$$= 1 - \frac{m \times c_p \times (T_4 - T_1)}{m \times c_p \times (T_3 - T_2)}$$

$$= 1 - \frac{T_4 - T_1}{T_3 - T_2} \qquad \text{(i)}$$

$$\text{Let } r_p = \text{pressure ratio} = \frac{p_2}{p_1} = \frac{p_3}{p_4}$$

$$\frac{T_2}{T_1} = \left\{\frac{p_2}{p_1}\right\}^{\frac{\gamma-1}{\gamma}} \qquad \therefore T_2 = T_1 \times r_p^{(\gamma-1)/\gamma}$$

$$\frac{T_3}{T_4} = \left\{\frac{p_3}{p_4}\right\}^{\frac{\gamma-1}{\gamma}} \qquad \therefore T_3 = T_4 \times r_p^{(\gamma-1)/\gamma}$$

$$T_3 - T_2 = r_p^{(\gamma-1)/\gamma}(T_4 - T_1)$$

Substituting this value of $T_3 - T_2$ into (i):

$$\text{Ideal thermal efficiency} = 1 - \frac{1}{r_p^{(\gamma-1)/\gamma}} \qquad \text{(ii)}$$

Showing that the ideal thermal efficiency depends upon the pressure ratio.

f Example. In a simple gas turbine plant working on the ideal constant pressure cycle, air is taken into the compressor at 1 bar, 16°C, and delivered at 5·5 bar. If the temperature at turbine inlet is 700°C, calculate (i) the temperature at the end of compression, (ii) temperature at exit from the turbine, (iii) the ideal thermal efficiency. Take $\gamma = 1{\cdot}4$.

$$\text{Pressure ratio } r_p = \frac{p_2}{p_1} = \frac{p_3}{p_4} = \frac{5{\cdot}4}{1} = 5{\cdot}4$$

$$\frac{\gamma - 1}{\gamma} = \frac{0{\cdot}4}{1{\cdot}4} = \frac{2}{7}$$

$$r_p^{(\gamma-1)/\gamma} = 5{\cdot}4^{2/7} = 1{\cdot}619$$

$$\frac{T_2}{T_1} = \left\{\frac{p_2}{p_1}\right\}^{\frac{\gamma-1}{\gamma}}$$

$$T_2 = 289 \times 1{\cdot}619 = 467{\cdot}8 \text{ K}$$

Temperature at end of compression
= 194·8°C. Ans. (i)

$$\frac{T_4}{T_3} = \left\{\frac{p_4}{p_3}\right\}^{\frac{\gamma-1}{\gamma}}$$

$$T_4 = \frac{973}{1{\cdot}619} = 601 \text{ K}$$

Temperature at end of expansion
= 328°C. Ans. (ii)

$$\text{Ideal thermal efficiency} = 1 - \frac{T_4 - T_1}{T_3 - T_2}$$

$$= 1 - \frac{601 - 289}{973 - 467{\cdot}8}$$

$$= 0{\cdot}3824 \text{ or } 38{\cdot}24\% \quad \text{Ans. (iii)}$$

Alternatively,

$$\text{Ideal thermal efficiency} = 1 - \frac{1}{r_p^{(\gamma-1)/\gamma}}$$

$$= 1 - \frac{1}{1{\cdot}619} = 0{\cdot}3823$$

HEAT EXCHANGER. By including a heat exchanger, some of the heat energy in the exhaust gases can be utilized by transferring it to the air before it enters the combustion chamber, resulting in less fuel being needed to raise the temperature of the air to the required turbine inlet temperature and therefore increasing the thermal efficiency. A typical arrangement is shown diagrammatically in Fig. 66.

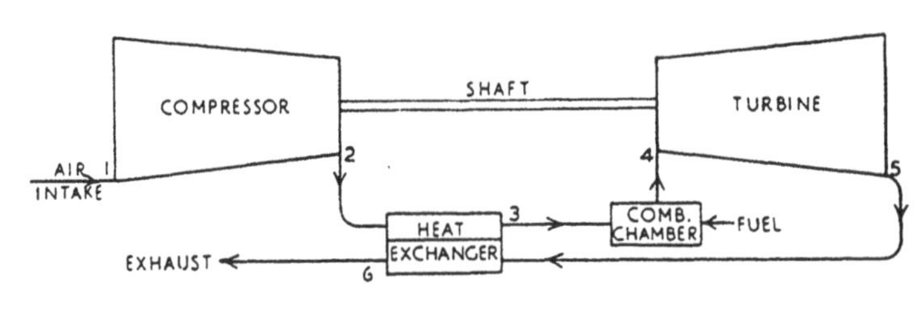

Fig. 66

The lowest temperature entering the heat exchanger is that of the compressed air T_2 and the highest temperature is that of the exhaust gases from the turbine T_5. The difference $T_5 - T_2$ is the overall temperature range for the exchanger. The air, on its passage through the exchanger is increased in temperature from T_2 to T_3, that is, an increase of $T_3 - T_2$. The ratio of the increase in air temperature to the overall temperature range is termed the *thermal ratio* or *effectiveness* of the heat exchanger, thus:

$$\text{Thermal ratio of heat exchanger} = \frac{T_3 - T_2}{T_5 - T_2}$$

ACTUAL CYCLE EFFICIENCY. The actual processes of compression and expansion can be estimated by applying the isentropic efficiency to the ideal isentropic processes. Isentropic efficiency is determined from enthalpy change, which, for a perfect gas, is given by:

$$\Delta h = mc_p \Delta T$$

The processes and cycle, for gas turbines are best illustrated, as in Fig. 67, by a T–s diagram.

f Example. An open cycle gas turbine unit has a pressure ratio of 6:1 and a maximum cycle temperature of 760°C. The isentropic efficiency of the compressor and turbine is 0·85. The minimum cycle temperature is 15°C.

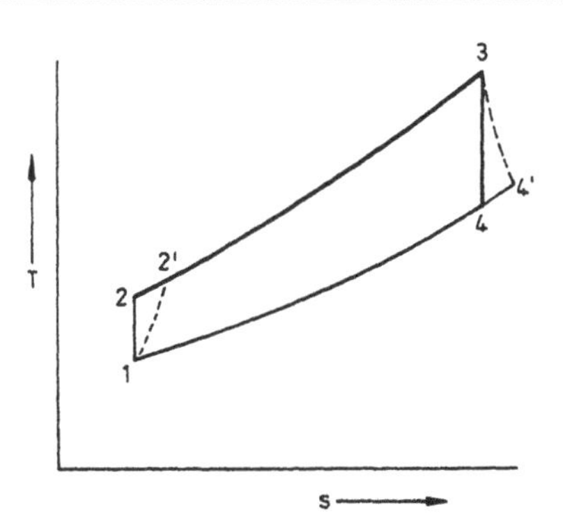

Fig. 67

(a) sketch the cycle on a temperature–entropy diagram;
(b) calculate the thermal efficiency of the unit.

For the working substance $c_p = 1005$ J/kg K $\gamma = 1{\cdot}4$.

Refer to Fig. 67: Ans. (a)

$$\frac{T_2}{T_1} = \left\{\frac{p_2}{p_1}\right\}^{\frac{\gamma-1}{\gamma}}$$

$$T_2 = 288 \times 6^{\frac{0\cdot4}{1\cdot4}}$$

$$= 480{\cdot}4 \text{ K}$$

$$T_2 - T_1 = 480{\cdot}4 - 288$$

$$= 192{\cdot}3 \text{ K}$$

$$T_2{}' - T_1 = \frac{192{\cdot}3}{0{\cdot}85}$$

$$= 226{\cdot}3 \text{ K}$$

$$T_2{}' = 226{\cdot}3 + 288$$

$$= 514{\cdot}3 \text{ K}$$

$$\frac{T_3}{T_4} = \left\{\frac{p_3}{p_4}\right\}^{\frac{\gamma-1}{\gamma}}$$

$$\frac{1033}{T_4} = 6^{\frac{0\cdot4}{1\cdot4}}$$

$$T_4 = 619{\cdot}3 \text{ K}$$

$$T_3 - T_4 = 1033 - 619{\cdot}3$$

$$= 413{\cdot}7 \text{ K}$$

$$T_3 - T_4{}' = 0{\cdot}85 \times 413{\cdot}7$$

$$= 351{\cdot}6 \text{ K}$$

$$T_4' = 681{\cdot}4 \text{ K}$$

$$\text{Thermal efficiency} = \frac{\text{Net work}}{\text{Heat supplied}} \times 100$$

$$= \frac{351{\cdot}6 - 226{\cdot}3}{1033 - 514{\cdot}3} \times 100$$

$$= 24{\cdot}16\% \quad \text{Ans. (b)}$$

Note: c_P and $\dot{m}$ on are common to each term in the above expression for thermal efficiency and cancel out leaving temperatures. c_P, although given, is not required.

TEST EXAMPLES 12

1. Dry saturated steam at 8 bar is expanded in turbine nozzles to a pressure of 5 bar 0·97 dry. Find (i) the velocity at exit. If the area at exit is 14·5 cm^2 find (ii) the mass flow of the steam in kg/s.

2. The velocity of the steam from the nozzles of a turbine is 450 m/s, the angle of the nozzles to blade motion is 20° and the blade entrance angle is 33°. Find (i) the linear velocity of the blades so that the steam enters without shock, and (ii) the rotational speed of the rotor in rev/s if the mean diameter of the blade ring is 660 mm.

3. At a certain stage of a reaction turbine, the steam leaves the guide blades and enters the moving blades at an absolute velocity of 243 m/s at an angle of 23° to the plane of rotation, and the blade velocity is 159 m/s. Fixed and moving blades have the same inlet and exit angles and the steam flow is 0·9 kg/s. Calculate (i) the inlet angle of the blades, (ii) the force on the blades, and (iii) the stage power.

4. A single stage impulse steam turbine has symmetrical rotor blades with inlet and outlet angles of 35°. The mean blade diameter is 600 mm, the turbine runs at 100 rev/s and the steam leaves the rotor in the axial direction:
  (a) Determine:
    (i) the nozzle angle;
    (ii) the velocity of steam leaving the nozzle;
    (iii) the power developed per kg of steam flow.
  (b) State the major assumption made to solve this problem.

*f*5. Dry saturated steam at 14 bar is expanded in a turbine nozzle to 10 bar, expansion following the law pV^n = constant, where the value of n is 1·135, calculate:
  (i) the dryness fraction of the steam at exit,
  (ii) the enthalpy drop through the nozzle per kg of steam,
  (iii) the velocity of discharge,
  (iv) the area of nozzle exit in mm^2 per kg of steam discharged per second.

*f*6. A perfect gas expands from 7 bar 150°C to atmospheric pressure isentropically through a convergent-divergent nozzle.

If the mass flow rate of the gas through the nozzle is 0·25 kg/s determine:
  (a) the velocity of the gas at the nozzle throat
  (b) the throat area.

Note. $\gamma = 1{\cdot}67$, $R = 2078{\cdot}5$ J/kg K and $c_p = 833{\cdot}9$ J/kg K for the gas.

Also $\dfrac{p_T}{p_1} = \left[\dfrac{2}{\gamma + 1}\right]^{\frac{\gamma}{\gamma - 1}}$ is the critical pressure ratio for a gas expanding in a convergent-divergent nozzle according to the law pV^γ = constant where p_T and p_1 are the pressures at throat and inlet respectively.

f 7. In an impulse turbine the theoretical enthalpy drop of the steam through the nozzles is 312·5 kJ/kg and 10% of this is lost in friction in the nozzles. The nozzle angle is 20°, the inlet angle of the blades is 35°, and the absolute velocity of the steam leaving the blades is 204 m/s in the direction of the axis of the turbine. Calculate on the basis of one kg of steam supplied per second:

(i) blade velocity so that there is no shock at steam entry,
(ii) blade angle at exit,
(iii) energy lost due to friction of the steam across the blades,
(iv) axial thrust,
(v) power supplied,
(vi) efficiency of the blading.

f 8. An engine is supplied with steam at a pressure of 15 bar and temperature 250°C, and the pressure of the exhaust is 0·16 bar. Assuming isentropic expansion, find (i) the dryness fraction of the steam after expansion, (ii) the Rankine efficiency.

f 9. Dry saturated steam at 20 bar is throttled to supply a turbine at 10 bar, this is followed by isentropic expansion to 1·4 bar. It is then reheated at constant pressure to 200°C and finally expanded isentropically to 0·4 bar. Using the *h–s* chart determine per kg of steam.

(i) changes in enthalpy during each stage.
(ii) the overall change in entropy.
(iii) the condition of the steam at the end of expansion.

f 10. Air is drawn into a gas turbine working on the constant pressure cycle, at 1 bar 21°C and compressed to 5·7 bar. The temperature at the end of heat supply is 680°C. Taking expansion and compression to be adiabatic where $c_v = 0{\cdot}718$ kJ/kg K, $c_p = 1{\cdot}005$ kJ/kg K, calculate (i) temperature at end of compression, (ii) temperature at exhaust, (iii) heat energy supplied per kg at constant pressure, (iv) increase in internal energy per kg from inlet to exhaust, (v) ideal thermal efficiency.

CHAPTER 13

BOILERS AND COMBUSTION

There are two types of boiler, (i) the fire-tube, in which the hot gases from the combustion of the fuel in the furnace pass through the tubes, while the water is around the outside of the tubes, and (ii) the water-tube boiler in which the water flows through the tubes while the hot gases pass around the outside of the tubes.

CAPACITY AND EQUIVALENT EVAPORATION

The *capacity* of a boiler is the mass of steam that can be produced by it in a given time, usually per hour. However, since the feed water temperature, and pressure and temperature of the steam varies with different plants, it is necessary for purposes of comparison, to refer the evaporative capacity to a common standard. This is the imaginary condition of assuming the feed water temperature as 100°C, to be converted into dry saturated steam at 100°C, and to reduce this to the mass of steam evaporated by unit mass of fuel burned in the furnace. This basis of comparison is termed the *equivalent evaporation, per kg of fuel, from and at* 100°C. For brevity, "per kg of fuel" is usually omitted.

Thus, if for example a boiler produces 60 Mg of steam per hour at 40 bar and 450°C, from feed water at 130°C, when 4800 kg of fuel are burned per hour, then:

Tables page 7, steam 40 bar 450°C, $h = 3330$
... ... 4, water at 130°C, $h = 546$

Heat energy transferred to each kg of steam
$$= 3330 - 546 = 2784 \text{ kJ}$$
Actual evaporative capacity is 60×10^3 kg/h therefore,
total heat energy transferred to steam
$$= 2784 \times 60 \times 10^3 \text{ kJ/h}$$

Now suppose the temperature of the feed water was 100°C and dry saturated steam at 100°C was produced, the heat energy required per kg of steam would be h_{fg} at 100°C which (tables page 2) is 2256·7 kJ, therefore the mass of steam that would be produced under these conditions of "from and at 100°C" would be:

$$\frac{\text{total heat energy transferred to steam per hour}}{\text{heat energy required for each kg}}$$

$$= \frac{2784 \times 60 \times 10^3}{2256 \cdot 7} = 74 \times 10^3 \text{ kg/h or } 74 \text{ Mg/h}$$

This is the capacity of the boiler from and at 100°C.

Since 4800 kg of fuel are burned per hour, the mass of steam that would be produced by each kg of fuel is,

$$\frac{74 \times 10^3}{4800} = 15 \cdot 42 \text{ kg steam/kg fuel}$$

This is the equivalent evaporation, per kg of fuel, from and at 100°C.

Summing up,

if m_s = actual mass of steam generated per unit mass of fuel burned

h_1 = enthalpy per kg of steam

h_w = enthalpy per kg of feed water

h_{fg} 100°C = enthalpy of evaporation per kg at 100°C then, equivalent evaporation from and at 100°C

$$= \frac{m_s (h_1 - h_w)}{h_{fg} \, 100°\text{C}}$$

BOILER EFFICIENCY

The efficiency of a boiler is the ratio of the heat energy transferred to the feed water in converting it into steam, to the heat energy supplied to the boiler by the combustion of the fuel.

The heat energy transferred to the water to produce steam is the difference between the enthalpy of the steam leaving the boiler and the enthalpy of the feed water entering the boiler, thus:

$$\text{mass of steam generated} \times (h_1 - h_w)$$

The heat energy supplied to the boiler is the energy released during combustion of the fuel, which is the product of the mass of fuel burned and its calorific value. The calorific value, as previously explained (Chapter 7) is the heat energy given off during complete combustion of unit mass of the fuel. Therefore,

$$\text{Boiler efficiency} = \frac{\text{Heat energy transferred to water and steam}}{\text{Heat energy supplied by fuel}}$$

$$= \frac{\text{mass of steam} \times (h_1 - h_w)}{\text{mass of fuel} \times \text{calorific value}}$$

or, if we let m_s represent the mass of steam generated per unit mass of fuel burned, then,

$$\text{Boiler efficiency} = \frac{m_s (h_1 - h_w)}{\text{calorific value of the fuel}}$$

Example. An oil-fired boiler working at a pressure of 15 bar generates 14·5 kg of steam per kg of fuel burned. The feed water temperature is 95°C and the steam leaves the boiler 0·98 dry. If the calorific value of the oil is 42 MJ/kg, calculate the thermal efficiency of the boiler and the equivalent evaporation from and at 100°C.

Tables page 4, steam 15 bar, $h_f = 845$ $h_{fg} = 1947$

... ... 2, water 95°C, $h_w = 398$

$$\text{Boiler steam, } h_1 = h_f + x h_{fg}$$
$$= 845 + 0{\cdot}98 \times 1947 = 2753$$

Heat energy transferred to steam per kg of fuel burned

$$= m_s (h_1 - h_w)$$
$$= 14{\cdot}5\,(2753 - 398)$$
$$= 14{\cdot}5 \times 2355 \text{ kJ}$$

Heat energy supplied to boiler per kg of fuel burned

$$= 42 \times 10^3 \text{ kJ}$$

$$\text{Boiler efficiency} = \frac{14{\cdot}5 \times 2355}{42 \times 10^3}$$
$$= 0{\cdot}8132 \text{ or } 81{\cdot}32\%. \quad \text{Ans. (i)}$$

Equivalent evaporation, per kg of fuel, from and at 100°C

$$= \frac{m_s (h_1 - h_w)}{h_{fg}\ 100^\circ\text{C}}$$
$$= \frac{14{\cdot}5 \times 2355}{2256{\cdot}7}$$
$$= 15{\cdot}13 \text{ kg steam/kg fuel.} \quad \text{Ans. (ii)}$$

FEED WATER

Every precaution must be taken to maintain water as pure as possible. One of the most common causes of contamination of the water is a leakage of sea water.

There is always a certain amount of loss in the steam circuit through the engines and this loss must be made up to maintain the working level of the water in the boiler. Make-up feed may be taken direct from the reserve fresh water tanks, or, distilled water may be produced by evaporating sea water in evaporators. It is usual to pass all make-up feed water through evaporators.

The quantity of dissolved solids present in water is expressed in parts of solids per million parts of water, abbreviated to *parts per million* and represented by p.p.m. This is the same ratio as the grammes of solids in one million grammes of water. Note that 10^6 g = 10^3 kg = 1 tonne and this is the mass of one cubic metre of pure water. For example, if there are 80 g of dissolved solids in one m^3 of water, it is expressed as 80 p.p.m. As a further example, 5 tonne of water having 150 p.p.m. dissolved solids, the total mass of dissolved solids is

$$5 \times 150 = 750 \text{ g.}$$

Equations dealing with the amount of dissolved solids in evaporators and boilers are dealt with on this basis. Thus, in boilers with a contaminated feed, the equation is built upon the simple principle:

$$\begin{array}{c}\text{initial mass [g] of}\\\text{solids in boiler}\end{array} + \begin{array}{c}\text{mass [g] of solids}\\\text{put in with feed}\end{array} = \begin{array}{c}\text{final mass [g] of}\\\text{solids in boiler}\end{array}$$

$$\begin{array}{c}\text{mass [t] of}\\\text{water in}\\\text{boiler}\end{array} \times \begin{array}{c}\text{initial}\\\text{p.p.m.}\end{array} + \begin{array}{c}\text{mass [t]}\\\text{of feed}\end{array} \times \begin{array}{c}\text{feed}\\\text{p.p.m.}\end{array} = \begin{array}{c}\text{mass [t] of}\\\text{water in}\\\text{boiler}\end{array} \times \begin{array}{c}\text{final}\\\text{p.p.m.}\end{array}$$

In evaporators where it is common to have a constant blowdown to maintain a steady pre-determined density of the water in the evaporator, the equation would be based upon

$$\text{mass of solids put in} = \text{mass of solids blown out}$$
$$\text{mass of feed} \times \text{feed p.p.m.} = \text{mass blown out} \times \text{blow out p.p.m.}$$

If the evaporated water contains a given amount of solids this would be included on the right hand side of the above equation.

Example. A boiler initially contains 4 tonne of water of 80 p.p.m. If the evaporation rate is 500 kg/h and the feed water contains 150 p.p.m. of dissolved solids, calculate (i) the density (p.p.m.) of the boiler water after 12 hours, and (ii) the time for the water to reach 2000 p.p.m. from the time of initial condition.

$$\begin{aligned}\text{Amount of feed} &= \text{evaporation rate}\\&= 500 \text{ kg/h} = 0{\cdot}5 \text{ tonne/h}\\&= 0{\cdot}5 \times 12 \text{ tonne in 12 hours}\end{aligned}$$

Solids in initially + solids put in = solids in finally

$$\text{water in boiler} \times \text{initial p.p.m.} + \text{amount of feed} \times \text{feed p.p.m.} = \text{water in boiler} \times \text{final p.p.m.}$$

$$4 \times 80 + 0{\cdot}5 \times 12 \times 150 = 4 \times \text{final density}$$

$$320 + 900 = 4 \times \text{final density}$$

$$\text{Final density} = \frac{1220}{4} = 305 \text{ p.p.m.} \quad \text{Ans. (i)}$$

Let t = time in hours for boiler water to reach 2000 p.p.m. then feed in t hours = $0{\cdot}5t$ tonne:

$$\text{water in boiler} \times \text{initial p.p.m.} + \text{amount of feed} \times \text{feed p.p.m.} = \text{water in boiler} \times \text{final p.p.m.}$$

$$4 \times 80 + 0{\cdot}5t \times 150 = 4 \times 2000$$

$$75t = 7680$$

$$t = 102{\cdot}4 \text{ hours} \quad \text{Ans. (ii)}$$

If an analysis of the feed water is given so that a distinction can be made with regard to the solids that remain in solution and those that precipitate to form scale when the water is heated, then the density in the evaporator or boiler is taken as that due to permanently soluble solids.

The density of sea water and the composition of the dissolved solids vary throughout different parts of the world. A typical sample could be taken as dissolved solids 32280 p.p.m. and mass analysis of the solid matter:

Sodium chloride	79·3%
Magnesium chloride	10·2%
Magnesium sulphate	6·1%
Calcium sulphate	3·8%
Calcium bicarbonate	0·6%

Scale is formed from magnesium sulphate, calcium sulphate and calcium bicarbonate. All scales are bad heat conductors and therefore hinder transfer of heat, the result being overheating of the metal heating surfaces with consequent loss of strength and possibility of collapse. The remainder of the solids may be regarded as permanently soluble solids.

PRINCIPLES OF COMBUSTION

Combustion of fuel is the chemical combination, at high temperature, of the combustible elements in the fuel with oxygen, heat energy being released in the process.

In furnaces of boilers and cylinders of internal combustion engines, the oxygen is obtained from an air supply, air being composed of approximately 23% oxygen and 77% nitrogen, by mass. The oxygen is the active element. Nitrogen, being an inert gas, takes no active part, it acts as a moderator, dilutes the products of combustion and, as it absorbs some of the heat energy produced it reduces the temperature of combustion.

The principal combustible elements in fuels are carbon and hydrogen, others may be present in small quantities, such as sulphur.

The air must be intimately mixed with the fuel and the amount of fuel which can be burned depends upon the quantity of air supplied. An excess over the theoretical minimum quantity of air for complete combustion is always necessary, the amount of excess depending upon the design of the combustion space and conditions under which the fuel is burned. If there is an insufficient air supply, combustion will not be complete, one indication of this being black smoke. If too much air is supplied, an unnecessary amount of heat energy will be carried away to waste. Each case represents a loss of efficiency.

Taking the composition of air by mass as 23% oxygen and 77% nitrogen, then 23 kg of oxygen will be obtained from 100 kg of air. In the same proportion, to obtain 1 kg of oxygen, the supply of air must be $\frac{100}{23}$ kg. Thus, the theoretical minimum mass of air required is $\frac{100}{23}$ times the mass of oxygen needed for complete combustion.

Elements are substances in their simplest form and consist of molecules which are identical. They are represented by a symbol, usually the first letter of their names except when necessary to distinguish between two whose names have the same first letter. The molecules of some elements consist of single atoms, such as carbon (C) and sulphur (S). Other elements consist of molecules of two atoms each and these are distinguished by the subscript 2, examples of these are hydrogen (H_2), nitrogen (N_2) and oxygen (O_2).

Compounds are chemical combinations of different elements and denoted by placing together the symbols of the elements which constitute the compound. Thus, each molecule of water (and steam) is composed of two atoms of hydrogen and one atom of oxygen, and is therefore written H_2O. Carbon dioxide is represented by CO_2 which signifies that each molecule of carbon dioxide is composed of one atom of carbon and two atoms of oxygen. The number of molecules, when more than one, is represented by placing that number in front, thus $3CO_2$ represents three molecules of carbon dioxide.

Atomic weights are pure numbers representing the relative masses of the atoms. For instance, if the atomic weight of hydrogen is 1 and the atomic weight of sulphur is 32, it means that the mass of one atom of sulphur is 32 times greater than the mass of one atom of hydrogen. These numbers are purely relative and neither the actual weight nor actual mass.

The *molecular weight* is the sum of the atomic weights of the atoms of which the molecules is composed.

A list of the relative masses of the substances involved in calculations on the combustion of fuels is given below, to the nearest whole number which is sufficiently accurate for practical purposes.

Substance	*Symbol*	*Atomic weight*	*Molecular weight*
ELEMENTS			
Hydrogen	H_2	1	$2 \times 1 = 2$
Carbon	C	12	$1 \times 12 = 12$
Nitrogen	N_2	14	$2 \times 14 = 28$
Oxygen	O_2	16	$2 \times 16 = 32$
Sulphur	S	32	$1 \times 32 = 32$
COMPOUNDS			
Water, Steam	H_2O		$2 \times 1 + 1 \times 16 = 18$
Carbon dioxide	CO_2		$1 \times 12 + 2 \times 16 = 44$
Sulphur dioxide	SO_2		$1 \times 32 + 2 \times 16 = 64$
Carbon monoxide	CO		$1 \times 12 + 1 \times 16 = 28$

STOICHIOMETRY. Is the determination of the exact proportion of elements to make pure chemical compounds. In examples that follow the expression "*stoichiometric air requirement*" is sometimes used. The stoichiometric air requirement is the exact amount of air required for complete combustion.

CALORIFIC VALUES. The calorific value (c.v.) of a substance is the amount of heat energy released during complete combustion of unit mass of that substance and, for fuels, it is usually expressed in megajoules of energy per kilogramme of mass [MJ/kg].

When hydrogen burns it combines with oxygen to form steam and the heat energy released is about 144 MJ per kg of hydrogen.

Carbon, if supplied with sufficient oxygen, will burn completely to carbon dioxide and in doing so, about 33·7 MJ of heat energy is released per kg of carbon. If there is a deficiency of oxygen, some or all of the carbon will burn to carbon monoxide and the heat energy released by each kg of carbon is then only about one-third of that when carbon dioxide is formed. Thus there is a great loss when carbon monoxide is produced due to an insufficient air supply to the fuel.

In the burning of sulphur, it chemically combines with oxygen to form sulphur dioxide and about 9·3 MJ of heat energy is released per kg of sulphur.

CHEMICAL EQUATIONS represent the proportions in which the elements combine and, by substituting the atomic weights, the relative mass of oxygen required for the burning of each combustible element can be calculated.

Combustion of hydrogen to steam,

$$2H_2 + O_2 = 2H_2O$$

inserting atomic weights,

$$\begin{aligned} 2 \times (2 \times 1) + 2 \times 16 &= 36 \\ 4\text{ kg } H_2 + 32\text{ kg } O_2 &= 36\text{ kg } H_2O \\ 1\text{ kg } H_2 + 8\text{ kg } O_2 &= 9\text{ kg } H_2O \end{aligned}$$

That is, 1 kg of hydrogen requires 8 kg of oxygen to burn it completely, and this produces 9 kg of steam.

Combustion of carbon to carbon dioxide,

$$\begin{aligned} C + O_2 &= CO_2 \\ 1 \times 12 + 2 \times 16 &= 44 \\ 12\text{ kg } C + 32\text{ kg } O_2 &= 44\text{ kg } CO_2 \\ 1\text{ kg } C + 2\tfrac{2}{3}\text{ kg } O_2 &= 3\tfrac{2}{3}\text{ kg } CO_2 \end{aligned}$$

Thus, 1 kg of carbon requires 2⅔ kg of oxygen to burn it completely and this produces 3⅔ kg of carbon dioxide.

Combustion of carbon to carbon monoxide,

$$\begin{aligned} 2C + O_2 &= 2CO \\ 2 \times 12 + 2 \times 16 &= 2 \times 28 \\ 24\text{ kg } C + 32\text{ kg } O_2 &= 56\text{ kg } CO \\ 1C + 1\tfrac{1}{3}\, O_2 &= 2\tfrac{1}{3}\, CO \end{aligned}$$

Thus, 1 kg of carbon if supplied with 1⅓ kg of oxygen gives incomplete combustion which produces 2⅓ kg of carbon monoxide.

Combustion of sulphur to sulphur dioxide,

$$\begin{aligned} S + O_2 &= SO_2 \\ 1 \times 32 + 2 \times 16 &= 64 \\ 32\,\text{kg S} + 32\,\text{kg } O_2 &= 64\,\text{kg } SO_2 \\ 1\,\text{kg S} + 1\,\text{kg } O_2 &= 2\,\text{kg } SO_2 \end{aligned}$$

Hence, 1 kg of sulphur requires 1 kg of oxygen to burn it and 2 kg of sulphur dioxide is produced.

The analysis of some fuels show that a little oxygen is present, in these cases it is usual to assume that this oxygen is combined with some of the hydrogen in the form of water and therefore this portion of hydrogen cannot be burned. We have seen above that hydrogen and oxygen combine in the proportion of 1 to 8, hence the amount of hydrogen which is not available for combustion is one-eighth of the mass of oxygen present in the fuel. The remaining hydrogen in the fuel is referred to as *available hydrogen*, thus,

$$\text{Available hydrogen} = H_2 - \frac{O_2}{8}$$

From the foregoing notes we can now write the expressions:

$$\text{Calorific value} = 33{\cdot}7\,C + 144\left(H_2 - \frac{O_2}{8}\right) + 9{\cdot}3\,S \text{ MJ/kg}$$

$$\text{Oxygen required} = 2\tfrac{2}{3}\,C + 8\left(H_2 - \frac{O_2}{8}\right) + S \text{ kg } O_2\text{/kg fuel}$$

$$\begin{aligned} \text{Minimum air reqd. (stoichiometric)} &= \frac{100}{23} \times \text{oxygen required} \\ &= \frac{100}{23}\left\{2\tfrac{2}{3}\,C + 8\left(H_2 - \frac{O_2}{8}\right) + S\right\} \text{kg air/kg fuel} \end{aligned}$$

Example. An oil fuel is composed of 86% carbon, 11% hydrogen, 2% oxygen and 1% impurities. Calculate the calorific value and stoichiometric mass of air required to burn 1 kg of the fuel, taking the calorific values of carbon and hydrogen as 33·7 and 144 MJ/kg respectively.

$$\begin{aligned} \text{Available hydrogen} &= H_2 - \frac{O_2}{8} \\ &= 0{\cdot}11 - \frac{0{\cdot}02}{8} = 0{\cdot}1075\,\text{kg} \end{aligned}$$

Heat energy from 0·86 kg of carbon

$$= 0{\cdot}86 \times 33{\cdot}7 = 28{\cdot}98 \text{ MJ}$$

Heat energy from 0·1075 kg of hydrogen

$$= 0{\cdot}1075 \times 144 = 15{\cdot}48 \text{ MJ}$$

Total heat energy per kg of fuel

$$= 28{\cdot}98 + 15{\cdot}48 = 44{\cdot}46 \text{ MJ/kg} \quad \text{Ans. (i)}$$

Oxygen required to burn the carbon

$$= 2\tfrac{2}{3} \text{ kg} \times 0{\cdot}86 = 2{\cdot}293 \text{ kg}$$

Oxygen required to burn the available hydrogen

$$= 8 \times 0{\cdot}1075 = 0{\cdot}86 \text{ kg}$$

$$\text{Total oxygen} = 2{\cdot}293 + 0{\cdot}86 = 3{\cdot}153 \text{ kg}$$

$$\begin{array}{l}\text{Air required} \\ \text{(stoichiometric)}\end{array} = \frac{100}{23} \times 3{\cdot}153 = 13{\cdot}71 \text{ kg air/kg fuel} \quad \text{Ans. (ii)}$$

HIGHER AND LOWER CALORIFIC VALUES. H_2O formed by the hydrogen in the fuel cannot exist as water in the high temperatures of boiler flue gases or the exhaust gases from internal combustion engines, it exists in the form of steam, and this steam passing away in the waste gases carries enthalpy of evaporation which is not available as heat energy to the boiler or engine.

Therefore, the theoretical calorific value of a fuel containing hydrogen, as calculated in the above example, is termed the *higher* (or *gross*) *calorific value* and, when the unavailable heat energy is subtracted from this, it is termed the *lower* (or *net*) *calorific value.* It has been recommended that the amount of heat energy to be considered as not available should be 2·442 MJ per kg of steam in the products of combustion. The amount of steam, as shown above, is nine times the mass of hydrogen in the fuel.

In the previous example the mass of hydrogen in each kilogramme of fuel is 0·11 kg, therefore there will be 9 × 0·11 = 0·99 kg of steam formed. The unavailable heat energy is therefore to be taken as 0·99 × 2·442 = 2·418.

Hence the lower calorific value of this fuel is:

$$44{\cdot}46 - 2{\cdot}418 = 42{\cdot}042 \text{ MJ/kg}$$

The calorific values of solid and liquid fuels can be found experimentally by means of a bomb calorimeter.

COMPOSITION OF FLUE GASES

An estimate of the analysis of the flue gases can be calculated from the composition of the fuel and mass of supply air as demonstrated in the following.

Example. An oil fuel is composed of 85% carbon, 12% hydrogen, 2% oxygen, and 1% incombustible solid matter. If the air supply is 50% in excess of the stoichiometric, find the mass of each product of combustion per kg of fuel burned and the percentage mass analysis of the flue gases.

$$\text{Stoichiometric air/kg fuel} = \frac{100}{23}\left\{2\tfrac{2}{3}\,C + 8\left(H_2 - \frac{O_2}{8}\right)\right\}$$

$$= \frac{100}{23}\left\{2\tfrac{2}{3}\,C + 8\,H_2 - O_2\right\}$$

$$= \frac{100}{23}\left\{2\tfrac{2}{3} \times 0{\cdot}85 + 8 \times 0{\cdot}12 - 0{\cdot}02\right\}$$

$$= \frac{100}{23} \times 3{\cdot}207 = 13{\cdot}94 \text{ kg air/kg fuel}$$

$$\text{Excess air} = 0{\cdot}5 \times 13{\cdot}94 = 6{\cdot}97 \text{ kg}$$

$$\text{Actual air supply} = 13{\cdot}94 + 6{\cdot}97 = 20{\cdot}91 \text{ kg air/kg fuel}$$

Mass of each product of combustion, Ans. (i):

Mass of CO_2 formed = $3\tfrac{2}{3} \times 0{\cdot}85 = 3{\cdot}117$ kg

Mass of H_2O formed = $9 \times 0{\cdot}12 = 1{\cdot}08$ kg

Mass of O_2 = excess of oxygen from the excess air

= 23% of $6{\cdot}97 = 1{\cdot}603$ kg

Mass of N_2 = mass of nitrogen in all the air supply

= 77% of $20{\cdot}91 = 16{\cdot}1$ kg

Total products of combustion

$= 3{\cdot}117 + 1{\cdot}08 + 1{\cdot}603 + 16{\cdot}1 = 21{\cdot}9$ kg/kg fuel

Alternatively, the total mass of products of combustion per kg fuel = mass of air supplied + (mass of fuel – incombustibles)

$= 20{\cdot}91 + (1 - 0{\cdot}01)$

$= 20{\cdot}91 + 0{\cdot}99 = 21{\cdot}9$ kg/kg fuel

Composition expressed as percentages, Ans. (ii):

$$CO_2 = \frac{3{\cdot}117}{21{\cdot}9} \times 100 = 14{\cdot}23\%$$

$$H_2O = \frac{1{\cdot}08}{21{\cdot}9} \times 100 = 4{\cdot}931\%$$

$$O_2 = \frac{1{\cdot}603}{21{\cdot}9} \times 100 = 7{\cdot}319\%$$

$$N_2 = \frac{16{\cdot}1}{21{\cdot}9} \times 100 = 73{\cdot}52\%$$

ORSAT APPARATUS

A volumetric analysis of the dry products of combustion can be made by the Orsat apparatus. This operates on the principle of passing a sample of the flue gases through a series of three bottles containing different solutions, each of which absorbs one of the constituents, and measuring each time the reduction in volume of the sample which is the volume of the constituent absorbed.

The solutions used for absorption are (i) caustic potash which can absorb CO_2, (ii) pyrogallic acid which can absorb CO_2 and O_2, (iii) cuprous chloride which can absorb CO_2, O_2 and CO. Therefore the first bottle through which the sample is passed must be that containing the caustic potash solution so that only CO_2 is absorbed, the remainder of the sample then contains O_2, CO and N_2. The sample is next passed through the bottle containing the pyrogallic acid solution which takes out the O_2, leaving CO and N_2. Finally, the sample is passed through the cuprous chloride solution which solution which absorbs the CO, and the remainder is assumed to be all N_2. The temperature and pressure are maintained constant throughout the operation.

It is convenient to draw in a sample of exactly 100 ml to be tested so that the percentage volumetric analysis is simply the volumes in ml.

CONVERSION FROM VOLUMETRIC TO MASS ANALYSIS. The volumetric analysis obtained by the Orsat apparatus can now be converted into a mass analysis. By Avogadro's law, equal volumes of any gas, at the same temperature and pressure, contain the same number of molecules, therefore, although the *mass* of a molecule of one gas is different to the mass of a molecule of another gas, the *volume* of each molecule is the same at the same temperature and pressure.

Hence the ratio of the volumes of each gas in the mixture of flue gases (N mols) can be converted into a ratio of masses by multiplying by the molecular weight of the gas (M), and from this the percentage composition by mass (m%) can be calculated.

Example. In a test on a sample of funnel gases the percentage composition by volume was found to be: $CO_2 = 8{\cdot}5\%$, $O_2 = 9{\cdot}5\%$, $CO = 3\%$ and the remainder 79% was assumed to be N_2. Convert these quantities into a mass analysis.

DFG	N	M	m	m%
CO_2	8·5	44	8·5 × 44 = 374	$\frac{374}{2974} \times 100 = 12{\cdot}58\%$
O_2	9·5	32	9·5 × 32 = 304	$\frac{304}{2974} \times 100 = 10{\cdot}22\%$
CO	3·0	28	3 × 28 = 84	$\frac{84}{2974} \times 100 = 2{\cdot}82\%$
N_2	79·0	28	79 × 28 = 2212	$\frac{2212}{2974} \times 100 = 74{\cdot}38\%$
			Total = 2974	Total = 100·00%

Example. The analysis of a sample of coal burned in the furnace of a boiler is 80% carbon, 5% hydrogen, 4% oxygen, and the remainder ash etc. Calculate (i) the stoichiometric air required per kg of coal, (ii) the actual mass of air if it is supplied with 70% excess, (iii) the percentage mass analysis of the products of combustion.

$$\text{Stoichiometric air} = \frac{100}{23}\left\{2\tfrac{2}{3}\,C + 8\left(H_2 - \frac{O_2}{8}\right)\right\}$$

$$= \frac{100}{23}\left\{2\tfrac{2}{3}\,C + 8\,H_2 - O_2\right\}$$

$$= \frac{100}{23}\left\{2\tfrac{2}{3} \times 0{\cdot}8 + 8 \times 0{\cdot}05 - 0{\cdot}04\right\}$$

$$= \frac{100}{23} \times 2{\cdot}493 = 10{\cdot}84 \text{ kg air/kg fuel} \quad \text{Ans. (i)}$$

$$\text{Excess air} = 0{\cdot}7 \times 10{\cdot}84 = 7{\cdot}59$$

$$\text{Actual air} = 10{\cdot}84 + 7{\cdot}59 = 18{\cdot}43 \text{ kg air/kg fuel} \quad \text{Ans. (ii)}$$

Mass products of combustion per kg of coal:

$$CO_2 = 3\tfrac{2}{3}\,C = 3\tfrac{2}{3} \times 0{\cdot}8 = 2{\cdot}933 \text{ kg}$$

$$9\,H_2 = 9 \times 0{\cdot}05 = 0{\cdot}45$$

$$O_2 = 23\% \text{ of excess air} = 0{\cdot}23 \times 7{\cdot}59 = 1{\cdot}746$$

$$N_2 = 77\% \text{ of all air} = 0{\cdot}77 \times 18{\cdot}43 = 14{\cdot}19$$

$$\text{Total} = 19{\cdot}319 \text{ kg}$$

% mass analysis Ans. (iii)

$$CO_2 = \frac{2{\cdot}933}{19{\cdot}319} \times 100 = 15{\cdot}18\%$$

$$H_2O = \frac{0{\cdot}45}{19{\cdot}319} \times 100 = 2{\cdot}33\%$$

$$O_2 = \frac{1{\cdot}746}{19{\cdot}319} \times 100 = 9{\cdot}04\%$$

$$N_2 = \frac{14{\cdot}19}{19{\cdot}319} \times 100 = 73{\cdot}45\%$$

Dry flue products are the total products less the amount of steam
$= 19{\cdot}319 - 0{\cdot}45 = 18{\cdot}869$ kg

f CONVERSION FROM MASS TO VOLUMETRIC ANALYSIS. Is sometimes required and can be obtained by reversing the above procedure. To illustrate, determine the percentage analysis of the dry flue gases by volume from the previous worked example.

DFG	m%	M	N	N%
CO_2	15·18	44	15·18 ÷ 44 = 0·345	$\frac{0{\cdot}345}{3{\cdot}2505} \times 100 = 10{\cdot}61\%$
O_2	9·04	32	9·04 ÷ 32 = 0·2825	$\frac{0{\cdot}2825}{3{\cdot}2505} \times 100 = 8{\cdot}69\%$
N_2	73·45	28	73·45 ÷ 28 = 2·623	$\frac{2{\cdot}623}{3{\cdot}2505} \times 100 = 80{\cdot}7\%$
			Total = 3·2505	Total = 100·00%

f HYDROCARBON FUELS C_xH_y. The group of fuels which have a chemical formula C_xH_y are called hydrocarbon fuels. Some examples are:

Methane CH_4
Propane C_3H_8
Butane C_4H_{10}

If we consider one kg of hydrocarbon fuel C_xH_y it will be necessary to obtain the masses of carbon and hydrogen contained in the one kg in order to evaluate the air required for combustion.

The mass of one mol of the fuel $= 12 \times x + 1 + y$ kg
$M = 12 \times x + 1 \times y$

$$\text{The fraction of hydrogen by mass} = \frac{1 \times y}{M} \text{ kg}$$

$$\text{and the fraction of carbon by mass} = \frac{12 \times x}{M} \text{ kg}$$

f Example. Determine the amount of air required for the complete combustion of 1 kg of propane, formula C_3H_8.

$$\text{Mass of 1 mol of the fuel} = 12 \times 3 + 1 \times 8 = 44 \text{ kg}$$

$$\text{Fraction of hydrogen by mass} = \frac{8}{44} = 0{\cdot}1818$$

$$\text{Fraction of carbon by mass} = \frac{12 \times 3}{44} = 0{\cdot}8182$$

$$= \Sigma 1{\cdot}0000$$

Hence 1 kg of C_3H_8 consists of 0·8182 kg of carbon and 0·1818 kg of hydrogen.

$$\text{Stoichiometric air} = \frac{100}{23}[0{\cdot}8182 \times 2\tfrac{2}{3} + 0{\cdot}1818 \times 8]$$

$$= 15{\cdot}81 \text{ kg/kg of fuel.}$$

Note: In 1 kg of a pure hydrocarbon fuel, if x is the mass of carbon then $(1 - x)$ is the mass of hydrogen.

f INCOMPLETE COMBUSTION. This may occur if insufficient air, or unsatisfactory combustion, results. The exhaust gas analysis will then include carbon monoxide. Balancing of the fuel-gas combustion equation allows analysis of the constituents and dry gas and air amount.

Balancing equations:

$$\frac{x}{12}C + \cdots\cdots \rightarrow aCO_2 + bCO + \cdots\cdots$$

C balance is $\frac{x}{12} = a + b$

Similarly for oxygen, hydrogen, etc. balances; these relations can be used to complete an otherwise unknown combustion equation; the mass of dry gases can be found from a C balance.

Relative mass of C in fuel

$$= \frac{12}{44}\{\text{Rel. mass (m) } CO_2\} + \frac{12}{28}\{\text{Rel. mass (m) CO}\} \text{ in gases}$$

$$= \frac{12}{44}\{\text{Rel. vol. (N) } CO_2 \times 44\} + \frac{12}{28}\{\text{Rel. vol. (N) CO} \times 28\}$$

$$= \{\text{Rel. vol. (N) } CO_2 + \text{Rel. vol. (N) CO}\}$$

f Example. The exhaust gas from an engine has a dry analysis by volume as follows:

CO_2 8·8%, CO 1·25%, O_2 6·9%, N_2 83·05%.

The fuel supplied to the engine has a mass analysis as follows:

C 84%, H_2 14%, O_2 2%

Determine the mass of air supplied per kg of fuel burnt.

Note: Air contains 20·7% oxygen by volume and 23% oxygen by mass. Atomic mass relationships: hydrogen = 1, carbon = 12, nitrogen = 14, oxygen = 16.

DFG	N	M	m
CO_2	8·8	44	387·2
CO	1·25	28	35
O_2	6·9	32	220·8
N_2	83·05	28	2325·4
			Total 2968·4

$$\text{Relative gas mass} = 2968{\cdot}4$$
$$\text{Relative C mass} = 12(8{\cdot}8 + 1{\cdot}25)$$
$$= 120{\cdot}6$$

i.e. 120·6 kg of C in 2968·4 kg of dry flue gases

$$0{\cdot}84 \text{ kg of C in } \frac{2968{\cdot}4 \times 0{\cdot}84}{120{\cdot}6} \text{ kg dry gases}$$

$$\text{Dry flue gases} = 20{\cdot}68 \text{ kg}$$
$$\text{Water vapour} = 9 \times 0{\cdot}14$$
$$= 1{\cdot}26 \text{ kg}$$
$$\text{Total gases} = 21{\cdot}94 \text{ kg}$$
$$\text{Mass of air supplied} = 20{\cdot}94 \text{ kg/kg fuel} \quad \text{Ans.}$$

TEST EXAMPLES 13

1. Steam at 30 bar, 375°C, is generated in a boiler at the rate of 30000 kg/h from feed water at 130°C. The fuel has a calorific value of 42 MJ/kg and the daily consumption is 53 tonne. Calculate (i) the boiler efficiency, (ii) the equivalent engine power if the overall efficiency of the plant is 0·13, (iii) the evaporative capacity of the boiler in kg/h from and at 100°C, (iv) the equivalent evaporation per kg of fuel from and at 100°C.

2. A boiler contains 3·5 tonne of water initially having 40 p.p.m. dissolved solids, and after 24 hours the dissolved solids in the water is 2500 p.p.m. If the feed rate is 875 kg/h, find the p.p.m. of dissolved solids contained in the feed water.

3. At its rated output a waste heat evaporator produces 10 tonne/day of fresh water containing 250 p.p.m. dissolved solids from sea water containing 31250 p.p.m. The dissolved solid content of the brine in the evaporator shell is maintained at 78125 p.p.m. Calculate the mass flow per day of:

(a) the sea water feed;
(b) the brine discharge.

4. The constituents of a fuel are 85% carbon, 13% hydrogen, and 2% oxygen. When burning this fuel in a boiler furnace the air supply is 50% in excess of the stoichiometric, inlet temperature of the air being 31°C and funnel temperature 280°C. Calculate (i) the calorific value of the fuel, (ii) mass of air supplied per kg of fuel, and (iii) the heat energy carried away to waste in the funnel gases expressed as a percentage of the heat energy supplied, taking the specific heat of the flue gases as 1·005 kJ/kg K.

5. A fuel consists of 84% carbon, 13% hydrogen, 2% oxygen, and the remainder incombustible solid matter. Calculate (i) the calorific value, (ii) the stoichiometric air required per kg of fuel, and (iii) an estimate of the mass analysis of the flue gases if 22 kg of air are supplied per kg of fuel burned.

6. The analysis of a fuel oil is 85·5% carbon, 11·9% hydrogen, 1·6% oxygen, and 1% impurities. Calculate the percentage CO_2 in the flue gases when (i) the quantity of air supplied is the stoichiometric, and when the excess air over the stoichiometric is (ii) 25%, (iii) 50%, and (iv) 75%.

f 7. The mass analysis of a fuel is 84% carbon, 14% hydrogen and 2% ash. It is burnt with 20% excess air relative to stoichiometric requirement. If 100 kg/h of this fuel is burnt in a boiler determine (i) the volumetric analysis of the dry flue gas (ii) the volumetric flow rate of the gas if its temperature and pressure are 250°C and 1 bar respectively.

Note: Air contains 23% oxygen by mass. $R_o = 8{\cdot}3143$ kJ/kg molK and atomic mass relationships are: oxygen 16, carbon 12, hydrogen 1.

f 8. The fuel oil supplied to a boiler has a mass analysis of 86% carbon, 12% hydrogen and 2% sulphur. The fuel is burned with an air/fuel ratio of 20:1.

Calculate:

(a) the mass analysis of the wet flue gases;
(b) the volumetric analysis of the wet flue gases.

f 9. Benzene fuel C_6H_6 is burnt in a boiler furnace under stoichiometric conditions of combustion. Calculate:

(a) the mass of air required for one kilogram of benzene.
(b) the mass analysis of the flue gases.
(c) the volumetric analysis of the dry flue gas.

f 10. The volumetric analysis of a dry flue gas is: carbon dioxide 10·8%, oxygen 7·2%, carbon monoxide 0·8% and the remainder nitrogen.

Calculate:

(a) the analysis of the dry flue gas by mass;
(b) the mass percentage of carbon and hydrogen in the fuel assuming it to be a pure hydrocarbon.

CHAPTER 14

REFRIGERATION

The natural transfer of heat energy is from a hot body to a colder body. The function of a refrigeration plant is to act as a heat pump and reverse this process.

Refrigerating machines can be divided into two classes, (i) those which require a supply of mechanical work which is the vapour compression system, and (ii) those which require a heat supply and work on the absorption system. The latter is more suited to small domestic use and only the vapour compression system is described here.

The refrigerating agent used in the circuit is a substance which will evaporate at low temperatures. The boiling and condensation points of a liquid depend upon the pressure exerted upon it, for example, if water is under atmospheric pressure it will evaporate at 100°C, if the pressure is 7 bar the water will not change into steam until its temperature is 165°C, at 14 bar the boiling point is 195°C and so on. The refrigerant used must evaporate at very low temperatures. The boiling point of carbon dioxide at atmospheric pressure is about – 78°C (note *minus* 78), by increasing the pressure the temperature at which liquid CO_2 will evaporate (or CO_2 vapour will condense) is raised accordingly so that any desired evaporation and condensation temperature can be attained, within certain limits, by subjecting it to the appropriate pressure.

Some of the agents employed as refrigerants and their more important characteristics are as follows:

Carbon dioxide (CO_2) is an inert vapour, non-poisonous, odourless, and has no corrosive action on the metals. Its natural boiling point is very low which means that it must be run at very high pressures to bring it to the conditions where it will evaporate and condense at the normal temperatures of a refrigerating machine. A further disadvantage is that its critical temperature is about 31°C which falls within the range of coolant temperatures. At its critical temperature and above, it is impossible to liquefy the vapour no matter to what pressure it is subjected. Virtually outdated.

Ammonia (NH_3). Is a poisonous vapour and therefore an ammonia machine should have a compartment of its own so that it can be sealed off in the case of a serious leakage; water will absorb ammonia and therefore a water spray is a good combatant against a leakage. Ammonia will corrode copper and copper alloys and therefore parts in contact with it should be made of such metals as nickel steel and monel metal. Its natural boiling point is about − 39°C therefore the pressures required are much lower than those required in a CO_2 machine. Largely outdated.

Freon-12 (CCl_2F_2). Requires low pressures. Is non-flammable, non-poisonous and non-corrosive. Physical properties make this a very desirable refrigerant - almost standard in use today. R12 is part of a large Freon group; others in use are R11, 22 and 502.

WORKING CYCLE

Fig. 68 illustrates diagrammatically the essential components of the vapour compression system, which consists of the Compressor driven by an electric motor, the Condenser which is circulated by the water, the Regulator (sometimes referred to as the

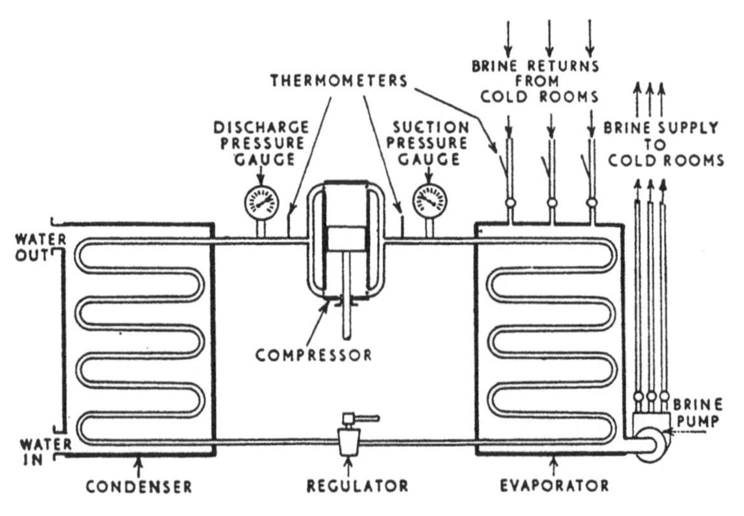

Fig. 68

expansion valve), and the Evaporator which is circulated by brine. It is a completely enclosed circuit, the same quantity of refrigerant passes continually through the system and it only requires to be charged when there are losses due to leakage. Any of the above agents may be used as the refrigerant, the difference in their working being only the pressures throughout the system, therefore in the following description the words "refrigerant", "the liquid" and "the vapour", etc. will be used.

The refrigerant is drawn as a vapour at low pressure from the evaporator into the compressor where it is compressed to a high pressure and delivered into the coils of the condenser in the state of a superheated vapour. As it passes through the condenser coils, the vapour is cooled and condensed into a liquid at approximately coolant inlet water temperature, the latent heat energy given up being absorbed by the coolant surrounding the coils. The liquid, still at a high pressure, passes along to the regulator, this is a valve just partially open to limit the flow through it. The pipe-line from the discharge side of the compressor to the regulating valve is under high pressure but from the regulating valve to the suction side of the compressor is at low pressure due to the regulating valve being just a little way open. As the liquid passes through the regulator from a region of high pressure to a region of low pressure, throttling occurs, and some of the liquid automatically changes itself into a vapour absorbing the required amount of heat energy to do so from the remainder of the liquid and causing it to fall to a low temperature, this temperature being regulated by the pressure so that the refrigerant enters the evaporator at a temperature lower than that of the brine. The liquid (more correctly a mixture of liquid and vapour) now passes through the coils of the evaporator where it receives heat to evaporate it before being taken into the compressor to go through the cycle again. The required heat absorbed by the liquid refrigerant in the coils of the evaporator to cause evaporation is extracted from the brine surrounding the coils, resulting in the brine being cooled to a low temperature (or directly from the air).

THE CIRCUIT OF THE REFRIGERANT

It is necessary to have a clear picture in mind when performing calculations involving the changes that take place on the refrigerant as it flows through the components of the circuit, therefore a simple diagrammatic sketch such as Fig. 69 should be made with points in

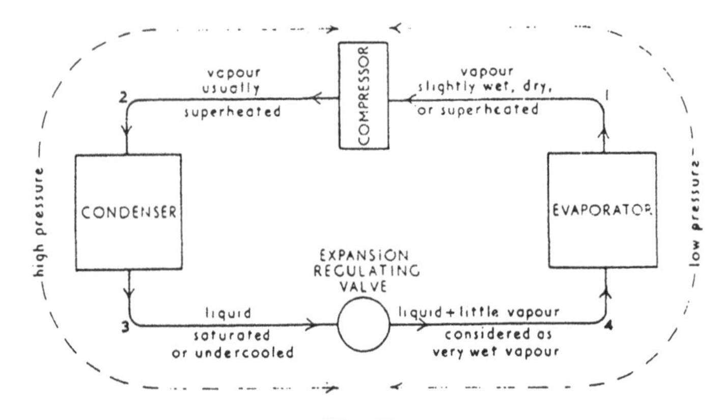

Fig. 69

the circuit suitably marked. Reference should be made to Fig. 69 when reading the following notes and examples.

Properties of some refrigerants are given in "Thermodynamic and Transport Properties of Fluids", SI Units, arranged by Y. R. Mayhew and G. F. C. Rogers, this booklet was referred to in previous Chapters dealing with properties of steam. In examples given here, extracts from these tables will be taken when appropriate, stating the page number.

Principles and calculations involving enthalpy of the refrigerant, dryness fraction, volume, effect of throttling, etc, follow the same pattern as for other vapours. However, whereas the enthalpy and entropy of steam are measured from the datum of water at 0°C, it is more convenient to choose a lower datum temperature for refrigerants to avoid negative quantities within the normal range of refrigeration. This datum is – 40°C, thus the enthalpy and entropy of the saturated liquid refrigerant are taken as zero at this temperature.

Note that when the value of h_{fg} is not given in the tables, it is found by subtraction: $h_{fg} = h_g - h_f$. Similarly for the value of s_{fg}.

Referring to Fig. 69:

The work done on the vapour in the compressor increases the specific enthalpy of the refrigerant from h_1 to h_2. Therefore the effective work done per second [kJ/s], which is the effective power [kW], is equal to the product of the mass flow [kg/s] of the refrigerant and its gain of specific enthalpy $(h_2 - h_1)$[kJ/kg].

In the condenser, heat transfer takes place from the refrigerant to the condenser circulating water, causing condensation of the refrigerant from a vapour to a liquid. The heat rejected by the refrigerant per kg is the change of specific enthalpy $h_2 - h_3$.

The effect of throttling the refrigerant through the expansion valve is to reduce the pressure of the saturated liquid with consequent reduction in temperature and the formation of a very small amount of vapour. During a throttling process there is no change of enthalpy, therefore h_4 is equal to h_3.

In passing through the evaporator coils, heat transfer takes place from the brine (or other surrounds) to the refrigerant, which cools the brine and causes the refrigerant to evaporate. The gain in specific enthalpy of the refrigerant is $h_1 - h_4$ and this is the *refrigerating effect* or *cooling effect* per kg of the refrigerant flowing through the circuit.

Example. In an NH_3 refrigerator, the ammonia leaves the evaporator and enters the compressor as dry saturated vapour at 2·68 bar, it leaves the compressor and enters the condenser at 8·57 bar with 50°C of superheat. it is condensed at constant pressure and leaves the condenser as saturated liquid. If the rate of flow of the refrigerant through the circuit is 0·45 kg/min calculate (i) the compressor power, (ii) the heat rejected to the condenser cooling water in kJ/s, and (iii) the refrigerating effect in kJ/s.

From tables page 12, NH_3:

2·68 bar, $h_g = 1430{\cdot}5$

8·57 bar, $h_f = 275{\cdot}1$ $\qquad h$ supht 50° = 1597·2

Mass flow of refrigerant

$$= \frac{0{\cdot}45}{60} = 0{\cdot}0075 \text{ kg/s}$$

Enthalpy gain per kg of refrigerant in compressor

$$= h_2 - h_1$$
$$= 1597{\cdot}2 - 1430{\cdot}5 = 166{\cdot}7 \text{ kJ/kg}$$

Enthalpy gain per second

$$= \text{mass flow [kg/s]} \times 166{\cdot}7 \text{ [kJ/kg]}$$
$$= 0{\cdot}0075 \times 166{\cdot}7 = 1{\cdot}25 \text{ kJ/s}$$

kJ/s = kW, therefore useful compressor power

$$= 1{\cdot}25 \text{ kW} \quad \text{Ans. (i)}$$

Enthalpy drop per kg of refrigerant through condenser

$$= h_2 - h_3$$
$$= 1597{\cdot}2 - 275{\cdot}1 = 1322{\cdot}1 \text{ kJ/kg}$$

$$\text{Heat rejected} = 0{\cdot}0075 \times 1322{\cdot}1$$
$$= 9{\cdot}915 \text{ kJ/s} \quad \text{Ans. (ii)}$$

Enthalpy gain per kg of refrigerant through evaporator

$$= h_1 - h_4$$
$$= 1430{\cdot}5 - 275{\cdot}1 = 1155{\cdot}4 \text{ kJ/kg}$$

(Note that the value of h_4 is the same as h_3 because there is no change of enthalpy in the throttling process through the expansion valve between the condenser exit and the evaporator inlet).

$$\text{Refrigerating effect} = 0{\cdot}0075 \times 1155{\cdot}4$$
$$= 8{\cdot}666 \text{ kJ/s} \quad \text{Ans. (iii)}$$

Example. In a Freon-12 refrigerator, the freon leaves the condenser as a saturated liquid at 20°C, and the evaporator temperature is – 10°C and the freon leaves the evaporator as a vapour 0·97 dry. Calculate (i) the dryness fraction at the evaporator inlet, (ii) the cooling effect per kg of refrigerant, and (iii) the volume flow of refrigerant entering the compressor if the mass flow is 0·1 kg/s.

From tables page 13, Freon-12:

20°C, $h_f = 54{\cdot}87$

– 10°C, $h_f = 26{\cdot}87$ $h_g = 183{\cdot}19$ $v_g = 0{\cdot}0766$

$h_{fg} = h_g - h_f = 183{\cdot}19 - 26{\cdot}87 = 156{\cdot}32$

Throttling process through expansion valve:

$$\text{Enthalpy after throttling } (h_4) = \text{Enthalpy before } (h_3)$$
$$h_{f8} + x_4 h_{fg4} = h_{f3}$$
$$26{\cdot}87 + x_4 \times 156{\cdot}32 = 54{\cdot}87$$
$$x_4 \times 156{\cdot}32 = 28$$
$$x_4 = 0{\cdot}1792 \quad \text{Ans. (i)}$$

Enthalpy gain of refrigerant through evaporator

$= \text{cooling effect} = h_1 - h_4$

$= (26{\cdot}87 + 0{\cdot}97 \times 156{\cdot}32) - (26{\cdot}87 + 0{\cdot}1792 \times 156{\cdot}32)$

$= (0{\cdot}97 - 0{\cdot}1792) \times 156{\cdot}32$

$= 0{\cdot}7908 \times 156{\cdot}32 = 123{\cdot}6$ kJ/kg Ans. (ii)

Alternatively, since $h_4 = h_3$:

cooling effect

$= (26{\cdot}87 + 0{\cdot}97 \times 156{\cdot}32) - 54{\cdot}87$

$= 123{\cdot}6$ kJ/kg

Specific volume of refrigerant leaving evaporator

$= 0{\cdot}97 \times 0{\cdot}0766$ m³/kg

Volume of refrigerant entering compressor per second [m³/kg]

$= 0{\cdot}1 \times 0{\cdot}97 \times 0{\cdot}0766$

$= 7{\cdot}43 \times 10^{-3}$ m³/s

or 7·43 litre/s Ans. (iii)

CAPACITY AND PERFORMANCE

The capacity of a refrigerating plant is usually expressed in terms of the tonnes of ice at 0°C that could be made in 24 hours from water at 0°C.

The *coefficient of performance* is the ratio of the refrigerating effect to the net work done on the refrigerant in the compressor:

$$\text{Coeff. of performance} = \frac{\text{heat extracted by refrigerant}}{\text{work done on refrigerant}}$$

$$= \frac{h_1 - h_4}{h_2 - h_1}$$

f In Chapter 8, the reversed Carnot cycle was explained as the ideal cycle for a refrigerator and the coefficient of performance of this cycle given as:

$$\frac{T_2}{T_1 - T_2}$$

where T_1 and T_2 are the highest and lowest absolute temperatures in the cycle. This is referred to as the *Carnot coefficient of performance*. Note however, that the notation is different to that used in the practical circuit of Fig. 68, to avoid confusion write subscripts H for highest and L for lowest:

$$\frac{T_L}{T_H - T_L}$$

f Example. A Freon-12 refrigerating machine operates on the ideal vapour compression cycle between the limits −15°C and 25°C. The vapour is dry saturated at the end of isentropic compression and there is no undercooling of the condensate in the condenser. Calculate (i) the dryness fraction at the suction of the compressor, (ii) the refrigerating effect per kg of refrigerant, (iii) the coefficient of performance, (iv) the Carnot coefficient of performance.

Tables Freon-12, page 13:

25°C, $h_f = 59{\cdot}7$ $h_g = 197{\cdot}73$ $s_g = 0{\cdot}6869$

−15°C, $h_f = 22{\cdot}33$ $h_g = 180{\cdot}97$

$h_{fg} = 180{\cdot}97 - 22{\cdot}33 = 158{\cdot}64$

$s_f = 0{\cdot}0906$ $s_g = 0{\cdot}7501$

$s_{fg} = 0{\cdot}7051 - 0{\cdot}0906 = 0{\cdot}6145$

Isentropic compression in compressor:

Entropy before (s_1) = Entropy after (s_2)

$$0{\cdot}0906 + x_1 \times 0{\cdot}6145 = 0{\cdot}6869$$
$$x_1 = 0{\cdot}9701 \quad \text{Ans. (i)}$$
$$h_1 = 22{\cdot}33 + 0{\cdot}9701 \times 158{\cdot}64$$
$$= 153{\cdot}9$$
$$\text{Refrigerating effect/kg} = h_1 - h_4$$

Since there is no undercooling, the condensate leaves the condenser as saturated liquid, hence $h_3 = 59{\cdot}7$, and $h_4 = h_3$ because there is no change of enthalpy during the throttling process, therefore:

$$\text{Refrigerating effect} = 153{\cdot}9 - 59{\cdot}7$$
$$= 94{\cdot}2 \text{ kJ/kg} \quad \text{Ans. (ii)}$$
$$\text{Work done in compressor/kg} = h_2 - h_1$$
$$= 197{\cdot}73 - 153{\cdot}9 = 43{\cdot}83 \text{ kJ/kg}$$

$$\text{Coefficient of performance} = \frac{\text{refrigerating effect}}{\text{work transfer}}$$
$$= \frac{h_1 - h_4}{h_2 - h_1} = \frac{94{\cdot}2}{43{\cdot}83}$$
$$= 2{\cdot}140 \quad \text{Ans. (iii)}$$

$$\text{Carnot coeff. of performance} = \frac{T_L}{T_H - T_L}$$

$$\text{where } T_H = 25 + 273 = 298 \text{ K}$$
$$T_L = -15 + 273 = 258 \text{ K}$$

$$\therefore \text{ Carnot c.o.p.} = \frac{258}{298 - 258} = 6{\cdot}45 \quad \text{Ans. (iv)}$$

T–s and *p–h* diagrams illustrate, and allow analysis of, the refrigeration circuit.

f Example. An ammonia refrigeration plant operates between temperature limits of – 20°C and 32°C. The refrigerant leaves the evaporator as saturated vapour and enters the expansion valve as saturated liquid. Assuming the isentropic efficiency of the compressor is 0·85 determine the coefficient of performance of the plant and sketch the cycle on *p–h* and *T–s* diagrams.

From tables

$$h_2 = 1469{\cdot}9 \text{ kJ/kg}$$
$$h_3 = h_4 = 332{\cdot}8 \text{ kJ/kg}$$

For isentropic compression $s_1 = s_2 = 4{\cdot}962$ kJ/kg K

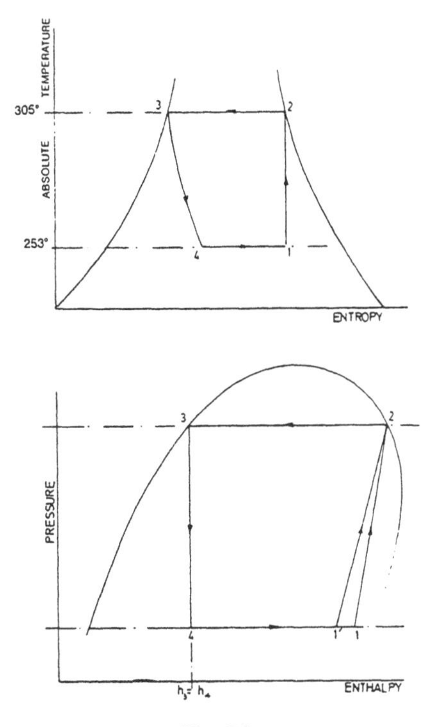

Fig. 70

at – 20°C

$$\begin{aligned} s_1 &= s_f + x(s_g - s_f) \\ 4{\cdot}962 &= 0{\cdot}368 + x(5{\cdot}623 - 0{\cdot}368) \\ x &= 0{\cdot}874 \text{ dry} \end{aligned}$$

also at – 20°C

$$\begin{aligned} h_1 &= h_f + x(h_g - h_f) \\ &= 89{\cdot}8 + 0{\cdot}874\,(1420 - 89{\cdot}8) \\ &= 1252{\cdot}4 \text{ kJ/kg} \end{aligned}$$

$$\begin{aligned} \text{Isentropic work done in compressor} &= h_2 - h_1 \\ &= 1469{\cdot}9 - 1252{\cdot}4 \\ &= 217{\cdot}5 \text{ kJ/kg} \end{aligned}$$

Actual work done in compressor × isentropic efficiency = isentropic work done in compressor

$$\therefore \text{ Actual work done in compressor} = \frac{217{\cdot}5}{0{\cdot}85} = 255{\cdot}9 \text{ kJ/kg}$$

$$\therefore\ h_1' = h_2 - 255{\cdot}9$$
$$= 1469{\cdot}9 - 255{\cdot}9 = 1214 \text{ kJ/kg}$$

$$\text{c.o.p.} = \frac{h_1' - h_4}{h_2 - h_1} = \frac{1214 - 332{\cdot}8}{1469{\cdot}9 - 1214}$$
$$= 3{\cdot}443 \quad \text{Ans.}$$

HEAT PUMPS If in Fig. 69 we consider the condenser as a heater where air, or water, is heated as the refrigerant condenses, the heated air or water being subsequently used for domestic purposes in accommodation spaces. We are then using the plant as a *heat pump*.

The coefficient of performance of the heat pump (or heat pump coefficient) is defined as the ratio:

$$\frac{\text{heat transfer in condenser}}{\text{work done on refrigerant}}$$

$$\textit{i.e.}\ \text{c.o.p.} = \frac{h_2 - h_3}{h_2 - h_1} \text{ for heat pump.}$$

TEST EXAMPLES 14

1. In a CO_2 refrigerating plant, the specific enthalpy of the refrigerant as it leaves the condenser is 135 kJ/kg, and as it leaves the evaporator it is 320 kJ/kg. If the mass flow of the refrigerant is 5 kg/min calculate the refrigerating effect per hour.

2. Freon-12 leaves the condenser of a refrigerating plant as a saturated liquid at 5·673 bar. The evaporator pressure is 1·509 bar and the refrigerant leaves the evaporator at this pressure and at a temperature of – 5°C. Calculate (i) the dryness fraction of the refrigerant as it enters the evaporator, (ii) the refrigerating effect per kg.

3. A refrigerating machine uses ammonia as the working fluid. It leaves the compressor as dry saturated vapour at 8·57 bar, passes through the condenser at this pressure and leaves as saturated liquid. The pressure in the evaporator is 1·902 bar and the ammonia leaves the evaporator 0·96 dry. If the rate of flow of the refrigerant through the circuit is 2 kg/min, calculate (i) the heat rejected in the condenser in kJ/min, (ii) the refrigerating effect in kJ/min, (iii) the volume taken into the compressor in m^3/min.

4. The dryness fractions of the CO_2 entering and leaving the evaporator of a refrigerating plant are 0·28 and 0·92 respectively. If the specific enthalpy of evaporation (h_{fg}) of CO_2 at the evaporator pressure is 290·7 kJ/kg, calculate (i) the refrigerating effect per kg, (ii) the mass of ice at – 5°C that would theoretically be made per day from water at 14°C when the mass flow of CO_2 through the machine is 0·5 kg/s. Take the values:

Specific heat of water = 4·2 kJ/kg K
Specific heat of ice = 2·04 kJ/kg K
Enthalpy of fusion of ice = 335 kJ/kg

5. A refrigerating machine is driven by a motor of output power 2·25 kW and 2·5 tonne of ice at –7°C is made per day from water at 18°C. Calculate the coefficient of performance of the machine and express its capacity in terms of tonnes of ice per 24 hours from and at 0°C, taking the following values:

Specific heat of water = 4·2 kJ/kg K
Specific heat of ice = 2·04 kJ/kg K
Enthalpy of fusion of ice = 335 kJ/kg

6. The refrigerant leaves the compressor and enters the condenser of a Freon-12 refrigerating plant at 5·673 bar and 50°C, and leaves the condenser as saturated liquid at the same pressure. At compressor suction the pressure is 1·826 bar and temperature 0°C. Calculate the coefficient of performance.

f 7. In an ammonia refrigerating plant the discharge from the compressor is 14·7 bar 63°C, the refrigerant leaves the condenser at this pressure as liquid with no undercooling. The suction pressure at the compressor is 2·077 bar and compression is isentropic. Calculate, for a mass flow of refrigerant of 0·15 kg/s, (i) the refrigerating effect, (ii) the work transfer in the compressor, (iii) the coefficient of performance.

f 8. A refrigeration plant using Freon-12 operates between 4·914 bar and 1·004 bar. The refrigerant is dry saturated on entry to the compressor and no undercooling takes place in the condenser. If compression is isentropic and losses are neglected, determine:

(a) the c.o.p.
(b) power input to compressor if the flow rate of refrigerant entering the compressor is 0·15 m^3/s

Sketch the cycle on temperature-entropy and pressure-enthalpy axes.

f 9. A heat pump using ammonia operates between saturation temperature limits of – 16°C and 40°C. The ammonia leaves the compressor after isentropic compression as dry saturated vapour and is then used to heat air from ambient, 13°C to 20°C which is circulated in an insulated room of internal volume 300 m^3. If the room requires 4 changes of air per hour determine the c.o.p. of the heat pump and the mass flow rate of ammonia.

Take for the air c_p = 1005 J/kg, R = 287 J/kgK and pressure 1·013 bar.

f 10. In a refrigeration plant Freon-12 leaves the condenser as a liquid at 25°C. The refrigerant leaves the evaporator as a dry saturated vapour at – 15°C and leaves the compressor at 6·516 bar and 40°C. The cooling load is 73·3 kW.

Calculate:

(a) the mass flow rate of refrigerant;
(b) the power of the compressor
(c) the condition of the refrigerant after the expansion valve.

SOLUTIONS TO TEST EXAMPLES 1

1. Mass of water lifted = 50×10^3 kg/h

Weight of water lifted = $50 \times 10^3 \times 9{\cdot}81$ N/h

$$\text{Power[W} = \text{J/s} = \text{N m/s]}$$

$$= \text{weight lifted per second [N/s]} \times \text{height [m]}$$

$$= \frac{50 \times 10^3 \times 9{\cdot}81 \times 8}{3600}$$

$$= 1090\text{ W} = 1{\cdot}09\text{ kW} \quad \text{Ans. (i)}$$

$$\text{Input power} = \frac{\text{output power}}{\text{efficiency}}$$

$$= \frac{1{\cdot}09}{0{\cdot}69} = 1{\cdot}58\text{ kW} \quad \text{Ans. (ii)}$$

$$\text{Energy [kW h]} = \text{power [kW]} \times \text{time [h]}$$

$$= 1{\cdot}58 \times 2 = 3{\cdot}16\text{ kW h} \quad \text{Ans. (iii)}$$

$$\text{Energy [MJ]} = \text{energy [kW h]} \times 3{\cdot}6\text{ [MJ/kW h]}$$

$$= 3{\cdot}16 \times 3{\cdot}6 = 11{\cdot}38\text{ MJ} \quad \text{Ans. (iv)}$$

2. A column of water 1 mm high exerts a pressure of $9{\cdot}81$ N/m^2, therefore 20 mm water is equivalent to:

$$20 \times 9{\cdot}81 = 196{\cdot}2\text{ N/m}^2 \quad \text{Ans. (ia)}$$

$$\text{One mbar} = 100\text{ N/m}^2$$

$$\therefore\ 1962{\cdot}2\text{ N/m}^2 = 1{\cdot}962\text{ mbar} \quad \text{Ans. (iia)}$$

A column of mercury 1 mm high exerts a pressure of $133{\cdot}3$ N/m^2, therefore 750 mmHg is equivalent to:

$$750 \times 133{\cdot}3 = 1 \times 10^5\text{ N/m}^2$$

$$= 100\text{ kN/m}^2 \quad \text{Ans. (ib)}$$

$$= 1\text{ bar} \quad \text{Ans. (iib)}$$

3. Abs. press. = (barometer mmHg − vac. gauge mmHg) × 133·3

$$= (757 - 715) \times 133{\cdot}3$$

$$= 42 \times 133{\cdot}3$$

$$= 5{\cdot}6 \times 10^3\text{ N/m}^2 = 5{\cdot}6\text{ kN/m}^2$$

$$= 0{\cdot}056\text{ bar} \quad \text{Ans.}$$

4. $C = (140 - 32) \times \frac{5}{9}$

$= 108 \times \frac{5}{9}$

$= 60°$ Ans. (i)

$C = (5 - 32) \times \frac{5}{9}$

$= -27 \times \frac{5}{9}$

$= -15°C$ Ans. (ii)

$C = (-31 - 32) \times \frac{5}{9}$

$= -63 \times \frac{5}{9}$

$= -35°C$ Ans. (iii)

$C = (-40 - 32) \times \frac{5}{9}$

$= -72 \times \frac{5}{9}$

$= -40°C$ Ans. (iv)

5. Area of tube bore $= 0{\cdot}7854 \times 0{\cdot}03^2\ m^2$

Area of 16 tubes $= 0{\cdot}7854 \times 0{\cdot}03^2 \times 16\ m^2$

Volume flow [m^3/s] = area [m^2] × velocity [m/s]

$= 0{\cdot}7854 \times 0.03^2 \times 16 \times 2$

$= 0{\cdot}02261\ m^3/s$

$1\ m^3 = 10^3$ litre

∴ Volume flow [l/s] $= 0{\cdot}02261 \times 10^3$

$= 22{\cdot}611$ l/s Ans. (i)

density = 0·85 g/ml = 0·85 kg/litre

Mass flow [kg/min] = vol. flow [litre/min] × density [kg/litre]

$= 22{\cdot}61 \times 60 \times 0{\cdot}85$

$= 1153$ kg/min Ans. (ii)

SOLUTIONS TO TEST EXAMPLES 2

1. Mechanical energy converted into heat energy per minute:

$$\begin{aligned} \text{Energy [kJ} &= \text{power [kW = kJ/s]} \times \text{time [s]} \\ &= 70 \times 60 \\ &= 4200\ \text{kJ} = 4{\cdot}2\ \text{MJ} \quad \text{Ans. (i)} \end{aligned}$$

All the heat energy being transferred to water:

$$\begin{aligned} Q\ [\text{kJ}] &= m\ [\text{kg}] \times c\ [\text{kJ/kg K}] \times (T_2 - T_1)\ [\text{K}] \\ 4200 &= m \times 4{\cdot}2 \times 10 \\ m &= 100\ \text{kg/min} \quad \text{Ans. (ii)} \end{aligned}$$

2. Friction force at effective radius of pads

$$\begin{aligned} &= \mu \times \text{force between surfaces} \\ &= 0.025 \times 240 = 6\ \text{kN} \end{aligned}$$

Work to overcome friction in one revolution

$$\begin{aligned} &= \text{friction force [kN]} \times \text{circumference [m]} \\ &= 6 \times 2\pi \times 0{\cdot}23\ \text{kN m} = \text{kJ} \end{aligned}$$

Power loss [kW = kJ/s] = work per rev [kJ] × rev/s

$$\begin{aligned} &= 6 \times 2\pi \times 0{\cdot}23 \times \frac{93}{60} \\ &= 13{\cdot}45\ \text{kW} \quad \text{Ans. (i)} \end{aligned}$$

Mechanical energy converted into heat energy per hour:

$$\begin{aligned} \text{Energy [kW h]} &= \text{power [kW]} \times \text{time [h]} \\ &= 13{\cdot}45 \times 1 = 13{\cdot}45\ \text{kW h} \\ 13{\cdot}45\ \text{kW h} \times 3{\cdot}6 &= 48{\cdot}42\ \text{MJ} \quad \text{Ans. (ii)} \end{aligned}$$

Quantity of heat energy transferred to oil per hour:

$$\begin{aligned} Q\ [\text{kJ}] &= m\ [\text{kg}] \times c\ [\text{kJ/kg K}] \times (\text{T}_2 - \text{T}_1)\ [\text{K}] \\ 48{\cdot}42 \times 10^3 &= m \times 2 \times 20 \\ m &= 1210{\cdot}5\ \text{kg/h} \quad \text{Ans.(iii)} \end{aligned}$$

3. Let θ = initial temperature of the copper
= temperature of the flue gases

$$\begin{aligned} \text{Heat lost by copper} &= \text{Heat gained by water} \\ m_c \times c_c \times \text{temp. fall}_c &= m_w \times c_w \times \text{temp. rise}_w \\ 1{\cdot}8 \times 0{\cdot}395 \times (\theta - 37{\cdot}2) &= 2{\cdot}27 \times 4{\cdot}2 \times (37{\cdot}2 - 20) \\ 0{\cdot}711\,\theta - 26{\cdot}44 &= 164 \\ 0{\cdot}711\,\theta &= 190{\cdot}44 \\ \theta &= 267{\cdot}9^\circ\text{C} \quad \text{Ans.} \end{aligned}$$

4. Mass of 2·3 litre of water = 2·3 kg

Let c [kJ/kg K] = specific heat of the iron

Heat lost by iron = Heat gained by water and vessel

$$2{\cdot}15 \times c \times (100 - 24{\cdot}4) = (2{\cdot}3 + 0{\cdot}18) \times 4{\cdot}2 \times (24{\cdot}4 - 17)$$

$$2{\cdot}15 \times c \times 75{\cdot}6 = 2{\cdot}48 \times 4{\cdot}2 \times 7{\cdot}4$$

$$c = \frac{2{\cdot}48 \times 4{\cdot}2 \times 7{\cdot}4}{2{\cdot}15 \times 75{\cdot}6}$$

$$= 0{\cdot}4742 \text{ kJ/kg K} \quad \text{Ans.}$$

5. Heat transferred to the ice to raise all of it in temperature from −5°C to 0°C

$$= 0{\cdot}5 \times 2{\cdot}04 \times 5 = 5{\cdot}1 \text{ kJ}$$

Heat transferred from the water and vessel in cooling from 17°C to 0°C

$$= (1{\cdot}8 + 0{\cdot}148) \times 4{\cdot}2 \times 17$$

$$= 1{\cdot}948 \times 4{\cdot}2 \times 17 = 139{\cdot}1 \text{ kJ}$$

Therefore, heat from water available to melt some of the ice

$$= 139{\cdot}1 - 5{\cdot}1 = 134 \text{ kJ}$$

Each kg of ice requires 335 kJ to melt it, therefore mass of ice melted by 134 kJ

$$= \frac{134}{335} = 0{\cdot}4 \text{ kg}$$

Hence,

$$\left.\begin{aligned} \text{Final mass of water} &= 1{\cdot}8 + 0{\cdot}4 = 2{\cdot}2 \text{ kg} \\ \text{Final mass of ice} &= 0{\cdot}5 - 0{\cdot}4 = 0{\cdot}1 \text{ kg} \end{aligned}\right\} \text{Ans.}$$

SOLUTIONS TO TEST EXAMPLES 3

1. Free expansion $= \alpha \times l \times (\theta_2 - \theta_1)$
$= 1{\cdot}25 \times 10^{-5} \times 3{\cdot}85 \times (260 - 18)$
$= 0{\cdot}01165$ m
$= 11{\cdot}65$ mm Ans.

2. Vol. of sphere $= \dfrac{\pi}{6} d^3 = \dfrac{\pi}{6} \times 0{\cdot}15^3 \text{m}^3$
Mass [kg] = volume [m³] × density kg/m³
$= \dfrac{\pi}{6} \times 0{\cdot}15^3 \times 7{\cdot}21 \times 10^3$
$= 12{\cdot}75$ kg
Q[kJ] $= m\,[\text{kg}] \times c\,[\text{kJ/kgK}] \times (T_2 - T_1)\,[\text{K}]$
$2110 = 12{\cdot}75 \times 0{\cdot}54 \times (T_2 - T_1)$
temp. rise $= \dfrac{2110}{12{\cdot}75 \times 0{\cdot}54} = 306{\cdot}7$ K
$= 306{\cdot}7$°C
Increase in diameter $= \alpha \times d \times (\theta_2 - \theta_1)$
$= 1{\cdot}12 \times 10^{-5} \times 150 \times 306{\cdot}7$
$= 0{\cdot}5152$ mm Ans.

3.

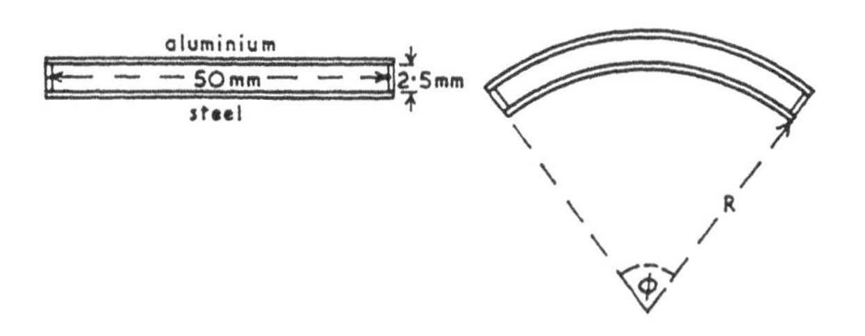

Free expansion $= \alpha \times l \times (\theta_2 - \theta_1)$
Heated length $= l + \alpha l\,(\theta_2 - \theta_1) = l\{1 + \alpha\,(\theta_2 - \theta_1)\}$
Heated length of aluminium strip
$= 50\,(1 + 2{\cdot}5 \times 10^{-5} \times 200)$
$= 50 \times 1{\cdot}005 = 50{\cdot}25$ mm
Heated length of steel strip
$= 50\,(1 + 1{\cdot}2 \times 10^{-5} \times 200)$
$= 50 \times 1{\cdot}0024 = 50{\cdot}12$ mm
Heated length of brass distance pieces
$= 2{\cdot}5\,(1 + 2 \times 10^{-5} \times 200)$

$$= 2{\cdot}5 \times 1{\cdot}004 = 2{\cdot}51 \text{ mm}$$

Let R = radius of curvature of steel

$R + 2{\cdot}51$ = ... aluminium

Subtended angle φ is common to both

$$\varphi \text{ [rad]} = \frac{\text{arc length [mm]}}{\text{radius [mm]}}$$

$$\therefore \quad \frac{50{\cdot}25}{R + 2{\cdot}51} = \frac{50{\cdot}12}{R}$$

$$50{\cdot}25R = 50{\cdot}12\,(R + 2{\cdot}51)$$
$$50{\cdot}25R = 50{\cdot}12R + 125{\cdot}8$$
$$0{\cdot}13R = 125{\cdot}8$$
$$R = 967{\cdot}7 \text{ mm}$$

Radius of steel = 967·7 mm Ans.

Radius of aluminium = 967·7 + 2·51

= 970·21 mm Ans.

4. Internal volume of pipe (= vol. of oil in pipe)

$$= 0{\cdot}7854 \times 0{\cdot}03^2 \times 13{\cdot}7 \text{ m}^3$$

Co-efficient of cubical expansion of steel pipe

$$\beta_p = 3 \times 1{\cdot}2 \times 10^{-5} = 3{\cdot}6 \times 10^{-5}/°\text{C}$$

Co-efficient of cubical expansion of oil

$$\beta_o = 9 \times 10^{-4} = 90 \times 10^{-5}/°\text{C}$$

Increase in volume of pipe = $\beta_p \times V \times (\theta_2 - \theta_1)$

Increase in volume of oil = $\beta_o \times V \times (\theta_2 - \theta_1)$

Oil overflow = $\beta_o V\,(\theta_2 - \theta_1) - \beta_p V\,(\theta_2 - \theta_1)$

$= (\beta_o - \beta_p) \times V \times (\theta_2 - \theta_1)$

$= (90 - 3{\cdot}6) \times 10^{-5} \times 0{\cdot}7854 \times 0{\cdot}03^2 \times 13{\cdot}7 \times 27$

$= 2{\cdot}259 \times 10^{-4} \text{ m}^3$

one cubic metre = 10^3 litres

$\therefore\ 2{\cdot}259 \times 10^{-4} \times 10^3 = 0{\cdot}2259$ litre Ans.

5. Free expansion = $\alpha \times l \times (\theta_2 - \theta_1)$

If fully restricted,

$$\text{Strain} = \frac{\text{change of length}}{\text{original length}} = \frac{\alpha l\,(\theta_2 - \theta_1)}{l} = \alpha\,(\theta_2 - \theta_1)$$

Stress = strain × E

$= 1{\cdot}12 \times 10^{-5} \times (220 - 17) \times 206 \times 10^9$

$= 1{\cdot}12 \times 203 \times 206 \times 10^4$

$= 4{\cdot}684 \times 10^8 \text{ N/m}^2 = 4684 \text{ MN/m}^2$

This compressive stress is to be relieved by exerting an initial tensile stress so that the compressive stress does not exceed 350 MN/m^2:

Initial tensile stress = 468·4 − 350
= 118·4 MN/m^2

SOLUTIONS TO TEST EXAMPLES 4

1. $$Q\,[\text{J}] = \frac{k\,[\text{W/m K}] \times A\,[\text{m}^2] \times t\,[\text{s}] \times (T_1 - T_2)\,[\text{K}]}{S\,[\text{m}]}$$

$$= \frac{150 \times 0{\cdot}7854 \times 0{\cdot}127^2 \times 60 \times 5}{0{\cdot}019}$$

$$= 3 \times 10^4\ \text{J} = 30\ \text{kJ} \quad \text{Ans.}$$

2. Total surface area of walls, ceiling and floor

$$= 2(4{\cdot}5 \times 2{\cdot}5 + 4 \times 2{\cdot}5 + 4{\cdot}5 \times 4)$$
$$= 78{\cdot}5\ \text{m}^2$$

$$Q = \frac{kAt\,(T_1 - T_2)}{S}$$

$$= \frac{5{\cdot}8 \times 10^{-2} \times 78{\cdot}5 \times 3600 \times 20}{0{\cdot}15}$$

$$= 2{\cdot}185 \times 10^6\ \text{J}$$
$$= 2{\cdot}185\ \text{MJ or } 2185\ \text{kJ} \quad \text{Ans.}$$

3. For each thickness:

$$Q = \frac{kAt \times \text{temp. diff.}}{S} \qquad \therefore\ \text{temp. diff.} = \frac{QS}{kAt}$$

For the three thicknesses:

$$T_D = \frac{Q}{AT}\left\{\frac{S_1}{k_1} + \frac{S_2}{k_2} + \frac{S_3}{k_3}\right\}$$

where $T_D = 20 - (-3) = 23°\text{C temp. diff.} = 23\ \text{K}$

$$\Sigma\left\{\frac{S}{k}\right\} = \frac{0{\cdot}03}{0{\cdot}2} + \frac{0{\cdot}168}{0{\cdot}042} + \frac{0{\cdot}035}{0{\cdot}2}$$
$$= 0{\cdot}15 + 4 + 0{\cdot}175$$
$$= 4{\cdot}325$$

Heat transfer per second per metre^2 of area:

$$T_D = \frac{Q}{At}\Sigma\left\{\frac{S}{k}\right\}$$

$$23 = \frac{Q}{1 \times 1} \times 4{\cdot}325$$

$$Q = \frac{23}{4 \cdot 325} = 5 \cdot 318 \text{ J} \quad \text{Ans. (i)}$$

Heat transfer per hour through whole side:

$$\begin{aligned} Q &= 5 \cdot 318 \times 6 \times 3 \cdot 7 \times 3600 \\ &= 4 \cdot 25 \times 10^5 \text{ J} \\ &= 425 \text{ kJ} \quad \text{Ans. (ii)} \end{aligned}$$

Temperature drop across outer wood wall:

$$\begin{aligned} &= \frac{QS_1}{k_1 At} \\ &= \frac{5 \cdot 318 \times 0 \cdot 03}{0 \cdot 2} \\ &= 0 \cdot 7977 \text{ K} \quad \text{say } 0 \cdot 8^\circ\text{C} \end{aligned}$$

Temperature at interface of cork and outer wood wall

$$= 20 - 0 \cdot 8 = 19 \cdot 2^\circ\text{C} \quad \text{Ans. (iiia)}$$

Temperature drop across cork:

$$= \frac{QS_2}{k_2 At}$$

$$= \frac{5 \cdot 318 \times 0 \cdot 168}{0 \cdot 042} = 21 \cdot 27 \text{ K} = 21 \cdot 27^\circ\text{C}$$

Temperature at interface of cork and inner wood wall

$$= 19 \cdot 2 - 21 \cdot 27 = -2 \cdot 07^\circ\text{C} \quad \text{Ans. (iiib)}$$

4.

$$\begin{aligned} Q &= 5 \cdot 67 \times 10^{-11} At \, (T_1^4 - T_2^4) \\ T_1 &= 488 \text{ K} \\ T_2 &= 318 \text{ K} \\ T_1^4 - T_2^4 &= 4 \cdot 647 \times 10^{10} \end{aligned}$$

$$Q = 5 \cdot 67 \times 10^{-11} \times 0 \cdot 7854 \times 0 \cdot 5^2 \times 3600 \times 4 \cdot 647 \times 10^{10}$$

$$= 1862 \text{ kJ} \quad \text{Ans.}$$

5.

$$\begin{aligned} \frac{1}{U} &= \frac{1}{h_G} + \frac{S_P}{k_P} + \frac{1}{h_A} \\ &= \frac{1}{31 \cdot 5} + \frac{0 \cdot 01}{50} + \frac{1}{32} \\ &= 0 \cdot 03175 + 0 \cdot 0002 + 0 \cdot 03125 \\ &= 0 \cdot 0632 \\ U &= 1/0 \cdot 0632 = 15 \cdot 82 \text{ W/m}^2 \text{ K} \quad \text{Ans. (i)} \\ Q &= UAtT_D \end{aligned}$$

$$Q = 15{\cdot}82 \times 1 \times 60 \times (280 - 35)$$
$$= 2{\cdot}327 \times 10^5 \text{ J} = 232{\cdot}7 \text{ kJ} \quad \text{Ans. (ii)}$$

6. Heat flow from outside atmosphere to exposed surface of outer wood wall, and also from exposed surface of inner wood wall to room atmosphere:

$Q = hAt \times$ temp. difference per m^2 of surface area per second,

$42 = 15 \times 1 \times 1 \times$ temp. difference

$$\text{temp. difference} = \frac{42}{15} = 2{\cdot}8°\text{C}$$

Temperature of exposed surface of outer wood

$$= 25 - 2{\cdot}8 = 22{\cdot}2°\text{C} \quad \text{Ans. (ia)}$$

Temperature of exposed surface of inner wood

$$= -20 + 2{\cdot}8 = -17{\cdot}2°\text{C} \quad \text{Ans. (ib)}$$

Heat flow through each wood wall:

$$Q = \frac{kAt \times \text{temp. difference}}{0{\cdot}03}$$

$$42 = \frac{0{\cdot}2 \times 1 \times 1 \times \text{temp. difference}}{S}$$

Temperature difference across each wood wall

$$= \frac{42 \times 0{\cdot}03}{0{\cdot}2} = 6{\cdot}3°\text{C}$$

Interface temperature of outer wood and cork

$$= 22{\cdot}2 - 6{\cdot}3 = 15{\cdot}9°\text{C} \quad \text{Ans. (iia)}$$

Interface temperature of inner wood and cork

$$= -17{\cdot}2 + 6{\cdot}3 = -10{\cdot}9°\text{C} \quad \text{Ans. (iib)}$$

Temperature difference across cork

$$= 15{\cdot}9 - (-10{\cdot}9) = 26{\cdot}8°\text{C}$$

Heat flow through cork:

$$Q = \frac{kAt \times \text{temp. difference}}{S}$$

$$S = \frac{0{\cdot}05 \times 1 \times 1 \times 26{\cdot}8}{42}$$

$$= 0{\cdot}0319 \text{ m} = 31{\cdot}9 \text{ mm} \quad \text{Ans. (iii)}$$

7. Let T_1 = abs. temp. of shell before lagging
= 503 K

T_2 = abs. temp. of surrounds before lagging
= 324 K

T_3 = abs. temp. of cleading after lagging
= 342 K

T_4 = abs. temp. of surrounds after lagging
$= 300\ \text{K}$

Heat radiated from shell before lagging

$$= 5{\cdot}67 \times 10^{-11}(T_1^4 - T_2^4)\ \text{kJ/m}^2\ \text{s}$$

Heat radiated from cleading after lagging

$$= 5{\cdot}67 \times 10^{-11}(T_3^4 - T_4^4)\ \text{kJ/m}^2$$

Heat saved by lagging

$$= 5{\cdot}67 \times 10^{-11}\{(T_1^4 - T_2^4) - (T_3^4 - T_4^4)\}\ \text{kJ/m}^2\ \text{s}$$
$$= 4{\cdot}74 \times 10^{10}$$

$$\text{Total surface area} = \text{area of ends} + \text{area of cylinder}$$
$$= \pi d^2 + \pi dl$$
$$= \pi d\,(d + l)$$
$$= \pi \times 1{\cdot}22\,(1{\cdot}22 + 4{\cdot}78)$$
$$= \pi \times 1{\cdot}22 \times 6$$
$$\text{Lagged area} = 0{\cdot}75 \times \pi \times 1{\cdot}22 \times 6$$
$$= 17{\cdot}26\ \text{m}^2$$

Total heat saved by lagging per hour

$$= 5{\cdot}67 \times 10^{-11} At\{(T_1^4 - T_2^4) - (T_3^4 - T_4^4)\}$$
$$= 5{\cdot}67 \times 10^{-11} \times 17{\cdot}26 \times 3600 \times 4{\cdot}74 \times 10^{10}$$
$$= 1{\cdot}67 \times 10^5\ \text{kJ}$$
$$= 167\ \text{MJ} \quad \text{Ans.}$$

8.

$$\text{For lagging } Q = \frac{2\pi k\,(T_1 - T_2)}{\ln\,(D/d)} \quad \text{W/m}$$

$$\text{For surface film } Q = \frac{hA\,(T_2 - T_3)}{L} \quad \text{W/m}$$
$$= h\pi D\,(T_2 - T_3)$$

$$Q \text{ for lagging} = Q \text{ for surface film}$$

$$\therefore \quad \frac{2\pi k\,(T_1 - T_2)}{\ln\,(D/d)} = h\pi D\,(T_2 - T_3)$$

$$i.e. \quad \frac{2\pi \times 0{\cdot}05\,(350 - T_2)}{\ln\left(\frac{340}{200}\right)} = 10 \times \pi \times \frac{340}{10^3}\,(T_2 - 15)$$

$$\text{From which } T_2 = 32{\cdot}6\,^\circ\text{C} \quad \text{Ans. (i)}$$
$$\therefore\ Q = hA\,(T_2 - T_3)$$
$$= 10 \times \pi \times \frac{340}{10^3} \times 20\,(32{\cdot}6 - 15)/10^3$$
$$= 3{\cdot}76\ \text{kJ/s} \quad \text{Ans. (ii)}$$

9.

$$\text{Let } T = \text{temperature of the cooling water at outlet}$$

$$\therefore \theta_1 = 410 - 10 \qquad \theta_2 = 130 - T$$
$$= 400$$

$$\therefore \theta_m = \frac{400 - (130 - T)}{\ln\left(\frac{400}{130 - T}\right)} \qquad (1)$$

$$\text{For the gas } Q = 0{\cdot}4 \times 1130 \times (410 - 130)/10^3$$
$$= 126{\cdot}56 \text{ kW}$$
$$\text{For the water } Q = 0{\cdot}5 \times 4190 \times (T - 10)/10^3$$
$$126{\cdot}56 = 0{\cdot}5 \times 4190 \times (T - 10)/10^3$$
$$\text{From which } T = 70{\cdot}4^\circ\text{C} \quad \text{substitute in (1)}$$

$$\therefore \theta_m = \frac{400 - (130 - 70{\cdot}4)}{\ln\left(\frac{400}{130 - 70{\cdot}4}\right)}$$

$$\theta_m = 178{\cdot}8^\circ\text{C}$$
$$\text{also } Q = UAt\theta_m$$

$$A = \frac{Q}{UAt\theta_m}$$

$$= \frac{126{\cdot}56 \times 10^3}{140 \times 1 \times 178{\cdot}8}$$

$$= 5{\cdot}06 \text{ m}^2 \quad \text{Ans.}$$

10. *Note*: This question illustrates one fluid being at constant temperature, *i.e.* the condensing steam

$$\text{Water mass flow} = 0{\cdot}7854 \times 0{\cdot}025^2 \times 0{\cdot}6 \times 10^3$$
$$= 0{\cdot}2945 \text{ kg/s}$$
$$Q = 0{\cdot}2945 \times 4{\cdot}18 \times 12$$
$$= 14{\cdot}77 \text{ kW}$$

$$\theta_m = \frac{27 - 15}{\ln\left(\frac{27}{15}\right)}$$

$$= 20{\cdot}41^\circ\text{C}$$
$$Q = UAt\theta_m$$

$$U = \frac{14770}{\pi \times 0{\cdot}025 \times 2{\cdot}75 \times 1 \times 20{\cdot}41}$$

Overall heat transfer coefficient

$$= 3350 \text{ W/m}^2\text{ K} \quad \text{Ans. (a)}$$

$$\frac{1}{U} = \frac{1}{h_o} + \frac{1}{h_i}$$

$$\frac{1}{3350} = \frac{1}{17000} + \frac{1}{h_i}$$

Inside surface heat transfer coefficient (i.e. between water and tube) is h_i:

$$= 4172 \text{ W/m}^2\text{ K} \quad \text{Ans. (b)}$$

SOLUTIONS TO TEST EXAMPLES 5

1.

$$p_1V_1 = p_2V_2$$
$$120 \times 1 = 960 \times V_2$$
$$V_2 = \frac{120 \times 1}{960} = 0{\cdot}125 \text{ m}^3 \quad \text{Ans. (i)}$$

$$p_1V_1 = p_2V_2$$
$$1{\cdot}05 \times V_1 = 42 \times 5{\cdot}6$$
$$V_1 = \frac{42 \times 5{\cdot}6}{1{\cdot}05} = 224 \text{ m}^3 \quad \text{Ans. (ii)}$$

2.

$$\frac{p_1V_1}{T_1} = \frac{p_2V_2}{T_2}$$
$$\frac{1350 \times 0{\cdot}2}{450} = \frac{250 \times 0{\cdot}9}{T_2}$$
$$T_2 = \frac{450 \times 250 \times 0{\cdot}9}{1350 \times 0{\cdot}2} = 375 \text{ K}$$
$$= 102^\circ\text{C} \quad \text{Ans.}$$

3.

$$pV = mRT$$
$$m_1 = \frac{pV}{RT}$$
$$= \frac{10 \times 100 \times 2}{287 \times 293}$$
$$= 0{\cdot}02378 \text{ kg}$$
$$m_2 = 0{\cdot}92 \times 0{\cdot}02378$$
$$= 0{\cdot}02188 \text{ kg}$$
$$T = \frac{pV}{mR}$$
$$= \frac{20 \times 100 \times 2}{0{\cdot}02188 \times 287}$$
$$= 636{\cdot}98 \text{ K}$$
$$\text{Temperature} = 363{\cdot}98^\circ\text{C} \quad \text{Ans.}$$

4. By Charles' law, when volume is constant,

$$\frac{p_1}{p_2} = \frac{T_1}{T_2}$$

$$p_2 = \frac{(3200 + 100) \times 308}{289}$$

$$= \frac{3300 \times 308}{289} = 3518 \text{ kN/m}^2 \text{ abs.}$$

$$\text{or, } 3518 - 100 = 3418 \text{ kN/m}^2 \text{ gauge} \quad \text{Ans. (i)}$$

$$Q\,[\text{kJ}] = m\,\text{kJ} \times c\,[\text{kJ/kg K}] \times (T_2 - T_1)\,[\text{K}]$$
$$= 20 \times 0{\cdot}718 \times (35 - 16)$$
$$= 272{\cdot}9 \text{ kJ} \quad \text{Ans. (ii)}$$

5. Volume of saloon = $12 \times 16{\cdot}5 \times 4 = 792 \text{ m}^3$
Volume of air to be supplied per hour
$= 2 \times 792 = 1584 \text{ m}^3$

Volume varies as the absolute temperature when the pressure is constant, therefore density varies inversely as the absolute temperature.

Density at 0°C and atmos. press. = $1{\cdot}293 \text{ kg/m}^3$
∴ density at 21°C and atmos. press.

$$= 1{\cdot}293 \times \frac{273}{294} = 1{\cdot}2 \text{ kg/m}^3$$

Mass of air to be supplied per hour
$= \text{volume } [\text{m}^3/\text{h}] \times \text{density } [\text{kg/m}^3]$
$= 1584 \times 1{\cdot}2 = 1901 \text{ kg/h}$

Heat extracted per hour

$$Q\,[\text{kJ/h}] = m\,[\text{kg/h}] \times c_p\,[\text{kJ/kg K}] \times (T_1 - T_2)\,[\text{K}]$$
$$= 1901 \times 1{\cdot}005 \times (30 - 21)$$
$$= 17190 \text{ kJ/h or } 17{\cdot}19 \text{ MJ/h} \quad \text{Ans. (i)}$$

Power [kW] = kilojoules per second

$$= \frac{17190}{3600} = 4{\cdot}775 \text{ kW} \quad \text{Ans. (ii)}$$

6. Negligible velocity, so kinetic energies are zero
Steady Flow Energy Equation:

$$h_1 + q = h_2 + w \quad \text{per unit mass}$$
$$w_{OUT} = h_1 - h_2 + q_{IN}$$
$$w_{OUT} = 1{\cdot}7 \times 0{\cdot}9 \times 480 - 1{\cdot}7 \times 10$$
$$= 717{\cdot}4 \text{ kW} \quad \text{Ans.}$$

7. At constant volume:

$$Q = mc_V (T_2 - T_1)$$
$$= 1{\cdot}36 \times 0{\cdot}718 \times (468 - 40)$$
$$= 417{\cdot}8 \text{ kJ} \quad \text{Ans. (a) (i)}$$

External work done = nil Ans. (a) (ii)

From the energy equation,

$$\begin{matrix}\text{Heat energy} \\ \text{supplied}\end{matrix} = \begin{matrix}\text{Increase in} \\ \text{internal energy}\end{matrix} + \begin{matrix}\text{External} \\ \text{work done}\end{matrix}$$

$$417{\cdot}8 = \text{Increase in internal energy} + 0$$

$\therefore$ Increase in internal energy = 417·8 kJ Ans. (a) (iii)

At constant pressure:

$$Q = mc_p (T_2 - T_1)$$
$$= 1{\cdot}36 \times 1{\cdot}005 \times (468 - 40)$$
$$= 584{\cdot}8 \text{ kJ} \quad \text{Ans. (b)(i)}$$

Internal energy depends only upon the temperature and is independent of changes in pressure and volume. Therefore the increase in internal energy is the same as before,

$$= 417{\cdot}8 \text{ kJ} \quad \text{Ans. (b) (iii)}$$

$$\begin{matrix}\text{Heat energy} \\ \text{supplied}\end{matrix} = \begin{matrix}\text{Increase in} \\ \text{internal energy}\end{matrix} + \begin{matrix}\text{External} \\ \text{work done}\end{matrix}$$

$$584{\cdot}8 = 417{\cdot}8 + \text{work done}$$
$$\text{Work done} = 584{\cdot}8 - 417{\cdot}8$$
$$= 167 \text{ kJ} \quad \text{Ans. (b) (ii)}$$

Note also:

$$R = c_P - c_V = 1{\cdot}005 - 0{\cdot}718 = 0{\cdot}287 \text{ kJ/kg K}$$
$$\text{Work done} = mR (T_2 - T_1)$$
$$= 1{\cdot}36 \times 0{\cdot}287 \times 428$$
$$= 167 \text{ kJ (as above)}$$

8. Ratio of partial pressures is equal to the ratio of partial volumes.

$$\left.\begin{aligned}\text{Partial press. of } CO_2 &= 0{\cdot}1 \times 1{\cdot}015 = 0{\cdot}1015 \text{ bar} \\ \text{Partial press. of } O_2 &= 0{\cdot}08 \times 1{\cdot}015 = 0{\cdot}0812 \text{ bar} \\ \text{Partial press. of } N_2 &= 0{\cdot}82 \times 1{\cdot}015 = 0{\cdot}8323 \text{ bar}\end{aligned}\right\} \text{Ans.}$$

$$pV = mRT \quad \therefore \; m = \frac{pV}{RT}$$

where p = partial pressure, and V = full volume

(or p = total pressure, and V = partial volume)

$$\text{Mass of } CO_2 = \frac{0{\cdot}1015 \times 10^2 \times 500 \times 10^{-6}}{0{\cdot}189 \times 293}$$

$$= 9{\cdot}162 \times 10^{-5} \text{ kg} = 0{\cdot}09162 \text{ g} \quad \text{Ans.}$$

$$\text{Mass of } O_2 = \frac{0{\cdot}0812 \times 10^2 \times 500 \times 10^{-6}}{0{\cdot}26 \times 293}$$

$$= 5{\cdot}33 \times 10^{-5} \text{ kg} = 0{\cdot}0533 \text{ g} \quad \text{Ans.}$$

$$\text{Mass of } N_2 = \frac{0{\cdot}8323 \times 10^2 \times 500 \times 10^{-6}}{0{\cdot}297 \times 293}$$

$$= 4{\cdot}782 \times 10^{-4} \text{ kg} = 0{\cdot}4782 \text{ g} \quad \text{Ans.}$$

9.

$$pv = RT \text{ and } R = \frac{R_0}{M}$$

$$\text{hence } pv = \frac{R_0}{M} \times T$$

$$\text{also } h = u + pv$$

$$\therefore \quad h = u + R_0\frac{T}{M}$$

$$\text{Hence } h_1 = u_1 + R_0\frac{T_1}{M} \quad \text{and } h_2 = u_2 + R_0\frac{T_2}{M}$$

$$u_1 - u_2 = 80$$

$$\therefore \quad h_1 - h_2 = 80 + \frac{R_0}{M}(T_1 - T_2)$$

$$= 80 + \frac{8{\cdot}314}{30}(200 - 100)$$

$$= 107{\cdot}71 \text{ kJ/kg}$$

Steady flow energy equation

$$h_1 + \tfrac{1}{2}c_1^2 + q = h_2 + \tfrac{1}{2}c_2^2 + w$$

$$\text{as } q \text{ and } w = 0$$

$$h_1 - h_2 = \tfrac{1}{2}(c_2^2 - c_1^2)$$

$$\text{hence } 107{\cdot}71 \times 10^3 = \tfrac{1}{2}(c_2^2 - 50^2)$$

$$c_2 = \sqrt{107{\cdot}71 \times 10^3 \times 2 + 50^2}$$

$$c_2 = 466{\cdot}82 \text{ m/s} \quad \text{Ans.}$$

10. 0·45 kg carbon monoxide $R = \frac{R_0}{M} = \frac{8{\cdot}314}{28} = 0{\cdot}2969$

0·23 kg oxygen $R = \frac{R_0}{M} = \frac{8{\cdot}314}{32} = 0{\cdot}2598$

0·77 kg nitrogen $R = \frac{R_0}{M} = \frac{8{\cdot}314}{28} = 0{\cdot}2969$

$$1{\cdot}45\,R = 0{\cdot}45 \times 0{\cdot}2969 + 0{\cdot}23 \times 0{\cdot}2598 + 0{\cdot}77 \times 0{\cdot}2969$$

$$1{\cdot}45 \text{ kg mixture } R = 0{\cdot}2911$$

$$pV = mRT$$

$$p = \frac{mRT}{V}$$

$$= \frac{1{\cdot}45 \times 0{\cdot}2911 \times 288}{0{\cdot}4}$$

$$= 303{\cdot}9 \text{ kN/m}^2$$

Total pressure = 3·039 bar Ans. (a)

$$pV = mRT$$

$$p = \frac{mRT}{V}$$

$$= \frac{0{\cdot}77 \times 0{\cdot}2911 \times 288}{0{\cdot}4}$$

$$= 164{\cdot}6 \text{ kN/m}^2$$

Partial pressure nitrogen = 1·646 bar Ans. (b)

SOLUTIONS TO TEST EXAMPLES 6

1. For polytropic expansion

$$p_1V_1^n = p_2V_2^n$$

$$2550 \times V_1^{1\cdot3} = 210 \times 0\cdot75^{1\cdot3}$$

$$V_1^{1\cdot3} = \frac{210 \times 0\cdot75^{1\cdot3}}{2550}$$

$$V_1 = 0\cdot75 \times \sqrt[1\cdot3]{\frac{210}{2550}}$$

$$= 0\cdot1099 \text{ m}^3 \quad \text{Ans.}$$

2. Ratio of compression $= \dfrac{8\cdot6}{1} = \dfrac{\text{initial volume}}{\text{final volume}} = \dfrac{V_2}{V_2}$

$$p_1V_1^n = p_2V_2^n$$

$$98 \times 8\cdot6^{1\cdot36} = p_2 \times 1^{1\cdot36}$$

$$p_2 = 98 \times 8\cdot6^{1\cdot36} = 1828 \text{ kN/m}^2 \quad \text{Ans. (i)}$$

$$\frac{p_1V_1}{T_1} = \frac{p_2V_2}{T_2}$$

$$\frac{98 \times 8\cdot6}{301} = \frac{1828 \times 1}{T_2}$$

$$T_2 = \frac{301 \times 1828}{98 \times 8\cdot6} = 653 \text{ K}$$

$$= 380°\text{C} \quad \text{Ans. (ii)}$$

3.

$$p_1V_1^n = p_2V_2^n$$

$$1750 \times 0\cdot05^n = 122\cdot5 \times 0\cdot375^n$$

$$\frac{1750}{122\cdot5} = \left\{\frac{0\cdot375}{0\cdot05}\right\}^n$$

$$n = 1\cdot32 \quad \text{Ans.}$$

4.

$$\frac{T_1}{T_2} = \left\{\frac{p_1}{p_2}\right\}^{\frac{n-1}{n}}$$

$$\frac{n-1}{n} = \frac{1\cdot35 - 1}{1\cdot35} = \frac{7}{27}$$

$$\frac{293}{T_2} = \left\{\frac{101 \cdot 3}{1420}\right\}^{\frac{7}{27}}$$

$$T_2 = 293 \times \left\{\frac{1420}{101 \cdot 3}\right\}^{\frac{7}{27}} = 581 \cdot 1 \text{ K}$$

$$581 \cdot 1 - 273 = 308 \cdot 1^\circ\text{C} \quad \text{Ans.}$$

5. For adiabatic expansion

$$\frac{T_1}{T_2} = \left\{\frac{V_2}{V_1}\right\}^{\gamma - 1}$$

$$\text{where } \gamma = \frac{c_P}{c_V} = \frac{1 \cdot 005}{0 \cdot 718} = 1 \cdot 4$$

$$\gamma - 1 = 1 \cdot 4 - 1 = 0 \cdot 4$$

$$\frac{339}{275} = \left\{\frac{V_2}{0 \cdot 014}\right\}^{0 \cdot 4}$$

$$V_2 = 0 \cdot 014 \times \sqrt[0 \cdot 4]{\frac{339}{275}}$$

$$= 0 \cdot 02362 \text{ m}^3 \quad \text{Ans.}$$

6.

$$\frac{T_1}{T_2} = \left\{\frac{p_1}{p_2}\right\}^{\frac{n-1}{n}}$$

$$\frac{305}{773} = \left\{\frac{117}{3655}\right\}^{\frac{n-1}{n}}$$

$$\text{or, } \frac{773}{305} = \left\{\frac{3655}{117}\right\}^{\frac{n-1}{n}}$$

$$\log\left\{\frac{773}{305}\right\} = \log\left\{\frac{3655}{117}\right\} \times \left\{\frac{n-1}{n}\right\}$$

$$0 \cdot 4039 = 1 \cdot 4947 \times \frac{n-1}{n}$$

$$1 \cdot 4947 = 1 \cdot 0908n$$

$$n = \frac{1 \cdot 4947}{1 \cdot 0908} = 1 \cdot 37$$

∴ law of compression is:

$$pV^{1\cdot37} = C \quad \text{Ans.}$$

7.

$$p_1V_1 = mRT_1$$

$$V_1 = \frac{1 \times 0{\cdot}287 \times 473}{20 \times 100}$$

$$= 0{\cdot}1142 \text{ m}^3$$

$$p_2V_2 = mRT_2$$

$$V_2 = \frac{1 \times 0{\cdot}287 \times 398}{10 \times 100}$$

$$= 0{\cdot}0679 \text{ m}^3$$

Work transfer (expansion) = Area under process straight line of pV diagram

= Mean pressure × change of volume

$= 100 \times 15\,(0{\cdot}1142 - 0{\cdot}0679)$

$= 69{\cdot}45$ kJ Ans. (a)

$c_V = 1005 - 287 = 718$ J/kg K

Change of internal energy $= mc_v\,(T_2 - T_1)$

$= 1 \times 0{\cdot}718\,(398 - 473)$

$= -53{\cdot}85$ kJ, i.e. decrease Ans. (b)

Heat transfer $= -53{\cdot}85 + 69{\cdot}45$

$= 15{\cdot}6$ kJ Ans. (c)

Change of enthalpy $= mc_P\,(T_2 - T_1)$

$= 1 \times 1{\cdot}005\,(398 - 473)$

$= -75{\cdot}375$ kJ, i.e. decrease Ans. (d)

8.

$$p_1V_1 = mRT_1$$

$$V_1 = \frac{1 \times 0{\cdot}287 \times 486}{1 \times 100}$$

$$= 1{\cdot}395 \text{ m}^3$$

$$p_2V_2 = m_2RT_2$$

$$m_2 = \frac{6 \times 100 \times 0{\cdot}2}{0{\cdot}287 \times 685}$$

$$= 0{\cdot}6103 \text{ kg}$$

No heat transfer takes place (insulated)
No external work is done
The process is at constant internal energy

$$m_1 c_V T_1 + m_2 c_V T_2 = m_3 c_V T_3$$
$$1 \times 486 + 0{\cdot}6103 \times 685 = 1{\cdot}6103 \times T_3$$
$$T_3 = 561{\cdot}4 \text{ K}$$

Final air temperature is 288·4°C Ans. (a)

$$p_3 V_3 = m_3 R T_3$$
$$p_3 = \frac{1{\cdot}6103 \times 0{\cdot}287 \times 561{\cdot}4}{1{\cdot}595}$$
$$= 162{\cdot}7 \text{ kN/m}^2$$

Final air pressure is 1·627 bar Ans. (b)

9.

$$pV = mRT$$
$$m = \frac{1 \times 100 \times 0{\cdot}1}{0{\cdot}287 \times 288}$$
$$= 0{\cdot}1210 \text{ kg}$$
$$\frac{T_2}{T_1} = \frac{p_2}{p_1} \quad \text{as } V \text{ constant}$$
$$\frac{773}{288} = \frac{p_2}{1}$$
$$p_2 = 2{\cdot}684 \text{ bar}$$
$$\frac{T_3}{T_2} = \left(\frac{p_3}{p_2}\right)^{\frac{n-1}{n}}$$
$$\frac{T_3}{773} = \left(\frac{1}{2{\cdot}684}\right)^{\frac{0{\cdot}5}{1{\cdot}5}}$$
$$T_3 = 556{\cdot}1 \text{ K}$$

Final temperature = 283·1°C Ans. (a)

$$\text{Work done} = \frac{mR\,(T_2 - T_3)}{n-1}$$
$$= \frac{0{\cdot}121 \times 0{\cdot}287\,(773 - 556{\cdot}1)}{0{\cdot}5}$$

$$= 15{\cdot}05 \text{ kJ} \quad \text{Ans. (b)}$$

$$\text{Heat supplied} = mc_v (T_2 - T_1) \text{ at constant volume}$$
$$= 0{\cdot}121 \times 0{\cdot}718\,(773 - 288)$$
$$= 42{\cdot}14 \text{ kJ} \quad \text{Ans. (c)}$$

$$\text{Heat transfer} = mc_v (T_3 - T_2)$$
$$= 0{\cdot}121 \times 0{\cdot}718\,(556{\cdot}1 - 773)$$
$$= -18{\cdot}84 + 15{\cdot}05$$
$$= -3{\cdot}79 \text{ kJ} \quad \text{Ans. (c)}$$

10.

$$T_1 = 783 \text{ K}$$
$$T_2 = 313 \text{ K}$$

$$\frac{p_1V_1}{T_1} = \frac{p_2V_2}{T_2}$$

$$\frac{36 \times 0{\cdot}125}{783} = \frac{p_2 \times 1{\cdot}5}{313}$$

$$p_2 = \frac{36 \times 0{\cdot}125 \times 313}{783 \times 1{\cdot}5}$$

$$= 1{\cdot}199 \text{ bar or } 119{\cdot}9 \text{ kN/m}^2 \quad \text{Ans. (i)}$$

$$p_1V_1^n = p_2V_2^n \quad \text{or} \quad \frac{T_1}{T_2} = \left\{\frac{V_2}{V_1}\right\}^{n-1}$$

$$36 \times 0{\cdot}125^n = 1{\cdot}199 \times 1{\cdot}5^n$$

$$\frac{36}{1{\cdot}199} = \left\{\frac{1{\cdot}5}{0{\cdot}125}\right\}^n$$

$$n = 1{\cdot}37 \quad \text{Ans. (ii)}$$
$$p_1V_1 = mRT_1$$

$$m = \frac{3600 \times 0{\cdot}125}{0{\cdot}284 \times 783} = 2{\cdot}023 \text{ kg} \quad \text{Ans. (iii)}$$

Increase in internal energy:

$$U_2 - U_1 = mc_V (T_2 - T_1)$$
$$= 2{\cdot}023 \times 0{\cdot}71 \times (313 - 783)$$
$$= 2{\cdot}023 \times 0{\cdot}71 \times (-470)$$
$$= -675{\cdot}1 \text{ kJ}$$

The minus sign indicates a *decrease* in internal energy.

Decrease in internal energy = 675·1 kJ Ans. (iv)

$$\text{Work done} = \frac{p_1V_1 - p_2V_2}{n-1} \text{ or } \frac{mR\,(T_1 - T_2)}{n-1}$$

$$= \frac{3600 \times 0{\cdot}125 - 119{\cdot}9 \times 1{\cdot}5}{1{\cdot}37 - 1}$$

$$= \frac{450 - 179{\cdot}9}{0{\cdot}37} = 730{\cdot}2 \text{ kJ} \quad \text{Ans. (v)}$$

$$\text{Heat supplied} = \begin{matrix}\text{Increase in}\\ \text{internal energy}\end{matrix} + \begin{matrix}\text{External}\\ \text{work done}\end{matrix}$$

$$= -675{\cdot}1 + 730{\cdot}2$$

$$= 55{\cdot}1 \text{ kJ} \quad \text{Ans. (vi)}$$

SOLUTIONS TO TEST EXAMPLES 7

1. Indicated mean effective pressure [kN/m²]

$$= \text{mean height of diagram [mm]} \times \text{spring scale [kN/m}^2\text{mm]}$$

$$= \frac{\text{area of diagram}}{\text{length of diagram}} \times \text{spring scale}$$

$$= \frac{390}{70} \times 1{\cdot}6 \times 10^2 = 891{\cdot}6 \text{ kN/m}^2$$

$$\text{ip} = p_m ALn \times 4 \text{ (for 4 cylinders)}$$

$$= 891{\cdot}6 \times 0{\cdot}7854 \times 0{\cdot}15^2 \times 0{\cdot}2 \times \frac{5{\cdot}5}{2} \times 4$$

$$= 34{\cdot}66 \text{ kW} \quad \text{Ans.}$$

2. $$\text{ip} = p_m ALn \times 8 \text{ (for 8 cylinders)}$$

$$= 1172 \times 0{\cdot}7854 \times 0{\cdot}75^2 \times 1{\cdot}125 \times \frac{110}{60 \times 2} \times 8$$

$$= 4272 \text{ kW} \quad \text{Ans. (i)}$$

$$\text{bp} = \text{ip} \times \text{mech. efficiency}$$
$$= 4272 \times 0{\cdot}86$$
$$= 3676 \text{ kW} \quad \text{Ans. (ii)}$$

3. $$\text{ip} = \frac{\text{bp}}{\text{mech. effic.}} = \frac{2250}{0{\cdot}84} = 2680 \text{ kW}$$

Let d = diameter of cylinders in metres

then L = $1{\cdot}25d$ metres

$$\text{ip} = p_m ALn \times 6$$
$$2680 = 10 \times 10^2 \times 0{\cdot}7854 \times d^2 \times 1{\cdot}25d \times 2 \times 6$$

$$d = \sqrt[3]{\frac{2680}{1000 \times 0{\cdot}7854 \times 1{\cdot}25 \times 2 \times 6}}$$

$$\left.\begin{aligned} &= 0{\cdot}6103 \text{ m} = 610{\cdot}3 \text{ mm} \\ L &= 1{\cdot}25 \times 610{\cdot}3 = 763 \text{ mm} \end{aligned}\right\} \text{Ans.}$$

4. Load on brake = $480 - 84 = 396$ N

Effective radius = ½ (1220 + 24) = 622 mm = 0·622 m

Brake power = T_ω

$$= 396 \times 0{\cdot}622 \times \frac{250 \times 2\pi}{60}$$

$$= 6449 \text{ W} = 6{\cdot}449 \text{ kW} \quad \text{Ans. (i)}$$

$$1 \text{ kWh} = 3{\cdot}6 \text{ MJ}$$

Heat carried away by water in one hour

$$= 0{\cdot}9 \times 6{\cdot}449 \times 3{\cdot}6 \times 10^3 \text{ kJ}$$

Heat received by water

$$= \text{mass} \times \text{spec. heat} \times \text{temp. rise}$$

$$Q \text{ [kJ]} = m \text{ [kg]} \times c_p \text{ [kJ/kg K]} \times \theta \text{ [K]}$$

$$= m \times 4{\cdot}2 \times 18$$

$$m \times 4{\cdot}2 \times 18 = 0{\cdot}9 \times 6{\cdot}449 \times 3{\cdot}6 \times 10^3$$

$$m = 276{\cdot}4 \text{ kg}$$

Mass of one litre of fresh water = one kg

Quantity in litres = 276·4 litres/h Ans. (ii)

5. Brake power [W] = Torque [Nm] × ω [rad/s]

When speed is 24·5 rev/s

bp = $T \times 2\pi \times 24{\cdot}5 = 153{\cdot}9\ T$ watts $= 0{\cdot}1539\ T$ kW

With all cylinders firing:

$$\text{bp} = 0{\cdot}1539 \times 193{\cdot}8 = 29{\cdot}84 \text{ kW} \quad \text{Ans. (i)}$$

$$\text{no. 1 cyl. cut out, bp} = 0{\cdot}1539 \times 130{\cdot}8 = 20{\cdot}13 \text{ kW}$$
$$\text{ip of no. 1 cyl.} = 29{\cdot}84 - 20{\cdot}13 = 9{\cdot}71 \text{ kW}$$
$$\text{no. 2 cyl. cut out, bp} = 0{\cdot}1539 \times 130{\cdot}2 = 20{\cdot}04 \text{ kW}$$
$$\text{ip of no. 2 cyl.} = 29{\cdot}84 - 20{\cdot}04 = 9{\cdot}8 \text{ kW}$$
$$\text{no. 3 cyl. cut out, bp} = 0{\cdot}1539 \times 129{\cdot}9 = 20{\cdot}0 \text{ kW}$$
$$\text{ip of no. 3 cyl.} = 29{\cdot}84 - 20 = 9{\cdot}84 \text{ kW}$$
$$\text{no. 4 cyl. cut out, bp} = 0{\cdot}1539 \times 131{\cdot}1 = 20{\cdot}18 \text{ kW}$$
$$\text{ip of no. 4 cyl.} = 29{\cdot}84 - 20{\cdot}18 = 9{\cdot}66 \text{ kW}$$

Total ip of engine

$$= 9{\cdot}71 + 9{\cdot}8 + 9{\cdot}84 + 9{\cdot}66 = 39{\cdot}01 \text{ kW} \quad \text{Ans. (ii)}$$

$$\text{Mech. effic.} = \frac{\text{brake power}}{\text{indicated power}}$$

$$= \frac{29{\cdot}84}{39{\cdot}01} = 0{\cdot}7649 = 76{\cdot}49\% \quad \text{Ans. (iii)}$$

6. $$\text{Brake thermal effic.} = \frac{3{\cdot}6 \text{ [MJ/kW h]}}{\text{kg fuel/brake kW h} \times \text{cal. value [MJ/kg]}}$$

$$= \frac{3{\cdot}6}{0{\cdot}255 \times 43{\cdot}5} = 0{\cdot}3245 \text{ or } 37{\cdot}45\% \quad \text{Ans. (ii)}$$

$$\text{Indicated thermal efficiency} = \frac{\text{brake thermal efficiency}}{\text{mechanical efficiency}}$$

$$= \frac{0{\cdot}3245}{0{\cdot}86} = 0{\cdot}3773 \text{ or } 37{\cdot}74\% \quad \text{Ans. (i)}$$

For each kg of fuel burned,
total mass of gases = 35 kg air + 1 kg fuel = 36 kg
Heat carried away in exhaust gases,

$$\begin{aligned} Q\,[\text{kJ}] &= m\,[\text{kg}] \times c_p\,[\text{kJ/kg K}] \times \theta\,[\text{K}] \\ &= 36 \times 1{\cdot}005 \times (393 - 26) \\ &= 1{\cdot}327 \times 10^4 \text{ kJ} \\ &= 13{\cdot}27 \text{ MJ/kg fuel} \\ \text{Heat supplied} &= 43{\cdot}5 \text{ MJ/kg fuel} \end{aligned}$$

∴ percentage heat in exhaust gases

$$= \frac{13{\cdot}27}{43{\cdot}5} \times 100 = 30{\cdot}52\% \quad \text{Ans. (iii)}$$

7.
$$\begin{aligned} \text{Ind. thermal effic.} &= 100\% - (31{\cdot}7 + 30{\cdot}8) \\ &= 37{\cdot}5\% \quad \text{Ans. (i)} \\ \text{Mechanical effic.} &= \frac{\text{bp}}{\text{ip}} = \frac{4060}{4960} \\ &= 0{\cdot}8185 \text{ or } 81{\cdot}85\% \quad \text{Ans. (ii)} \\ \text{Overall effic.} &= \text{ind. thermal effic.} \times \text{mech. effic.} \\ &= 0{\cdot}375 \times 0{\cdot}8185 \\ &= 0{\cdot}307 \text{ or } 30{\cdot}7\% \quad \text{Ans. (iii)} \end{aligned}$$

Specific fuel consumption (indicated)

$$= \frac{27 \times 10^3}{24 \times 4960} = 0{\cdot}2269 \text{ kg/ind. kW h} \quad \text{Ans. (iv)}$$

$$\begin{aligned} \text{Indicated thermal effic.} &= \frac{3{\cdot}6\ [\text{MJ/kW h}]}{\text{kg fuel/ind. kW h} \times \text{cal. val. [MJ/kg]}} \\ 0{\cdot}375 &= \frac{3{\cdot}6}{0{\cdot}2269 \times \text{cal. value}} \\ \text{Calorific value} &= \frac{3{\cdot}6}{0{\cdot}375 \times 0{\cdot}2269} \\ &= 42{\cdot}32 \text{ MJ/kg} \quad \text{Ans. (v)} \end{aligned}$$

8. BEFORE ALTERATION:

$$\begin{aligned} V_1 &= 87{\cdot}5 + 12{\cdot}5 = 100 \\ V_2 &= 12{\cdot}5 \\ p_1 V_1^{1{\cdot}35} &= p_2 V_2^{1{\cdot}35} \end{aligned}$$

$$0.97 \times 100^{1\cdot35} = p_2 \times 12{\cdot}5^{1\cdot35}$$

$$p_2 = 0{\cdot}97 \times \left\{\frac{100}{12{\cdot}5}\right\}^{1\cdot35}$$

$$= 0{\cdot}97 \times 8^{1\cdot35} = 16{\cdot}07 \text{ bar} \quad \text{Ans. (i)}$$

AFTER ALTERATION:

$$V_1 = 87{\cdot}5 + 10 = 97{\cdot}5$$

$$V_2 = 10$$

$$p_1 V_1^{1\cdot35} = p_2 V_2^{1\cdot35}$$

$$0.97 \times 97{\cdot}51^{1\cdot35} = p_2 \times 10^{1\cdot35}$$

$$p_2 = 0{\cdot}97 \times 9{\cdot}75^{1\cdot35} = 20{\cdot}99 \text{ bar} \quad \text{Ans. (ii)}$$

9. Due to the "hit and miss" governor cutting off the supply of gas, the number of power strokes per second is the number of explosions per second, which is $123 \div 60 = 2{\cdot}05$

$$\text{ip} = p_m ALn$$

$$= 3{\cdot}93 \times 10^2 \times 0{\cdot}7854 \times 0{\cdot}18^2 \times 0{\cdot}3 \times 2{\cdot}05$$

$$= 6{\cdot}152 \text{ kW} \quad \text{Ans. (i)}$$

$$\text{Mech. effic.} = \frac{\text{brake power}}{\text{indicated power}}$$

$$= \frac{4{\cdot}33}{6{\cdot}152} = 0{\cdot}7039 = 70{\cdot}39\% \quad \text{Ans. (ii)}$$

Indicated thermal efficiency

$$= \frac{\text{heat equivalent of work done in cyl. [kJ/h]}}{\text{heat supplied [kJ/h]}}$$

$$= \frac{6{\cdot}152 \times 3600}{3{\cdot}1 \times 17{\cdot}6 \times 10^3}$$

$$= 0{\cdot}4059 = 40{\cdot}59\% \quad \text{Ans. (iii)}$$

$$\text{Brake thermal effic.} = \text{ind. thermal effic.} \times \text{mech. effic.}$$

$$= 0{\cdot}4059 \times 0{\cdot}7039$$

$$= 0{\cdot}2858 = 28{\cdot}58\% \quad \text{Ans. (iv)}$$

10. Mean piston speed $= 2\,Ln$

$$6 = 2 \times 1{\cdot}5 \times n$$

$$n = 2 \text{ rev/s}$$

$$\text{brake power} = p_m ALn$$
$$= 7 \times 100 \times 0{\cdot}7854 \times 0{\cdot}75^2 \times 1{\cdot}5 \times 2$$

$$\text{brake power} = 927{\cdot}8 \text{ kW}$$

$$\frac{\text{power 1}}{\text{swept volume 1}} = \frac{\text{power 2}}{\text{swept volume 2}}$$

$$\frac{370}{0{\cdot}7854 \times D^2 \times 2{\cdot}0D} = \frac{927{\cdot}8}{0{\cdot}7854 \times 0{\cdot}75^2 \times 1{\cdot}5}$$

$$D^3 = \frac{370 \times 0{\cdot}5625 \times 1{\cdot}5}{927{\cdot}8 \times 2{\cdot}0}$$

$$D = 0{\cdot}552 \text{ m}$$

$$\text{cylinder bore} = 0{\cdot}552 \text{ m} \quad \text{Ans. (a)}$$

$$\text{brake power} = p_m ALn$$
$$370 = p_m \times 100 \times 0{\cdot}7854 \times 0{\cdot}552^2 \times 1{\cdot}104 \times 2$$

$$p_m = 7{\cdot}002 \text{ bar}$$

$$\text{brake mean effective pressure} = 7{\cdot}002 \text{ bar} \quad \text{Ans. (b)}$$

SOLUTIONS TO TEST EXAMPLES 8

1.

$$p_1V_1^{1\cdot35} = p_2V_2^{1\cdot35}$$
$$1 \times 8\cdot5^{1\cdot35} = p_2 \times 1^{1\cdot35}$$
$$p_2 = 8\cdot5^{1\cdot35} = 17\cdot97 \text{ bar} \quad \text{Ans. (i)}$$

$$\frac{p_1V_1}{T_1} = \frac{p_2V_2}{T_2}$$

$$\frac{1 \times 8\cdot5}{313} = \frac{17\cdot97 \times 1}{T_2}$$

$$T_2 = \frac{313 \times 17\cdot97}{8\cdot5} = 661\cdot8 \text{ K}$$

$$= 388\cdot8°\text{C} \quad \text{Ans. (ii)}$$

$$\frac{T_3}{T_2} = \frac{p_3}{p_2}$$

$$T_3 = \frac{661\cdot8 \times 31}{17\cdot97} = 1141 \text{ K}$$

$$= 868°\text{C} \quad \text{Ans. (iii)}$$

2.

$$\frac{T_2}{T_1} = \left(\frac{V_1}{V_2}\right)^{\gamma - 1}$$ (Refer to Fig. 26, p–V diagram)

$$T_2 = 323 \times 5^{0\cdot4}$$
$$= 615 \text{ K}$$

$$\text{Heat supplied} = c_v (T_3 - T_2)$$
$$930 = 0\cdot717 (T_3 - 615)$$
$$T_3 = 1912 \text{ K}$$
$$\text{Maximum temp.} = 1639°\text{C} \quad \text{Ans. (a)}$$

$$\frac{T_3}{T_4} = \left(\frac{V_4}{V_3}\right)^{\gamma-1}$$

$$T_4 = \frac{1912}{5^{0\cdot4}}$$

$$T_4 = 1004 \text{ K}$$

$$\text{Work done} = \text{heat supplied} - \text{heat rejected}$$
$$= c_V [(T_3 - T_2) - (T_4 - T_1)]$$
$$= 0\cdot717 [(1912 - 615) - (1004 - 323)]$$
$$= 441\cdot7 \text{ kJ/kg of working fluid} \quad \text{Ans. (b)}$$

Ideal thermal efficiency of cycle $= \dfrac{441{\cdot}7}{0{\cdot}717 \times 1297} \times 100$

$= 47{\cdot}5\%$ Ans. (c)

3.

$$\gamma = \frac{c_P}{c_V} = \frac{1005}{718} = 1{\cdot}4$$

$$\frac{T_2}{T_1} = \left(\frac{V_1}{V_2}\right)^{\gamma-1}$$

$$T_2 = 330 \times 8^{0\cdot4} = 758 \text{ K}$$

Energy added $= c_v\,(T_3 - T_2)$

$1250 = 0{\cdot}718\,(T_3 - 758)$

$T_3 = 2499$ K

Max. temperature $= 2216$°C Ans. (a) (i)

$$p_1V_1^{\gamma} = p_2V_2^{\gamma}$$

$$p_2 = 8^{1\cdot4} \times 1 = 18{\cdot}37 \text{ bar}$$

$$\frac{p_2}{T_2} = \frac{p_3}{T_3}$$

$$p_3 = 2499 \times \frac{18{\cdot}37}{758}$$

Max. pressure $= 60{\cdot}56$ bar Ans. (a)(ii)

For pressure–volume diagram, refer to Fig. 26 Ans. (b)(i)
For temperature–entropy diagram, refer to Fig. 49 Ans. (b)(ii)

4. Refer to Fig. 28 (p–V diagram)

$$\gamma = \frac{c_P}{c_V} = \frac{1005}{718} = 1{\cdot}4$$

$$p_2 = 1 \times (16)^{1\cdot4}$$

Max. pressure = 48·5 bar Ans. (a)

$$\frac{T_2}{T_1} = \left(\frac{V_1}{V_2}\right)^{\gamma-1}$$

$$T_2 = 330 \times 16^{0\cdot4}$$
$$= 1000 \text{ K}$$

$$\text{Energy added} = c_P (T_3 - T_2)$$
$$1250 = 1{\cdot}005\,(T_3 - 1000)$$
$$T_3 = 2244 \text{ K}$$

Max. temperature = 1961°C Ans. (b)

$$\frac{V_3}{V_2} = \frac{T_3}{T_2}$$

$$V_3 = 1 \times \frac{2244}{1000}$$

$$= 2{\cdot}244$$

$$\frac{T_4}{T_3} = \left(\frac{V_3}{V_4}\right)^{\gamma-1}$$

$$T_4 = 2244 \times \left(\frac{2{\cdot}244}{16}\right)^{0\cdot4}$$

$$T_4 = 1023 \text{ K}$$

$$\frac{p_4}{p_1} = \frac{T_4}{T_1}$$

$$p_4 = 1 \times \frac{1023}{330}$$

$$= 3{\cdot}1 \text{ bar}$$

$$\text{Energy removed} = c_v (T_4 - T_1)$$
$$= 0{\cdot}718\,(1023 - 330)$$
$$= 497{\cdot}6 \text{ kJ/kg}$$

Cycle efficiency (A.S.E.) $= \dfrac{1250 - 497{\cdot}6}{1250} \times 100$

$$= 60{\cdot}19\%$$

$$\text{m.e.p.} = \frac{p_2 (V_3 - V_2) + \dfrac{p_3V_3 - p_4V_4}{\gamma - 1} - \dfrac{p_2V_2 - p_1V_1}{\gamma - 1}}{V_1 - V_2}$$

$$= \frac{48{\cdot}5\,(2{\cdot}244 - 1) + \dfrac{48{\cdot}5 \times 2{\cdot}244 - 3{\cdot}1 \times 16}{0{\cdot}4} - \dfrac{48{\cdot}5 \times 1 - 1 \times 16}{0{\cdot}4}}{16 - 1}$$

$$= \frac{1}{15}(60{\cdot}33 + 148{\cdot}1 - 81{\cdot}25)$$

$= 8{\cdot}479$ bar Ans. (d)

5. Referring to Fig. 30, data given:

$p_1 = 1 \quad p_3 = p_4 = 41$
$V_1 = V_5 = 10{\cdot}7 \quad V_2 = V_3 = 1$
$T_1 = 305$
$T_4 = 1866$

$$\gamma = \frac{c_P}{c_V} = \frac{1{\cdot}005}{0{\cdot}718} = 1{\cdot}4$$

COMPRESSION PERIOD:

$$p_1 V_1^\gamma = p_2 V_2^\gamma \qquad p_2 = \frac{1 \times 10{\cdot}7^{1{\cdot}4}}{1^{1{\cdot}4}}$$

$$= 27{\cdot}62 \text{ bar}$$

$$\frac{p_1 V_1}{T_1} = \frac{p_2 V_2}{T_2} \qquad T_2 = \frac{305 \times 27{\cdot}62 \times 1}{1 \times 10{\cdot}7}$$

$$= 787{\cdot}2 \text{ K}$$

$$= 514{\cdot}2\text{°C}$$

PART COMBUSTION AT CONSTANT VOLUME:

Absolute temperature varies as absolute pressure,

$$\therefore\ T_3 = T_2 \times \frac{p_3}{p_2} = \frac{787{\cdot}2 \times 41}{27{\cdot}62} = 1169 \text{ K}$$

$$= 896\text{°C}$$

PART COMBUSTION AT CONSTANT PRESSURE:

Volume varies as absolute temperature,

$$\therefore\ V_4 = V_3 \times \frac{T_4}{T_3} = \frac{1 \times 1866}{1169} = 1{\cdot}596$$

EXPANSION PERIOD:

$$p_4V_4^{1\cdot4} = p_5V_5^{1\cdot4}$$

$$p_5 = \frac{41 \times 1{\cdot}596^{1\cdot4}}{10{\cdot}7^{1\cdot4}} = 2{\cdot}858 \text{ bar}$$

$$\text{Since } V_5 = V_1, \text{ then } \frac{T_5}{T_1} = \frac{p_5}{p_1}$$

$$T_5 = \frac{305 \times 2{\cdot}858}{1} = 871{\cdot}6 \text{ K}$$

$$= 598{\cdot}6°\text{C}$$

Hence, pressures and temperatures required:

$$\left.\begin{aligned} p_2 = 27{\cdot}62 \text{ bar}, \ \theta_2 &= 514{\cdot}2°\text{C} \\ \theta_3 &= 896°\text{C} \\ p_5 = 2{\cdot}858 \text{ bar}, \ \theta_5 &= 598{\cdot}6°\text{C} \end{aligned}\right\} \text{ Ans. (i)}$$

Ideal thermal efficiency

$$= 1 - \frac{\text{heat rejected}}{\text{heat supplied}}$$

$$= 1 - \frac{c_V(T_5 - T_1)}{c_V(T_3 - T_2) + c_P(T_4 - T_3)}$$

$$= 1 - \frac{(T_5 - T_1)}{(T_3 - T_2) + \gamma(T_4 - T_3)}$$

$$= 1 - \frac{871{\cdot}6 - 305}{(1169 - 787{\cdot}2) + 1{\cdot}4\,(1866 - 1169)}$$

$$= 1 - 0{\cdot}4175$$

$$= 0{\cdot}5825 = 58{\cdot}25\% \quad \text{Ans. (ii)}$$

Note: Compression ratio and maximum pressures are low for this example compared to modern practice.

6. Cycle efficiency $= \dfrac{T_1 - T_2}{T_1} \times 100$

$$= \frac{1300 - 300}{1300} \times 100$$

$$= 76{\cdot}9\% \quad \text{Ans. (a)}$$

Refer to Fig. 48 (temperature–entropy) of Chapter 11.

Also the description of practical constraints to the Carnot cycle – leading on to Fig. 61 (temperature–entropy) for the vapour (steam) plant and the Rankine cycle (Chapter 12).

Difficulties with compressor for wet vapour Ans. (b)(i)

Substitution of feed pump to handle condensed liquid Ans. (b)(ii)

7.

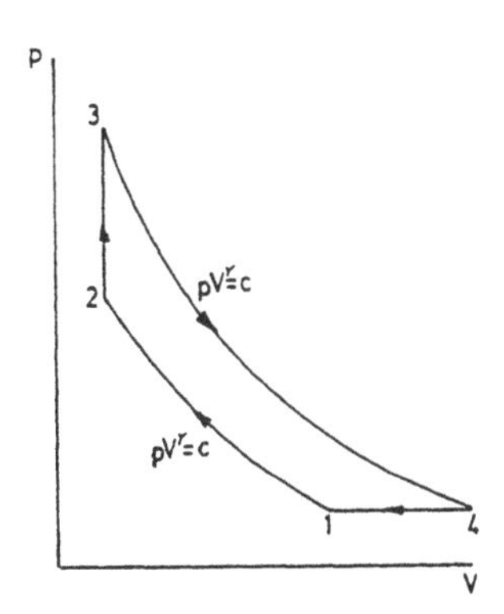

$$\gamma = \frac{c_P}{c_V} = \frac{1000}{678} = 1{\cdot}475$$

$$\frac{T_1}{T_2} = \left(\frac{p_1}{p_2}\right)^{\frac{\gamma-1}{\gamma}}$$

$$T_2 = 333 \times 4{\cdot}5^{\frac{0{\cdot}475}{1{\cdot}475}} = 540{\cdot}5 \text{ K}$$

$$\frac{p_3}{T_3} = \frac{p_2}{T_2}$$

$$T_3 = \frac{1{\cdot}35}{1} \times 540{\cdot}5 = 729 \text{ K}$$

$$\frac{T_3}{T_4} = \left(\frac{p_3}{p_4}\right)^{\frac{\gamma-1}{\gamma}}, \quad p_3 = 4{\cdot}5 \times 1{\cdot}35 = 6{\cdot}075$$

$$T_4 = 729\left(\frac{1}{6{\cdot}075}\right)^{\frac{0{\cdot}475}{1{\cdot}475}} = 407{\cdot}8 \text{ K}$$

Thermal efficiency of the cycle

$$= 1 - \frac{c_P (T_4 - T_1)}{c_V (T_3 - T_2)} \times 100\%$$

$$= 1 - 1{\cdot}475 \frac{(407{\cdot}8 - 333)}{(729 - 540{\cdot}5)} \times 100\%$$

$= 41{\cdot}5\%$ Ans.

8.

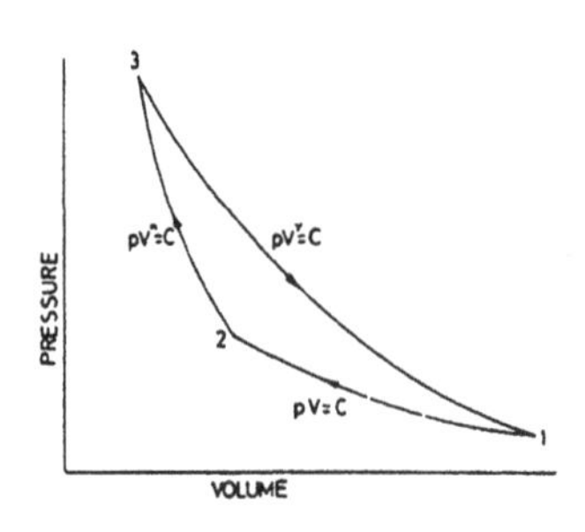

$$\frac{T_3}{T_1} = \left(\frac{p_3}{p_1}\right)^{\frac{\gamma-1}{\gamma}}$$

$$T_3 = 288 \times 4{\cdot}5^{\frac{0.4}{1.4}} = 428 \text{ K}$$

$$\frac{T_3}{T_2} = \left(\frac{p_3}{p_2}\right)^{\frac{n-1}{n}} \quad \text{and } T_1 = T_2$$

$$\therefore \frac{428}{288} = 2^{\frac{n-1}{n}}$$

$$\log 1{\cdot}486 = \frac{n-1}{n} \times \log 2$$

$$0{\cdot}172 \times n = (n-1) \times 0{\cdot}3010$$

$$n = 2{\cdot}33$$

From 1 to 2 Work transfer $= -mRT_1 \ln r$ $\qquad r = p_2/p_1$

$= -287 \times 288 \times \ln 2/10^3$

$= -57{\cdot}29$ kJ/kg Ans. (a)(i)

From 1 to 2 Heat transfer $= -57{\cdot}29$ kJ/kg Ans. (b)(i)

From 2 to 3 Work transfer $= -\dfrac{mR\,(T_3 - T_2)}{n-1}$

$$= -\frac{287\,(428-288)}{10^3\,(2{\cdot}33-1)}$$

$= -30{\cdot}2$ kJ/kg Ans. (a) (ii)

From 2 to 3 Heat transfer $= \dfrac{\gamma - n}{\gamma - 1} \times (-30{\cdot}2)$

$$= \left(\frac{1{\cdot}4 - 2{\cdot}33}{1{\cdot}4 - 1}\right) \times (-30{\cdot}2)$$

$= 70{\cdot}3$ kJ/kg Ans. (b) (ii)

From 3 to 1 Work transfer $= \dfrac{mR\,(T_3 - T_1)}{\gamma - 1}$

$$= \frac{287\,(428-288)}{10^3\,(1{\cdot}4-1)}$$

$= 100{\cdot}5$ kJ/kg Ans. (a) (iii)

From 3 to 1 Heat transfer $= 0$ Ans. (b) (iii)

9.

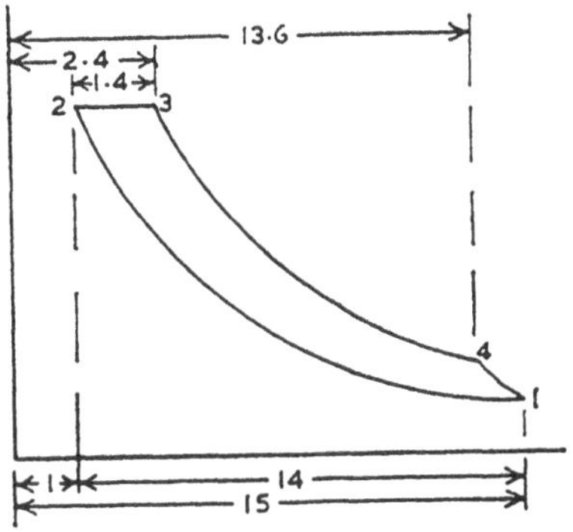

Let $V_1 = 15$, and $V_2 = 1$

then, stroke volume $= 15 - 1 = 14$

Fuel burning period $= \dfrac{1}{10} \times 14 = 1{\cdot}4$

$\therefore V_3 = 1{\cdot}4 + 1 = 2{\cdot}4$

$$V_4 = 1 + \frac{9}{10} \times 14 = 13{\cdot}6$$

$$T_1 = 314\text{K}$$

COMPRESSION PERIOD:

$$\frac{T_2}{T_1} = \left\{\frac{V_1}{V_2}\right\}^{n-1}$$

$$T_2 = 314 \times 15^{0{\cdot}34} = 788{\cdot}4 \text{ K}$$

$\therefore$ Temperature at end of compression

$= 515{\cdot}4°\text{C}$ Ans. (i)

BURNING PERIOD:

$$\frac{T_3}{T_2} = \frac{V_3}{V_2}$$

$$T_3 = 788{\cdot}4 \times 2{\cdot}4 = 1892 \text{ K}$$

$\therefore$ Temperature at end of combustion

$= 1619°\text{C}$ Ans. (ii)

EXPANSION PERIOD:

$$\frac{T_4}{T_3} = \left\{\frac{V_3}{V_4}\right\}^{n-1}$$

$$T_4 = 1892 \times \left\{\frac{2{\cdot}4}{13{\cdot}6}\right\}^{0{\cdot}34} = 1049 \text{ K}$$

$\therefore$ Temperature at end of expansion

$= 776°\text{C}$ Ans. (iii)

Note: This is not an ideal cycle

10. Refer to Fig. 28 (p–V diagram)

$$p_1V_1^n = p_2V_2^n$$

$$p_2 = 1 \times 12^{1{\cdot}36}$$

$$= 29{\cdot}38 \text{ bar}$$

$$\frac{T_2}{T_1} = \left(\frac{V_1}{V_2}\right)^{n-1}$$

$$T_2 = 353 \times 12^{0{\cdot}36}$$

$$= 863{\cdot}4 \text{ K}$$

$$\frac{T_3}{T_2} = \frac{V_3}{V_2}$$

$$V_3 = \frac{1923}{863{\cdot}4}$$

$$= 2{\cdot}227$$

$$p_3V_3^n = p_4V_4^n$$

$$p_4 = 29{\cdot}38 \times \left(\frac{2{\cdot}227}{12}\right)^{1{\cdot}4}$$

$$= 2{\cdot}780 \text{ bar}$$

$$\text{m.e.p.} = \frac{p_2(V_3 - V_2) + \dfrac{p_3V_3 - p_4V_4}{n-1} - \dfrac{p_2V_2 - p_1V_1}{n-1}}{V_1 - V_2}$$

$$= \frac{29{\cdot}38\,(2{\cdot}227 - 1) + \dfrac{29{\cdot}38 \times 2{\cdot}227 - 2{\cdot}78 \times 12}{0{\cdot}4} - \dfrac{29{\cdot}38 \times 1 - 1 \times 12}{0{\cdot}36}}{12 - 1}$$

$$= \frac{1}{11}(36{\cdot}05 + 80{\cdot}18 - 48{\cdot}28)$$

$$= 6{\cdot}177 \text{ bar} \quad \text{Ans. (a)}$$

$$\text{Power} = pm\,ALn$$

$$= \frac{6{\cdot}177 \times 0{\cdot}034 \times 200}{60}$$

$$= 70{\cdot}0 \text{ kW} \quad \text{Ans. (b)}$$

Note: This is not an ideal cycle

SOLUTIONS TO TEST EXAMPLES 9

1. $$\text{Clearance length [mm]} = \frac{\text{clearance volume [mm}^3\text{]}}{\text{area of cylinder [mm}^2\text{]}}$$

$$= \frac{900 \times 10^3}{0{\cdot}7854 \times 250^2} = 18{\cdot}33 \text{ mm}$$

$$V_1 = 350 + 18{\cdot}33 = 368{\cdot}33 \text{ mm}$$
$$p_1 V_1^{1{\cdot}25} = p_2 V_2^{1{\cdot}25}$$
$$0{\cdot}986 \times 368{\cdot}3^{1{\cdot}25} = 4{\cdot}1 \times V_2^{1{\cdot}25}$$
$$V_2 = 368{\cdot}3 \times \sqrt[1{\cdot}25]{\frac{0{\cdot}986}{4{\cdot}1}}$$
$$= 117{\cdot}8 \text{ mm}$$

$$\text{Compression period} = V_1 - V_2$$
$$= 368{\cdot}3 - 117{\cdot}8$$
$$= 250{\cdot}5 \text{ mm} \quad \text{Ans.}$$

2. Refer to Fig. 38 Ans. (a)

$$p_3 V_3^n = p_4 V_4^n$$
$$\left(\frac{p_3}{p_4}\right)^{\frac{1}{n}} = \frac{V_4}{V_3}$$
$$V_4 = \sqrt[1{\cdot}3]{8} \times 0{\cdot}08$$
$$= 0{\cdot}3961 \text{ m}^3$$

$$\text{Suction volume} = 1{\cdot}08 - 0{\cdot}3961$$
$$= 0{\cdot}6839 \text{ m}^3$$

$$\text{Volumetric effic.} = \frac{0{\cdot}6839}{1} \times 100$$
$$= 68{\cdot}39\% \quad \text{Ans. (b)(i)}$$

$$\text{Free air delivery} = 0{\cdot}7854 \times 0{\cdot}152^2 \times 0{\cdot}105 \times 0{\cdot}6839 \times 12$$
$$= 0{\cdot}01564 \text{ m}^3\text{/s} \quad \text{Ans. (b)(ii)}$$

3. Refer to Fig. 38 Ans. (a)

$$pV = mRT$$
$$1 \times 100 \times V_1 = 0{\cdot}0035 \times 0{\cdot}287 \times 288$$
$$V_1 = 0{\cdot}002893 \text{ m}^3$$

$$\text{Stroke volume} = V_1 - V_3$$
$$= 0{\cdot}7854 \times 0{\cdot}14^2 \times 0{\cdot}18$$
$$= 0{\cdot}002771 \text{ m}^3$$

$$\text{Clearance volume} = V_3$$
$$= 0{\cdot}002893 - 0{\cdot}002771$$
$$= 0{\cdot}000122 \text{ m}^3 \quad \text{Ans. (b) (i)}$$

$$p_3V_3{}^n = p_4V_4{}^n$$
$$8{\cdot}5 \times 0{\cdot}000122^{1\cdot32} = 1 \times V_4{}^{1\cdot32}$$
$$V_4 = 0{\cdot}0006166 \text{ m}^3$$

$$V_1 - V_4 = 0{\cdot}002893 - 0{\cdot}0006166$$
$$= 0{\cdot}0022764 \text{ m}^3$$

$$\text{Volumetric effic.} = \frac{0{\cdot}002276}{0{\cdot}002771} \times 100$$
$$= 82{\cdot}14\% \quad \text{Ans. (b)(ii)}$$

4. Refer to Fig. 36 Ans. (a)

$$\text{Stroke volume} = 0{\cdot}7854 \times 0{\cdot}1^2 \times 0{\cdot}12$$
$$= 0{\cdot}0009424 \text{ m}^3$$

Let N be rev/min

$$\text{Suction volume/min} = 3 \times N \times 0{\cdot}0009424$$
$$= \text{Free air delivery/min}$$

$$N = \frac{1{\cdot}2}{3 \times 0{\cdot}0009424}$$

$$\text{Operating speed} = 424{\cdot}4 \text{ rev/min} \quad \text{Ans. (b) (i)}$$

$$p_1V_1{}^n = p_2V_2{}^n$$
$$1 \times 0{\cdot}0009424^{1\cdot3} = 6{\cdot}6 \times V_2{}^{1\cdot3}$$
$$V_2 = 0{\cdot}0002203 \text{ m}^3$$

$$\text{Work/cycle} = \frac{n}{n-1}(p_2V_2 - p_1V_1) \times 3$$

$$= \frac{3 \times 1{\cdot}3 \times 10^2}{0{\cdot}3}(6{\cdot}6 \times 0{\cdot}0002203 - 1 \times 0{\cdot}0009424)$$

$$= 0{\cdot}6651 \text{ kJ}$$

$$\text{Work/s} = 0{\cdot}6651 \times \frac{424{\cdot}4}{60}$$

$$= 4{\cdot}704 \text{ kW}$$

Input power $= \dfrac{4{\cdot}704}{0{\cdot}9}$

$= 5{\cdot}227$ kW Ans. (b) (ii)

5. Volume of free air (at 1·01 bar from the atmosphere) to make 22 m^3 at 31·01 bar abs., at the same temperature

$$= 22 \times \frac{31{\cdot}01}{1{\cdot}01} \text{ m}^3$$

Volume of free air to make 22 m^3 at 20·01 bar abs.

$$= 22 \times \frac{22{\cdot}01}{1{\cdot}01} \text{ m}^3$$

$\therefore$ Volume of free air to supply the difference

$$= \frac{22}{1{\cdot}01}(31{\cdot}01 - 20{\cdot}01)$$

$$= \frac{22 \times 11}{1{\cdot}01} = 239{\cdot}6 \text{ m}^3 \quad . \; . \; . \; \text{(i)}$$

Volume of free air dealt with by compressor per minute
$= 0{\cdot}7854\,(0{\cdot}35^2 - 0{\cdot}075^2) \times 0{\cdot}3 \times 0{\cdot}92 \times 170$
$= 4{\cdot}306$ m^3/min . . . (ii)

$\therefore$ Time $= \dfrac{239{\cdot}6}{4{\cdot}306} = 55{\cdot}64$ minutes Ans.

6. Referring to Fig. 38

$$\text{Work/cycle} = \frac{n}{n-1}\{(p_2V_2 - p_1V_1) - (p_3V_3 - p_4V_4)\}$$

$$= \frac{n}{n-1}\{(p_2V_2 - p_1V_1) - (p_2V_3 - p_1V_4)\}$$

$$= \frac{n}{n-1}\{p_2(V_2 - V_3) - p_1(V_1 - V_4)\}$$

Piston swept volume $(V_1 - V_3)$

$$= 0{\cdot}7854 \times 0{\cdot}2^2 \times 0{\cdot}23 = 7{\cdot}226 \times 10^{-3} \text{ m}^3$$

Working in litres of volume and bars of pressure:

$V_3 = 364$ cm$^3 = 0{\cdot}364$ litre
$V_1 =$ piston swept vol. + clearance vol.
$= 7{\cdot}226 + 0{\cdot}364 = 7{\cdot}59$

$$p_1V_1^n = p_2V_2^n$$
$$1 \times 7{\cdot}59^{1{\cdot}28} = 5 \times V_2^{1{\cdot}28}$$
$$V_2 = \frac{7{\cdot}59}{\sqrt[1{\cdot}28]{5}} = 2{\cdot}158$$
$$p_3V_3^n = p_4V_4^n$$
$$5 \times 0{\cdot}364^{1{\cdot}28} = 1 \times V_4^{1{\cdot}28}$$
$$V_4 = 0{\cdot}364 \times \sqrt[1{\cdot}28]{5} = 1{\cdot}28$$

Expressing pressure in kN/m² (1 bar = 10² kN/m²) and volumes in m³ (1 litre = 10⁻³ m³) to obtain work in kJ:

$$\text{Work/cycle} = \frac{n}{n-1}\ \{p_2(V_2 - V_3) - p_1(V_1 - V_4)\}$$
$$= \frac{1{\cdot}28}{0{\cdot}28} \times 10^2 \times 10^{-3}\ \{5(2{\cdot}158 - 0{\cdot}363) - 1(7{\cdot}59 - 1{\cdot}28)\}$$
$$= 1{\cdot}216 \text{ kJ}$$

Indicated power [kW] = 1·216 [kJ/cycle] × 2 [cycle/s]
= 2·432 kW Ans. (i)

$$\text{Mean indicated press. [kN/m}^2] = \frac{\text{area of diagram [kJ]}}{\text{length of diagram [m}^3]}$$
$$= \frac{1{\cdot}216}{7{\cdot}226 \times 10^{-3}} = 168{\cdot}3 \text{ kN/m}^2 = 1{\cdot}683 \text{ bar} \quad \text{Ans. (ii)}$$

$$\text{Vol. effic.} = \frac{\text{volume drawn in per stroke}}{\text{piston swept volume}}$$
$$= \frac{V_1 - V_4}{V_1 - V_3} = \frac{7{\cdot}59 - 1{\cdot}28}{7{\cdot}226}$$
$$= 0{\cdot}8732 \text{ or } 87{\cdot}32\% \quad \text{Ans. (iii)}$$

7. Referring to Fig. 38

From pV/T = constant, volume rate of air induced at suction pressure and temperature

$$= \frac{5 \times 1{\cdot}013 \times 297}{60 \times 0{\cdot}98 \times 289} = 0{\cdot}08853 \text{ m}^3\text{/s}$$

$$\text{Work/cycle} = \frac{n}{n-1}\, p_1(V_1 - V_4)\left[\left\{\frac{p_2}{p_1}\right\}^{\frac{n-1}{n}} - 1\right]$$

The above expression is the work per cycle when $(V_1 - V_4)$ is the volume drawn in per cycle. Similarly, if $(V_1 - V_4)$ is taken as the volume drawn in per second, the expression will give work per second, which is power, thus:

$$\text{Power [kW]} = \text{work per second [kJ/s = kN m/s]}$$

$$\frac{n}{n-1} = \frac{1{\cdot}25}{0{\cdot}25} = 5 \qquad \frac{n-1}{n} = \frac{1}{5}$$

$$V_1 - V_4 = 0{\cdot}08853 \text{ m}^3/\text{s}$$
$$p_1 = 0{\cdot}98 \times 10^2 \text{ kN/m}^2$$
$$\frac{p_2}{p_1} = \text{pressure ratio} = 4{\cdot}55$$

$$\text{Power} = \frac{n}{n-1} p_1(V_1 - V_4)\left[\left\{\frac{p_2}{p_1}\right\}^{\frac{n-1}{n}} - 1\right]$$
$$= 5 \times 0{\cdot}98 \times 10^2 \times 0{\cdot}08853(4{\cdot}55^{\frac{1}{5}} - 1)$$
$$= 15{\cdot}36 \text{ kW} \quad \text{Ans. (i)}$$

Let stroke volume $(V_1 - V_3)$ on Fig. 38 be represented by unity, then $V_3 = 0{\cdot}05$ and $V_1 = 1{\cdot}05$

$$\text{From } p_3V_3^n = p_4V_4^n$$

$$V_4 = 0{\cdot}05 \times \sqrt[1{\cdot}25]{4{\cdot}55} = 0{\cdot}1681$$

$$\text{Suction period} = V_1 - V_4$$
$$= 1{\cdot}05 - 0{\cdot}1681 = 0{\cdot}8819$$

$$\text{Vol. effic.} = \frac{\text{vol. drawn in per stroke}}{\text{stroke volume}}$$
$$= \frac{0{\cdot}8819}{1} = 0{\cdot}8819 \quad \text{Ans. (ii)}$$

Actual volume of air drawn in per stroke [m³]

$$= \frac{0{\cdot}088\,53 \text{ [m}^3\text{/s]}}{2 \times 8 \text{ [strokes/s]}}$$

$$\text{Piston swept vol.} = \frac{\text{induced volume}}{\text{volumetric efficiency}}$$
$$= \frac{0{\cdot}088\,53}{2 \times 8 \times 0{\cdot}8819}$$

Let d [m] = cyl. diameter, stroke = $1{\cdot}2d$

Piston swept vol. = $0{\cdot}7854d^2 \times 1{\cdot}2d$

$$d = \sqrt[3]{\frac{0{\cdot}08853}{0{\cdot}7854 \times 1{\cdot}2 \times 2 \times 8 \times 0{\cdot}8819}}$$

$$= 0{\cdot}1881 \text{ m}$$

$$\left.\begin{array}{r}\text{Diameter of cylinder} = 188{\cdot}1 \text{ mm}\\ \text{Stroke} = 1{\cdot}2 \times 188{\cdot}1 = 255{\cdot}7 \text{ mm}\end{array}\right\} \text{Ans. (iii)}$$

8.
$$\dot{m} = \frac{p\dot{V}}{RT}$$

$$\dot{m} = \frac{10^5 \times 17}{287 \times 306 \times 60}$$

$$\dot{m} = 0{\cdot}323 \text{ kg/s}$$

For minimum work $p_2 = \sqrt{p_3 p_1}$

$$p_2 = \sqrt{16 \times 1}$$

$$p_2 = 4\text{bar}$$

$$\frac{T_1}{T_2} = \left(\frac{p_1}{p_2}\right)^{\frac{n\ 1}{n}}$$

$$\therefore\ T_2 = T_1\left(\frac{p_2}{p_1}\right)^{\frac{n-1}{n}}$$

$$= 306\,(4)^{\frac{0{\cdot}3}{1{\cdot}3}}$$

$$T_2 = 421{\cdot}4 \text{ K or } 148{\cdot}4\text{°C}$$

$$\text{First stage power} = \frac{n}{n-1}\dot{m}R\,(T_2 - T_1)$$

$$= \frac{1{\cdot}3}{0{\cdot}3} \times 0{\cdot}323 \times 287\,(148{\cdot}4 - 33)/10^3$$

$$= 46{\cdot}37 \text{ kW} \quad \text{Ans. (a)}$$

Heat rejected to intercooler

$$= \dot{m}\,c_p\,(T_2 - T_1)$$

$$= 0{\cdot}323 \times 1005\,(148{\cdot}4 - 33) \times 60/10^6$$

$$= 2{\cdot}25 \text{ MJ/min} \quad \text{Ans. (b)}$$

9. From $\frac{pV}{T}$ = constant, volume rate of air at suction pressure and temperature

$$= \frac{1{\cdot}013 \times 0{\cdot}6083 \times 300}{0{\cdot}97 \times 288} = 0{\cdot}6617 \text{ m}^3/\text{s}$$

Let stroke volume $V_1 - V_3$ in Fig. 38 equal unity.

$$\text{Then } V_3 = 0{\cdot}06,\ V_1 = 1{\cdot}06$$

$$p_3V_3^n = p_4V_4^n$$

$$V_4 = 0{\cdot}06 \times \sqrt[1{\cdot}32]{\frac{4{\cdot}85}{0{\cdot}97}} = 0{\cdot}2031$$

$$\text{Actual volume of air drawn in per stroke} = \frac{0{\cdot}6617}{2 \times 5}\left[\frac{\text{m}^3/\text{s}}{\text{stroke/s}}\right]$$

$$= 0{\cdot}06617 \text{ m}^3$$

$$\text{and } \frac{V_1 - V_4}{V_1 - V_3} = \frac{0{\cdot}06617}{\text{piston swept volume}}$$

$$\therefore \quad \text{Piston swept volume} = \frac{0{\cdot}06617}{1{\cdot}06 - 0{\cdot}2031} = 0{\cdot}07812 \text{ m}^3$$

$$\text{also, Piston swept volume} = 0{\cdot}7854\, d^2 \times d$$

d = cylinder diameter and stroke

$$\therefore\ 0{\cdot}07812 = 0{\cdot}7854\, d^3$$

$$d = 0{\cdot}463 \text{ m} \quad \text{Ans. (a)}$$

Isothermal efficiency

$$= \frac{\ln r}{\frac{n}{n-1}\left[r^{\frac{n-1}{n}} - 1\right]}$$

$$= \frac{\ln 5}{\frac{1{\cdot}32}{0{\cdot}32}\left[5^{\frac{0{\cdot}32}{1{\cdot}32}} - 1\right]}$$

$$= 0{\cdot}818 \text{ or } 81{\cdot}8\% \quad \text{Ans. (b)}$$

10. Let r = stage pressure ratio for minimum work condition
Let s = number of stages of compression

$$\frac{p_2}{p_1} = \left(\frac{T_2}{T_1}\right)^{\frac{n}{n-1}}$$

$$r = \left(\frac{368}{308}\right)^{\frac{1\cdot3}{0\cdot3}}$$

$$r = 2\cdot162$$

$$\text{now } p_2 = rp_1$$

$$p_3 = rp_2 = r^2p_1$$

$$p_4 = rp_3 = r^3p_1 \text{ etc.}$$

$$\therefore\ p_{s-1} = r^sp_1$$

$$\text{hence } 100 = 2\cdot162^s \times 1$$

$$s = 5\cdot97 \text{ say 6 stages} \quad \text{Ans. (a)}$$

$$\text{Power input per stage} = \frac{n}{n-1}\dot{m}R\,(T_2 - T_1)$$

$$= \frac{1\cdot3}{0\cdot3} \times 0\cdot1 \times 0\cdot287\,(368 - 308)$$

$$= 7\cdot45 \text{ kW}$$

$$\text{Compressor power input} = 7\cdot45 \times 6 \text{ stages}$$

$$= 44\cdot7 \text{ kW} \quad \text{Ans. (b)}$$

SOLUTIONS TO TEST EXAMPLES 10

1. Tables page 2, water at 80°C, $h = 334{\cdot}9$

   ... ... 4, steam at 9 bar:

$$h = h_f + xh_{fg}$$
$$= 743 + 0{\cdot}96 \times 2031 = 2693$$

Heat energy transferred = change in enthalpy
$$= 2693 - 334{\cdot}9$$
$$= 2358{\cdot}1 \text{ kJ/kg} \quad \text{Ans.}$$

2. Tables page 7, 30 bar 350°C, $h = 3117$

   ... ... 3, 0·06 bar dryness 0·88:

$$h = h_f + xh_{fg}$$
$$= 152 + 0{\cdot}88 \times 2415 = 2277$$

Enthalpy drop per kg = $3117 - 2277$
$$= 840 \text{ kJ/kg} \quad \text{Ans. (i)}$$

At 0·5 kg of steam per second, total change in enthalpy in the steam through the turbine per second
$$= 0{\cdot}5 \times 840 = 420 \text{ kJ/s}$$

kilojoules per second = kilowatts

∴ Power equivalent = 420 kW Ans. (ii)

3. Tables page 7, 20 bar 350°C, $h = 3138$, $v = 0{\cdot}1386$

   ... ... 4, 20 bar 0·98 dry:

$$h = h_f + xh_{fg}$$
$$= 909 + 0{\cdot}98 \times 1890 = 2761 \text{ kJ/kg}$$
$$v = xv_g$$
$$= 0{\cdot}98 \times 0{\cdot}09957$$
$$= 0{\cdot}09757 \text{ m}^3\text{/kg}$$

Heat energy supplied to steam in superheaters
$$= 3138 - 2761 = 377 \text{ kJ/kg} \quad \text{Ans. (i)}$$

% increase in specific volume

$$= \frac{0{\cdot}1386 - 0{\cdot}09757}{0{\cdot}09757} \times 100$$

$$= 42{\cdot}06\% \quad \text{Ans. (ii)}$$

4. Tables page 4, 4 bar, $v_g = 0{\cdot}4623$

   8 bar, $v_g = 0{\cdot}2403$

   For 8 bar, dryness 0·94,

   $v = xv_g = 0{\cdot}94 \times 0{\cdot}2403 = 0{\cdot}2258 \text{ m}^3\text{/kg}$

$$p_1 v_1^n = p_2 v_2^n$$

$$8 \times 0{\cdot}2258^{1{\cdot}12} = 4 \times v_2^{1{\cdot}12}$$

$$v_2 = 0{\cdot}2258 \times \sqrt[1{\cdot}12]{2} = 0{\cdot}4194 \text{ m}^3\text{/kg}$$

0·4194 m³/kg is the specific volume of wet steam at 4 bar, let its dryness fraction = x:

$$v = xv_g$$

$$0{\cdot}4194 = x \times 0{\cdot}4623$$

$$x = 0{\cdot}9072 \quad \text{Ans.}$$

5. Tables page 4, steam 2·4 bar, h_g = 2715
... ... 2, water 42°C, h = 175·8
... ... 3, water 99·6°C, h = 417

See Fig. 41. Consider one kg of steam from boiler, let x kg be tapped off low pressure turbine to heater, then $(1 - x)$ kg passes through condenser and as water to hotwell.

Enthalpy of heating steam and water entering heater = Enthalpy of water leaving heater

$$x \times 2715 + (1 - x) \times 175{\cdot}8 = 1 \times 417$$

$$2715x + 175{\cdot}8 - 175{\cdot}8x = 417$$

$$2539{\cdot}2x = 241{\cdot}2$$

$$x = 0{\cdot}095$$

$$\therefore \ \% \text{ of steam tapped off} = 9{\cdot}5\% \quad \text{Ans.}$$

6. Tables page 4, steam 16 bar, $h_f = 859$, $h_{fg} = 1935$
steam 8 bar, $h_f = 721$, $h_{fg} = 2048$

Enthalpy after throttling = Enthalpy before

$$721 + x \times 2048 = 859 + 0{\cdot}98 \times 1935$$

$$x \times 2048 = 2034$$

$$x = 0{\cdot}9931 \quad \text{Ans.}$$

7.

$$t_s = 165°\text{C}$$

$$p = 7 \text{ bar}$$

$$v_{g1} = 0{\cdot}2728 \text{ m}^3\text{/kg}$$

} tables, page 4

$$V_1 = 0{\cdot}75 \times 0{\cdot}2728 \times 0{\cdot}2$$

$$= 0{\cdot}0409 \text{ m}^3$$

$$V_2 = 0{\cdot}0818 \text{ m}^3$$

$$v_{g2} = \frac{0{\cdot}0818}{0{\cdot}2}$$

$$= 0{\cdot}4090 \text{ m}^3\text{/kg}$$

$$t = 355°\text{C}$$

i.e. using tables, page 7, interpolating at 7 bar with v_g 0·4090

Final temperature = 355°C Ans. (a)

$$\text{Work done} = p\,(V_2 - V_1)$$
$$= 7 \times 100\,(0{\cdot}0818 - 0{\cdot}0409)$$
$$= 28{\cdot}64 \text{ kJ} \quad \text{Ans. (b)}$$

$$\text{Heat energy transfer} = H_1 - H_2$$
$$= 0{\cdot}2\,[h_g - (h_f + xh_{fg})]$$
$$= 0{\cdot}2\,[3175 - (697 + 0{\cdot}75 \times 2067)]$$
$$= 185{\cdot}5 \text{ kJ} \quad \text{Ans. (c)}$$

8. Tables page 4, 15 bar, $h_f = 845$, $h_{fg} = 1947$
1·1 bar, $t_s = 102{\cdot}3$, $h_g = 2680$

Dryness fraction by separator,

$$x_1 = \frac{m_2}{m_2 + m_1} = \frac{10}{10{\cdot}55} = 0{\cdot}9479$$

Dryness fraction by throttling calorimeter:

Enthalpy before throttling = Enthalpy after

$$845 + x_2 \times 1947 = 2680 + 2(111 - 102{\cdot}3)$$
$$x_2 \times 1947 = 1852{\cdot}4$$
$$x_2 = 0{\cdot}9513$$

Dryness fraction of steam:

$$x = x_1 \times x_2$$
$$= 0{\cdot}9479 \times 0{\cdot}9513$$
$$= 0{\cdot}9018 \quad \text{Ans.}$$

9. 1st Case: Absolute press. = 1·9 + 1 = 2·9 bar
Tables page 4, when temp. of steam is 130°C,
$p = 2{\cdot}7$ bar, $v_g = 0{\cdot}6686 \text{ m}^3/\text{kg}$

$$\text{Mass of steam} = \frac{4{\cdot}25}{0{\cdot}6686} = 6{\cdot}356 \text{ kg} \quad \text{Ans. (a)(i)}$$

Air pressure = total press. – steam press.
= 2·9 – 2·7 = 0·2 bar = 20 kN/m²

$$pV = mRT$$

$$\therefore \quad m = \frac{20 \times 4{\cdot}25}{0{\cdot}287 \times 403}$$
$$= 0{\cdot}7348 \text{ kg} \quad \text{Ans. (b) (i)}$$

2nd Case: Absolute press. = 6·25 + 1 = 7·25 bar

Tables page 4, when temp. of steam is 165°C,
$p = 7$ bar, $v_g = 0{\cdot}2728$ m³/kg

$$\text{Mass of steam} = \frac{4{\cdot}25}{0{\cdot}2728} = 15{\cdot}58 \text{ kg} \quad \text{Ans. (a)(ii)}$$

$$\begin{aligned}\text{Air pressure} &= \text{total press.} - \text{steam press.}\\ &= 7{\cdot}25 - 7 = 0{\cdot}25 \text{ bar} = 25 \text{ kN/m}^2\end{aligned}$$

$pV = mRT$

$$\begin{aligned}\therefore \quad m &= \frac{25 \times 4{\cdot}25}{0{\cdot}287 \times 438}\\ &= 0{\cdot}845 \text{ kg} \quad \text{Ans. (b) (ii)}\end{aligned}$$

10. $p_{wv} = 0{\cdot}02337$ bar

Using tables, page 2, t_s 20°C

$$\begin{aligned}p_A &= 1 - 0{\cdot}02337\\ &= 0{\cdot}97663 \text{ bar}\end{aligned}$$

$$\frac{p_N}{p_O} = \frac{V_N}{V_O} \quad \text{using partial volumes}$$

$$\frac{p_N}{0{\cdot}97663 - p_N} = \frac{0{\cdot}79}{0{\cdot}21}$$

$$\begin{aligned}p_N &= 3{\cdot}674 - 3{\cdot}762\, p_N\\ p_N &= 0{\cdot}7715 \text{ bar}\\ p_O &= 0{\cdot}97663 - 0{\cdot}7715\\ &= 0{\cdot}2051 \text{ bar}\end{aligned}$$

$$\text{Partial pressures} = \left\{\begin{matrix}0{\cdot}2051 \text{ bar, oxygen}\\ 0{\cdot}7715 \text{ bar, nitrogen}\\ 0{\cdot}02337 \text{ bar, water vapour}\end{matrix}\right\} \quad \text{Ans. (a)}$$

$$\begin{aligned}p_A v &= R_A T\\ v &= \frac{0{\cdot}287 \times 293}{100 \times 0{\cdot}97663}\\ &= 0{\cdot}861 \text{ m}^3\text{/kg of dry air}\\ v_g &= 57{\cdot}84 \text{ m}^3\text{/kg}\end{aligned}$$

using tables, page 2, t_s 20°C

$$\begin{aligned}\text{kg of water vapour} &= \frac{0{\cdot}861}{57{\cdot}84}\\ &= 0{\cdot}01488 \text{ per kg of dry air}\end{aligned}$$

% Absolute humidity = 1·488 Ans. (b)

SOLUTIONS TO TEST EXAMPLES 11

1. Tables page 4, 17 bar, $s_f = 2{\cdot}372 \quad s_{fg} = 4{\cdot}028$

$$s = 2{\cdot}372 + 0{\cdot}95 \times 4{\cdot}028$$
$$= 6{\cdot}198 \text{ kJ/kgK} \quad \text{Ans.}$$

2. Tables page 4, 195°C (14 bar) $s_f = 2{\cdot}284 \; s_{fg} = 4{\cdot}185$

$$s = 2{\cdot}284 + 0{\cdot}9 \times 4{\cdot}185$$
$$= 6{\cdot}05 \text{ kJ/kg} \quad \text{Ans.}$$

3. Tables page 4, 5·5 bar, $s_g = 6{\cdot}790$
   Tables page 3, 0·2 bar, $s_f = 0{\cdot}832$, $s_{fg} = 7{\cdot}075$

Entropy after expansion = Entropy before

$$0{\cdot}832 + x \times 7{\cdot}075 = 6{\cdot}79$$
$$x \times 7{\cdot}075 = 5{\cdot}958$$
$$x = 0{\cdot}8422 \quad \text{Ans.}$$

4. Tables page 7, by interpolation:
   15 bar 350°C $s = 7{\cdot}102$
   20 bar 350°C $s = 6{\cdot}957$

For 5 bar increase $s = 0{\cdot}145$ decrease

For 2 bar increase $s = \frac{2}{5} \times 0{\cdot}145 = 0{\cdot}058$ decrease

∴ 17 bar 350°C, $s = 7{\cdot}102 - 0{\cdot}058 = 7{\cdot}044$

Page 4, 1·7 bar, $s_f = 1{\cdot}475 \quad s_{fg} = 5{\cdot}707$

Entropy after expansion = Entropy before

$$1{\cdot}475 + x \times 5{\cdot}707 = 7{\cdot}044$$
$$x \times 5{\cdot}707 = 5{\cdot}569$$
$$x = 0{\cdot}9759 \quad \text{Ans.}$$

5. Tables page 4, 22 bar, $h_g = 2801$, $s_g = 6{\cdot}305$
   7 bar, $h_g = 2764$
   1.4 bar, $s_f = 1{\cdot}411$, $s_{fg} = 5{\cdot}835$

THROTTLING PROCESS:

Enthalpy after = Enthalpy before
∴ Enthalpy at 7 bar = 2801

Throttled steam is therefore superheated.
Tables page 7, interpolating:

Enthalpy of superheat = $2801 - 2764 = 37$
At 7 bar, $h = 2846$ for 200°C

$$h = 2764 \text{ for } 165°\text{C (sat. temp.)}$$
$$\text{increase } h = 82 \text{ for } 35°\text{C increase}$$
$$\text{difference in temp. for } h = 37, = \frac{37}{82} \times 35 = 15{\cdot}8°\text{C}$$

$\therefore$ Degree of superheat at 7 bar = 15·8°C Ans. (i)

Entropy of steam at 7 bar with 15·8°C of superheat:

$$7 \text{ bar } 200°\text{C} \quad s = 6{\cdot}888$$
$$165°\text{C} \quad s = 6{\cdot}709$$
$$\text{for increase } 35°\text{C} \quad s = 0{\cdot}179 \text{ increase}$$
$$\ldots \quad \ldots \quad 15{\cdot}8°\text{C} \quad s = \frac{15{\cdot}8}{35} \times 0{\cdot}179 = 0{\cdot}0808$$

$\therefore$ Entropy at 7 bar, 15·8°C superheat

$$= 6{\cdot}709 + 0{\cdot}0808 = 6{\cdot}7898$$
$$\text{Increase in entropy} = 6{\cdot}7898 - 6{\cdot}305$$
$$= 0{\cdot}4848 \text{ kJ/kg K} \quad \text{Ans. (ii)}$$

ISENTROPIC EXPANSION from 7 to 1·4 bar:

$$\text{Entropy after} = \text{Entropy before}$$
$$1{\cdot}411 + x \times 5{\cdot}835 = 6{\cdot}7898$$
$$x \times 5{\cdot}835 = 5{\cdot}3788$$
$$x = 0{\cdot}9217 \quad \text{Ans. (iii)}$$

SOLUTIONS TO TEST EXAMPLES 12

1. Tables page 4, 8 bar, $h_g = 2769$
5 bar, $h_f = 640 \quad h_{fg} = 2109$
$v_g = 0{\cdot}3748$

$$\text{Enthalpy drop} = 2769 - (640 + 0{\cdot}97 \times 2109) = 83 \text{ kJ/kg}$$

$$\text{Velocity} = \sqrt{2 \times 83 \times 10^3} = 407{\cdot}4 \text{ m/s} \quad \text{Ans. (i)}$$

Spec. vol. of steam at exit = $0{\cdot}97 \times 0{\cdot}3748$ m³/kg

$$\text{Mass flow [kg/s]} = \frac{\text{volume flow [m}^3\text{/s]}}{\text{spec. vol. [m}^3\text{/kg]}}$$

$$= \frac{\text{area [m}^2] \times \text{velocity [m/s]}}{\text{spec. vol. [m}^3\text{/kg]}}$$

$$= \frac{14{\cdot}5 \times 10^{-4} \times 407{\cdot}4}{0{\cdot}97 \times 0{\cdot}3748}$$

$$= 1{\cdot}625 \text{ kg/s} \quad \text{Ans. (ii)}$$

2. Referring to Fig. 57:

Angle between u and $v_{r1} = 180° - \beta_1 = 180° - 33° = 147°$

Angle between v_1 and $v_{r1} = \beta_1 - \alpha_1 = 33° - 20° = 13°$

By sine rule:

$$\frac{u}{\sin 13°} = \frac{v_1}{\sin 147°}$$

$$u = \frac{450 \times \sin 13°}{\sin 147°}$$

$$\text{Rotational speed [rev/s]} = \frac{\text{linear velocity [m/s]}}{\text{circumference [m]}}$$

$$= \frac{185{\cdot}9}{\pi \times 0{\cdot}66} = 89{\cdot}64 \text{ rev/s} \quad \text{Ans. (ii)}$$

3. Referring to Fig. 60:

$$v_{a1} = v_1 \sin \alpha_1 = 243 \times \sin 23^\circ = 94{\cdot}95 \text{ m/s}$$
$$v_{w1} = v_1 \cos \alpha_1 = 243 \times \cos 23^\circ = 223{\cdot}7 \text{ m/s}$$
$$x = v_{w1} - u = 223{\cdot}7 - 159 = 64{\cdot}7 \text{ m/s}$$
$$\tan \beta_1 = \frac{v_{a1}}{x} = \frac{94{\cdot}95}{64{\cdot}7} = 1{\cdot}468$$

Blade inlet angle = 55° 44′ Ans. (i)

Since the combined vector diagram of inlet and exit velocities is symmetrical, $v_{w2} = x$,

$$v_w = v_{w1} + v_{w2} = 223{\cdot}7 + 64{\cdot}7 = 288{\cdot}4 \text{ m/s}$$
$$\text{Force on blades} = \dot{m}v_w$$
$$= 0{\cdot}9 \times 288{\cdot}4 = 259{\cdot}5 \text{ N} \quad \text{Ans. (ii)}$$
$$\text{Power [W]} = \text{force [N]} \times \text{velocity [m/s]}$$
$$= 259{\cdot}5 \times 159$$
$$= 4{\cdot}127 \times 10^4 \text{ W} = 41{\cdot}27 \text{ kW} \quad \text{Ans. (iii)}$$

4.

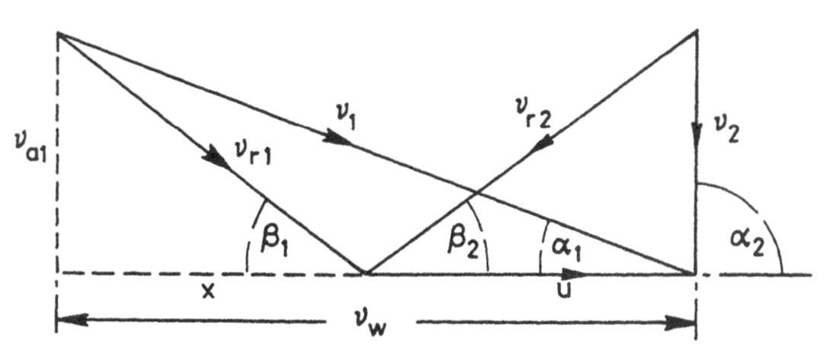

$$u = \pi \times 0{\cdot}6 \times 100$$
$$= 188{\cdot}5 \text{ m/s}$$
$$v_{r2} = \frac{188{\cdot}5}{\cos 35^\circ}$$
$$= 230{\cdot}1 \text{ m/s}$$

Let $v_{r1} = v_{r2}$ (see assumption following)

$$v_{a1} = 230{\cdot}1 \times \sin 35^\circ$$
$$= 132 \text{ m/s}$$
$$v_w = 2 \times 188{\cdot}5$$
$$= 377 \text{ m/s}$$

$$\tan \alpha_1 = \frac{132}{377}$$

$$\alpha_1 = 19{\cdot}36°$$

Nozzle angle $= 19{\cdot}36°$ Ans. (a)(i)

$$v_1 = \frac{377}{\cos \alpha_1}$$

$= 399{\cdot}6$ m/s Ans. (a)(ii)

Power $= \dot{m} v_w u$

$= 1 \times 377 \times 188{\cdot}5 \times 10^{-3}$

$= 71{\cdot}06$ kW Ans. (a)(iii)

Assumed – no frictional losses across the blades Ans. (b)

5. From h-s chart $h_1 = 2795$ kJ/kg, $s_1 = 6{\cdot}27$ kJ/kg K
$h_2 = 2795$ kJ/kg
$h_3 = 2455$ kJ/kg
$h_4 = 2870$ kJ/kg
$h_5 = 2615$ kJ/kg, $s_5 = 7{\cdot}63$ kJ/kg K

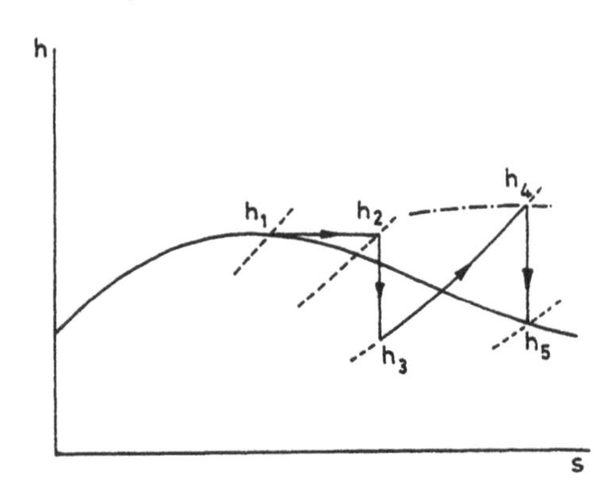

Changes in enthalpy during each stage, final – initial
Stage 1 $h_2 - h_1 = 0$
Stage 2 $h_3 - h_2 = -340$ kJ/kg
Stage 3 $h_4 - h_3 = 415$ kJ/kg
Stage 4 $h_5 - h_4 = -255$ kJ/kg Ans.

Overall changes in entropy $= s_5 - s_1 = 1{\cdot}36$ kJ/kg K Ans.

Condition of steam at end of final expansion is *dry saturated.* Ans.

6.
$$p_T = p_1\left[\frac{2}{\gamma+1}\right]^{\frac{\gamma}{\gamma-1}}$$

$$= 7\left[\frac{2}{1{\cdot}67+1}\right]^{\frac{1{\cdot}67}{1{\cdot}67-1}}$$

$$= 3{\cdot}407 \text{ bar}$$

$$T_T = T_1\left[\frac{p_T}{p_1}\right]^{\frac{\gamma-1}{\gamma}}$$

$$T_T = 423\left[\frac{3{\cdot}407}{7}\right]^{\frac{1{\cdot}67-1}{1{\cdot}67}}$$

$$= 316{\cdot}9 \text{ K or } 43{\cdot}87^\circ\text{C}$$

Enthalpy drop to throat $= c_p[T_1 - T_T]$

i.e. $h = 833{\cdot}9[423 - 316{\cdot}87]$

$h = 88501{\cdot}8$ J/kg

Velocity of gas at throat $= \sqrt{2h}$

$v_T = \sqrt{2 \times 88501{\cdot}8}$

$v_T = 420{\cdot}7$ m/s Ans.

$$p_T \dot{V}_T = \dot{m}RT_T$$

$$\dot{V}_T = \frac{\dot{m}RT_T}{p_T}$$

$$= \frac{0{\cdot}25 \times 2078{\cdot}5 \times 316{\cdot}9}{3{\cdot}407 \times 10^5}$$

$$= 0{\cdot}4833 \text{ m}^3/\text{s}$$

Throat area $= \dfrac{\dot{V}_T}{v_T} = \dfrac{0{\cdot}4833}{420{\cdot}7}$

$= 0{\cdot}001149 \text{ m}^2$ or $11{\cdot}49 \text{ cm}^2$ Ans.

7. Velocity at nozzle exit v_1

$$= \sqrt{2 \times 312{\cdot}5 \times 10^3 \times 0{\cdot}9}$$
$$= 750 \text{ m/s}$$

Ref. Fig. 57:

$$v_{a1} = v_1 \sin \alpha_1 = 750 \times \sin 20° = 256{\cdot}5 \text{ m/s}$$
$$v_{w1} = v_1 \cos \alpha_1 = 750 \times \cos 20° = 704{\cdot}9 \text{ m/s}$$
$$x = \frac{v_{a1}}{\tan \beta_1} = \frac{256{\cdot}5}{\tan 35°} = 366{\cdot}4 \text{ m/s}$$

$$\text{Blade velocity} = u = v_{w1} - x$$
$$= 704{\cdot}9 - 366{\cdot}4$$
$$= 388{\cdot}5 \text{ m/s} \quad \text{Ans. (i)}$$

Absolute velocity of exit steam is in the direction of the turbine axis, therefore $\alpha_2 = 90°$

$$\tan \beta_2 = \frac{v_2}{u} = \frac{204}{338{\cdot}5} = 0{\cdot}6027$$

Exit angle of blades = 31° 5′ Ans. (ii)

$$v_{r1} = \frac{v_{a1}}{\sin \beta_1} = \frac{256{\cdot}5}{\sin 35°} = 447{\cdot}2 \text{ m/s}$$

$$v_{r2} = \frac{v_2}{\sin \beta_2} = \frac{204}{\sin 31°\ 5'} = 395{\cdot}2 \text{ m/s}$$

Loss of kinetic energy of steam across the blades

$$= \tfrac{1}{2} \dot{m}(v_{r1}^2 - v_{r2}^2)$$
$$= \tfrac{1}{2} \times 1 \times (447{\cdot}2^2 - 395{\cdot}2^2)$$
$$= 21\,900 \text{ J} = 21{\cdot}9 \text{ kJ/kg steam} \quad \text{Ans. (iii)}$$

$$\text{Axial thrust} = \dot{m}(v_{a1} - v_{a2})$$
$$= 1 \times (256{\cdot}5 - 204)$$
$$= 52{\cdot}5 \text{ N/kg of steam} \quad \text{Ans. (iv)}$$
$$\text{Power} = \dot{m} v_w u$$

Since the steam leaves the turbine axially, that is, at 90° to the blade movement, there is no velocity of whirl at exit, $v_{w2} = 0$ $\therefore$ $v_w = v_{w1}$

$$\text{Power} = 1 \times 704{\cdot}9 \times 338{\cdot}5$$
$$= 2{\cdot}386 \times 10^5 \text{ W}$$
$$= 238{\cdot}6 \text{ kW/kg of steam} \quad \text{Ans. (v)}$$

$$\text{Diagram efficiency} = \frac{\text{work done on blades}}{\text{work supplied}}$$

$$= \frac{\dot{m}v_w u\ [\text{J/s}]}{\frac{1}{2}\dot{m}v_1^2\ [\text{J/s}]} = \frac{2uv_w}{v_1^2}$$

$$= \frac{2 \times 338{\cdot}5 \times 704{\cdot}9}{750^2}$$

$$= 0{\cdot}8484 \text{ or } 84{\cdot}84\% \quad \text{Ans. (vi)}$$

8. Tables page 7, 15 bar 250°C, $h = 2925$ $s = 6{\cdot}711$
page 3, 0·16 bar, $h_f = 232$ $s_f = 0{\cdot}772$
$h_{fg} = 2369$ $s_{fg} = 7{\cdot}213$

$$\text{Entropy after expansion} = \text{Entropy before}$$
$$0{\cdot}772 + x \times 7{\cdot}213 = 6{\cdot}711$$
$$x \times 7{\cdot}213 = 5{\cdot}939$$
$$\therefore \text{ dryness fraction } x = 0{\cdot}8235 \quad \text{Ans. (i)}$$

$$\text{At } 0{\cdot}16 \text{ bar}, h = h_f + xh_{fg}$$
$$= 232 + 0{\cdot}8235 \times 2369 = 2183$$

$$\text{Rankine efficiency} = \frac{h_1 - h_2}{h_1 - h_{f2}}$$

$$= \frac{2925 - 2183}{2925 - 232} = \frac{742}{2693}$$

$$= 0{\cdot}2755 \text{ or } 27{\cdot}55\% \quad \text{Ans. (ii)}$$

9. Tables page 4, 14 bar, $h_g = 2790$ $v_g = 0{\cdot}1408$
10 bar, $h_f = 763$ $h_{fg} = 2015$
$v_g = 0{\cdot}1944$

$$p_1 v_1^{1{\cdot}135} = p_2 v_2^{1{\cdot}135}$$
$$14 \times 0{\cdot}1408^{1{\cdot}135} = 10 \times v_2^{1{\cdot}135}$$
$$v_2 = 0{\cdot}1408 \times \sqrt[1{\cdot}35]{1{\cdot}4} = 0{\cdot}1893 \text{ m}^3\text{/kg}$$

Spec. vol. of dry steam at 10 bar is 0·1944 m³/kg therefore,

$$\text{dryness after expansion} = \frac{0{\cdot}1893}{0{\cdot}1944} = 0{\cdot}974 \quad \text{Ans. (i)}$$

$$\text{Spec. enthalpy drop} = 2790 - (763 + 0{\cdot}974 \times 2015)$$
$$= 64 \text{ kJ/kg} \quad \text{Ans. (ii)}$$

$$\text{Velocity} = \sqrt{2 \times 64 \times 10^3} = 357{\cdot}8 \text{ m/s} \quad \text{Ans. (iii)}$$

Volume flow [m³/s] = area [m²] × velocity [m/s]

For mass flow of 1 kg/s:

$$\text{Area [m}^2\text{]} = \frac{0{\cdot}1893}{357{\cdot}8} = 5{\cdot}293 \times 10^{-4}\ \text{m}^2$$

$$5{\cdot}293 \times 10^{-4} \times 10^6 = 529{\cdot}3\ \text{mm}^2 \quad \text{Ans. (iv)}$$

10. Referring to Fig. 65:

$$\gamma = \frac{c_P}{c_V} = \frac{1{\cdot}005}{0{\cdot}718} = 1{\cdot}4$$

$$r_p^{(\gamma-1)/\gamma} = 5{\cdot}7^{0{\cdot}4/1{\cdot}4} = 1{\cdot}644$$

$$\frac{T_2}{T_1} = \left\{\frac{p_2}{p_1}\right\}^{\frac{\gamma-1}{\gamma}}$$

$$T_2 = 294 \times 1{\cdot}644 = 483{\cdot}3\ \text{K}$$

Temperature at end of compression = 210·3°C Ans. (i)

$$\frac{T_4}{T_3} = \left\{\frac{p_4}{p_3}\right\}^{\frac{\gamma-1}{\gamma}}$$

$$T_4 = \frac{953}{1{\cdot}644} = 579{\cdot}7\ \text{K}$$

Alternatively, $\frac{T_4}{T_3} = \frac{T_1}{T_2}$ because pressure ratios are equal

$$\therefore \quad T_4 = \frac{953 \times 294}{483{\cdot}3} = 579{\cdot}7\ \text{K}$$

Temperature at end of expansion = 306·7°C Ans. (ii)

Heat energy supplied per kg

$$= m \times c_P \times (T_3 - T_2)$$
$$= 1 \times 1{\cdot}005 \times (953 - 483{\cdot}3)$$
$$= 472{\cdot}1\ \text{kJ/kg} \quad \text{Ans. (iii)}$$

Increase in internal energy per kg from inlet to exhaust

$$= m \times c_V \times (T_4 - T_1)$$
$$= 1 \times 0{\cdot}718 \times (579{\cdot}7 - 294)$$
$$= 205{\cdot}1\ \text{kJ/kg} \quad \text{Ans. (iv)}$$

$$\text{Ideal thermal effic.} = 1 - \frac{1}{r_p^{(\gamma-1)/\gamma}}$$

$$= 1 - \frac{1}{1 \cdot 644} = 0 \cdot 3918 \quad \text{Ans. (v)}$$

Alternatively,

$$\text{Thermal effic.} = 1 - \frac{T_4 - T_1}{T_3 - T_2}$$

$$= 1 - \frac{579 \cdot 7 - 294}{953 - 483 \cdot 3} = 0 \cdot 3919$$

SOLUTIONS TO TEST EXAMPLES 13

1. Tables page 4, water 130°C, $h = 546$
   ... ... 7, steam 30 bar 375°C, by interpolation,

$$\begin{aligned} h \text{ at 30 bar } 400^\circ\text{C} &= 3231 \\ h \text{ at 30 bar } 350^\circ\text{C} &= 3117 \\ \text{difference for } 50^\circ\text{C} &= 114 \\ \text{difference for } 25^\circ\text{C} &= 57 \end{aligned}$$

$\therefore$ h at 30 bar 375°C = 3117 + 57 = 3174

Heat energy transferred to steam per hour

$$= 30\,000 \times (3174 - 546) = 30\,000 \times 2628 \text{ kJ/h}$$

$$\text{Hourly fuel consumption} = \frac{53 \times 10^3}{24} = 2209 \text{ kg/h}$$

Heat energy supplied by fuel per hour

$$= 2209 \times 42 \times 10^3 \text{ kJ/h}$$

$$\text{Efficiency} = \frac{30\,000 \times 2628}{2209 \times 42 \times 10^3}$$

$$= 0{\cdot}85 \text{ or } 85\% \quad \text{Ans. (i)}$$

Heat energy supplied to plant by fuel [kJ/s = kW]

$$= \frac{2209 \times 42 \times 10^3}{3600}$$

Energy converted into engine power

$$= 0{\cdot}13 \times \text{energy supplied}$$

$$= \frac{0{\cdot}13 \times 2209 \times 42 \times 10^3}{3600} = 3349 \text{ kW} \quad \text{Ans. (ii)}$$

Tables page 2, h_{fg} at 100°C = 2256·7
Evaporative capacity from and at 100°C

$$= \frac{30\,000 \times 2628}{2256{\cdot}7} = 34920 \text{ kg/h} \quad \text{Ans. (iii)}$$

Equivalent evaporation, per kg fuel, from and at 100°C

$$= \frac{34920}{2209} = 15{\cdot}81 \text{ kg steam/kg fuel} \quad \text{Ans. (iv)}$$

2. Solids in initially + solids put in = solids in finally

$$\text{water in boiler} \times \text{initial p.p.m.} + \text{amount of feed} \times \text{feed p.p.m.} = \text{water in boiler} \times \text{final p.p.m.}$$

$$3{\cdot}5 \times 40 + 0{\cdot}875 \times 24 \times \text{feed p.p.m.} = 3{\cdot}5 \times 2500$$
$$21 \times \text{feed p.p.m.} = 8610$$
$$\text{feed p.p.m.} = 410 \quad \text{Ans.}$$

3. mass of solids put in = mass of solids blown out + mass of solids in evaporated output

mass of feed × feed p.p.m. = mass of blow out × blow out p.p.m. + mass evaporated × evaporated p.p.m.

Let x be mass flow per day of sea water feed
then $(x - 10)$ is mass flow per day of brine discharge

$$x \times 31\,250 = (x - 10) \times 78125 + 10 \times 250$$
$$x = 16{\cdot}613$$
$$(x - 10) = 6{\cdot}613$$

Mass flow of sea water feed = 16·613 tonne/day Ans. (a)
Mass flow of brine discharge = 6·613 tonne/day Ans. (b)

4. Per kg of fuel:

$$\text{Available hydrogen} = H_2 - \frac{O_2}{8}$$
$$= 0{\cdot}13 - \frac{0{\cdot}02}{8} = 0{\cdot}1275 \text{ kg}$$

$$\text{Cal. value} = 33{\cdot}7\,C + 144\left(H_2 - \frac{O_2}{8}\right)$$
$$= 33{\cdot}7 \times 0{\cdot}85 + 144 \times 0{\cdot}1275$$
$$= 28{\cdot}64 + 18{\cdot}36 = 47 \text{ MJ/kg} \quad \text{Ans. (i)}$$
$$\text{Stoichiometric air} = \frac{100}{23}\left\{2\tfrac{2}{3}\,C + 8\left(H_2 - \frac{O_2}{8}\right)\right\}$$
$$= \frac{100}{23}\{2\tfrac{2}{3} \times 0{\cdot}85 + 8 \times 0{\cdot}1275\}$$
$$= \frac{100}{23} \times 3{\cdot}287 = 14{\cdot}29 \text{ kg air/kg fuel}$$
$$\text{Actual air} = 1{\cdot}5 \times 14{\cdot}29 = 21{\cdot}44 \text{ kg air/kg fuel} \quad \text{Ans. (ii)}$$

Products of combustion per kg of fuel burned
$$= 21{\cdot}44 + 1 \text{ kg fuel} = 22{\cdot}44 \text{ kg}$$

Heat energy carried away
$$= \text{mass} \times \text{spec. heat} \times \text{temp. rise}$$
$$= 22{\cdot}44 \times 1{\cdot}005 \times (553 - 304) = 5614 \text{ kJ/kg fuel}$$

as a percentage of the heat energy supplied

$$= \frac{5614}{47 \times 10^3} \times 100 = 11{\cdot}95\ \% \quad \text{Ans. (iii)}$$

5. Available hydrogen $= H_2 - \frac{O_2}{8}$

$$= 0{\cdot}13 - \frac{0{\cdot}02}{8} = 0{\cdot}1275 \text{ kg}$$

Cal. value $= 33{\cdot}7 \times 0{\cdot}84 + 144 \times 0{\cdot}1275$
$= 28{\cdot}31 + 18{\cdot}36 = 46{\cdot}67$ MJ/kg Ans. (i)

Stoichiometric air $= \frac{100}{23} \times$ oxygen required

$$= \frac{100}{23}\{2\tfrac{2}{3} \times 0{\cdot}84 + 8 \times 0{\cdot}1275\}$$

$$= \frac{100}{23} \times 3{\cdot}26 = 14{\cdot}18 \text{ kg air/kg fuel} \quad \text{Ans. (ii)}$$

Per kg fuel burned:
Mass of gases in the products from
22 kg air + (1 kg fuel – 0·01 kg incombustibles)
= 22·99 kg
Mass of oxygen in 22 kg of air
$= 0{\cdot}23 \times 22 = 5{\cdot}06$ kg
Surplus oxygen $= 5{\cdot}06 - 3{\cdot}26 = 1{\cdot}8$ kg
Mass of nitrogen in 22 kg of air
$= 0{\cdot}77 \times 22 = 16{\cdot}94$ kg
CO_2 formed $= 3\tfrac{2}{3} \times 0{\cdot}84 = 3{\cdot}08$ kg
H_2O formed $= 9 \times 0{\cdot}13 = 1{\cdot}17$ kg
% composition of gases by mass Ans. (iii):

$$CO_2 = \frac{3{\cdot}08}{22{\cdot}99} \times 100 = 13{\cdot}4\%$$

$$H_2O = \frac{1{\cdot}17}{22{\cdot}99} \times 100 = 5{\cdot}09\%$$

$$O_2 = \frac{1{\cdot}8}{22{\cdot}99} \times 100 = 7{\cdot}83\%$$

$$N_2 = \frac{16{\cdot}94}{22{\cdot}99} \times 100 = 73{\cdot}68\%$$

6. Per kg of fuel burned:

$$\text{Stoichiometric air} = \frac{100}{23}\left\{2\tfrac{2}{3}\,C + 8\left(H_2 - \frac{O_2}{8}\right)\right\}$$

$$= \frac{100}{23}\{2\tfrac{2}{3} \times 0{\cdot}85 + 8 \times 0{\cdot}1275\}$$

$$= \frac{100}{23} \times 3{\cdot}216 = 13{\cdot}98 \text{ kg air/kg fuel}$$

$$CO_2 = 3\tfrac{2}{3} \times 0{\cdot}855 = 3{\cdot}135 \text{ kg}$$

When air supply is stoichiometric:
Mass of products of combustion

$$= 13{\cdot}98 \text{ kg air} + (1 \text{ kg fuel} - 0{\cdot}01 \text{ kg impurities})$$
$$= 13{\cdot}98 + 0{\cdot}99 = 14{\cdot}97 \text{ kg}$$

$$\%\,CO_2 = \frac{3{\cdot}135}{14{\cdot}97} \times 100 = 20{\cdot}94\% \quad \text{Ans. (i)}$$

When air supply is 25% excess:
Mass of products of combustion

$$= 1{\cdot}25 \times 13{\cdot}98 \text{ kg air} + 0{\cdot}99 = 18{\cdot}47 \text{ kg}$$

$$\%\,CO_2 = \frac{3{\cdot}135}{18{\cdot}47} \times 100 = 16{\cdot}97\% \quad \text{Ans. (ii)}$$

When air supply is 50% excess:
Mass of products of combustion

$$= 1{\cdot}5 \times 13{\cdot}98 + 0{\cdot}99 = 21{\cdot}96 \text{ kg}$$

$$\%CO_2 = \frac{3{\cdot}135}{21{\cdot}96} \times 100 = 14{\cdot}28\% \quad \text{Ans. (iii)}$$

When air supply is 75% excess:
Mass of products of combustion

$$= 1{\cdot}75 \times 13{\cdot}98 + 0{\cdot}99 = 25{\cdot}45 \text{ kg}$$

$$\%CO_2 = \frac{3{\cdot}135}{25{\cdot}45} \times 100 = 12{\cdot}31\% \quad \text{Ans. (iv)}$$

7. $$\text{Stoichiometric air} = \frac{100}{23}[\,2\tfrac{2}{3} \times 0{\cdot}84 + 0{\cdot}14 \times 8]$$

$$= 14{\cdot}6 \text{ kg/kg of fuel}$$

$$\text{Actual air supplied} = 14{\cdot}6 \times 1{\cdot}2$$
$$= 17{\cdot}52 \text{ kg/kg of fuel}$$

i.e. $17{\cdot}52 \times 0{\cdot}23 = 4{\cdot}03$ kg of O_2

$17{\cdot}52 - 4{\cdot}03 = 13{\cdot}49$ kg of N_2

CO_2, $3{\cdot}667 \times 0{\cdot}84 = 3{\cdot}08$ kg/kg of fuel

$$i.e.\ 2{\cdot}667 \times 0{\cdot}84 = 2{\cdot}24\ O_2$$

$$H_2O,\ 0{\cdot}14 \times 9 = 1{\cdot}26 \text{ kg/kg of fuel}$$

$$i.e.\ 0{\cdot}14 \times 8 = \underline{1{\cdot}12\ O_2}$$

$$\underline{3{\cdot}36\ O_2}$$

$$O_2,\ 4{\cdot}03 - 3{\cdot}36 = 0{\cdot}67 \text{ kg/kg of fuel.} \quad \text{Ans.}$$

DFG	m	M	N	N%
CO_2	3·08	44	3·08 ÷ 44 = 0·07	12·22
O_2	0·67	32	0·67 ÷ 32 = 0·021	3·65
N_2	13·49	28	13·49 ÷ 28 = 0·482	84·13
			Total = 0·573	100
H_2O	1·26	18	1·26 ÷ 18 = 0·07	

Total mass of gases (wet and dry) = 3·08 + 0·67 + 13·49 + 1·26

= 18·5 kg/kg of fuel

Alternatively 17·52 + 0·98 = 18·5 kg/kg of fuel

Mass flow rate of gases = $\dot{m}$ = 100 × 18·5 = 1850 kg/h

$$\text{From } pV = mRT,\ R = \frac{R_0}{M}$$

$$pV = \frac{m}{M} R_0 T$$

$$\frac{m}{M} = 0{\cdot}573 + 0{\cdot}07 = 0{\cdot}643$$

$$\therefore \quad V = \frac{0{\cdot}643 \times 8{\cdot}3143 \times 523}{1 \times 10^2}$$

$$= 27{\cdot}95 \text{ m}^3\text{/kg of fuel}$$

$$\dot{V} = 27{\cdot}95 \times 100$$

$$= 2795 \text{ m}^3\text{/h} \quad \text{Ans.}$$

8. Stoichiometric air $= \frac{100}{23}(2\tfrac{2}{3}C + 8H + S)$

$= 4{\cdot}348(2{\cdot}667 \times 0{\cdot}86 + 8 \times 0{\cdot}12 + 0{\cdot}02)$

= 14·23 kg/kg fuel

Actual air = 20·00 kg/kg fuel

Excess air = 5·77 kg/kg fuel

Mass products of combustion per kg fuel:

$CO_2 = 3\tfrac{2}{3} \times 0{\cdot}86 = 3{\cdot}154$ kg

$H_2O = 9 \times 0{\cdot}12 = 1{\cdot}08$

$$\begin{aligned} SO_2 &= 2 \times 0{\cdot}02 &&= 0{\cdot}04 \\ \text{Excess } O_2 &= 0{\cdot}23 \times 5{\cdot}77 &&= 1{\cdot}329 \\ \text{All } N_2 &= 0{\cdot}77 \times 20 &&= \underline{15{\cdot}40} \\ & \text{Total} &&= 21{\cdot}003 \text{ kg} \end{aligned}$$

% mass analysis of the wet flue gases:

$$CO_2 = \frac{315{\cdot}4}{21} = 15{\cdot}01$$

$$H_2O = \frac{108}{21} = 5{\cdot}14 \qquad \text{Ans. (a)}$$

$$SO_2 = \frac{4}{21} = 0{\cdot}19$$

$$O_2 = \frac{132{\cdot}9}{21} = 6{\cdot}32$$

$$N_2 = \frac{1540}{21} = 73{\cdot}33$$

% volume analysis of the wet flue gases:

DFG	*m*%	*M*	N	N%
CO_2	15·01	44	0·341	9·89
H_2O	5·14	18	0·286	8·30
SO_2	0·19	64	0·003	0·09
O_2	6·32	32	0·198	5·73
N_2	73·33	28	2·619	76·00
			3·447	

Ans. (b)

9. Mass of 1 mol of the fuel $= 12 \times 6 + 1 \times 6$
$= 78$ kg

$$H_2 \text{ fraction by mass} = \frac{6}{78} = 0{\cdot}0769$$

$$C \text{ fraction by mass} = \frac{72}{78} = \underline{0{\cdot}9231}$$

$$1{\cdot}0000$$

$$\text{Stoichiometric air} = \frac{100}{23}(2{\cdot}667 \times 0{\cdot}9231 + 0{\cdot}0769 \times 8)$$

$$= 13{\cdot}38 \text{ kg/kg fuel} \qquad \text{Ans. (a)}$$

$$\text{Mass of gas} = 14{\cdot}38 \text{ kg/kg fuel}$$

$$CO_2 = 3{\cdot}667 \times 0{\cdot}9231 = 3{\cdot}385$$
$$H_2O = 0{\cdot}0769 \times 9 = 0{\cdot}692$$
$$N_2 = 0{\cdot}77 \times 13{\cdot}38 = 10{\cdot}303$$
$$14{\cdot}380$$

$$CO_2 = \frac{3{\cdot}385}{14{\cdot}38} \times 100 = 23{\cdot}54\%$$

$$H_2O = \frac{0{\cdot}692}{14{\cdot}38} \times 100 = 4{\cdot}81\% \qquad \text{Ans. (b)}$$

$$N_2 = \frac{10{\cdot}303}{14{\cdot}38} \times 100 = 71{\cdot}65\%$$

$$DFG = 14{\cdot}38 - 0{\cdot}692 = 13{\cdot}688 \text{ kg/kg fuel}$$

$$CO_2 = \frac{3{\cdot}385}{13{\cdot}688} \times 100 = 24{\cdot}73\%$$

$$N_2 = \frac{10{\cdot}303}{13{\cdot}688} \times 100 = 75{\cdot}27\%$$

DFG	*m*%	*M*	N	N%
CO_2	24·73	44	0·562	17·29
N_2	75·27	28	2·688	82·71
			Total 3·250	

Ans. (c)

10.

DFG	N	*M*	*m*	*m*%
CO_2	10·8	44	475·2	15·83
CO	0·8	28	22·4	0·75
O_2	7·2	32	230·4	7·68
N_2	81·2	28	2273·6	75·75
			Total 3001·6	

Let x be C mass
then $(1 - x)$ is H_2 mass

$$\text{Relative gas mass} = 3001{\cdot}6$$
$$\text{Relative C mass} = 12\,(10{\cdot}8 + 0{\cdot}8)$$
$$= 139{\cdot}2$$
$$\text{Dry flue gas mass} = \frac{3001{\cdot}6 \times x}{139{\cdot}2}$$
$$= 21{\cdot}56x \text{ kg}$$
$$\text{Water vapour} = 9 - 9x \text{ kg}$$
$$\text{Total gases} = 9 - 12{\cdot}56x \text{ kg}$$

$$\text{Mass of air supplied} = 8 - 12{\cdot}56 \text{ kg/kg fuel} \quad (1)$$
$$\text{Mass of } N_2 \text{ supplied} = \frac{2273 \times x}{139{\cdot}2}$$
$$= 16{\cdot}33 \text{ kg}$$
$$\text{Mass of air supplied} = \frac{16{\cdot}33 \times 100}{77}$$
$$= 21{\cdot}21x \text{ kg/kg fuel} \quad (2)$$

From (1) and (2):

$$8 - 12{\cdot}56x = 21{\cdot}21x$$
$$x = 0{\cdot}925$$
$$(1 - x) = 0{\cdot}075$$
$$\text{Mass percentage of C} = 92{\cdot}5\% \text{ in the fuel}$$
$$\text{Mass percentage of } H_2 = 7{\cdot}5\% \text{ in the fuel} \qquad \text{Ans.}$$

SOLUTIONS TO TEST EXAMPLES 14

Refer to Fig. 69 for all solutions

1. Specific enthalpy gain of refrigerant through evaporator

$$= h_1 - h_4 = 320 - 135 = 185 \text{ kJ/kg}$$

(h_4 being equal to h_3 because there is no change of enthalpy in the throttling process through the expansion valve)

$$\begin{aligned} \text{Refrig. effect [kJ/h]} &= \text{mass flow [kg/h]} \times (h_1 - h_4) \text{ [kJ/kg]} \\ &= 5 \times 60 \times 185 \\ &= 5{\cdot}55 \times 10^4 \text{ kJ/h or } 55{\cdot}5 \text{ MJ/h} \quad \text{Ans.} \end{aligned}$$

2. From tables page 13, Freon-12,

5·673 bar $\quad h_f = 54{\cdot}87$
1·509 bar $\quad t_s = -20°\text{C} \quad h_f = 17{\cdot}82 \quad h_g = 178{\cdot}73$
$h_{fg} = h_g - h_f = 178{\cdot}73 - 17{\cdot}82 = 160{\cdot}91$

Since the saturation temperature at 1·509 bar is –20°C and the refrigerant at this pressure leaves the evaporator at –5°C, it is superheated by 15°.

$$h \text{ at } 1{\cdot}509 \text{ bar superheated } 15° = 187{\cdot}75$$

Throttling between condenser exit and evaporator inlet:

$$\begin{aligned} \text{Enthalpy after } (h_4) &= \text{Enthalpy before } (h_3) \\ 17{\cdot}82 + x_4 \times 160{\cdot}91 &= 54{\cdot}87 \\ x_4 \times 160{\cdot}91 &= 37{\cdot}05 \\ x_4 &= 0{\cdot}2303 \quad \text{Ans. (i)} \end{aligned}$$

$$\begin{aligned} \text{Refrig. effect/kg} &= h_1 - h_4 = h_1 - h_3 \\ &= 187{\cdot}75 - 54{\cdot}87 \\ &= 132{\cdot}88 \text{ kJ/kg} \quad \text{Ans. (ii)} \end{aligned}$$

3. From tables page 12, NH_3,

8·57 bar $\quad h_f = 275{\cdot}1 \quad h_g = 1462{\cdot}6$
1·902 bar $\quad h_f = 89{\cdot}8 \quad h_g = 1420{\cdot}0 \quad v_g = 0{\cdot}6237$
$\quad h_{fg} = 1420 - 89{\cdot}8 = 1330{\cdot}2$

Specific enthalpy drop through condenser

$$= h_2 - h_3 = 1462{\cdot}6 - 275{\cdot}1 = 1187{\cdot}5 \text{ kJ/kg}$$

Heat rejected in condenser (by 2 kg)

$$= 2 \times 1187{\cdot}5 = 2375 \text{ kJ/min} \quad \text{Ans. (i)}$$

Specific enthalpy gain in evaporator

$$h_1 - h_4 = h_1 - h_3$$
$$= (89{\cdot}8 + 0{\cdot}96 \times 1330{\cdot}2) - 275{\cdot}1$$
$$= 1366{\cdot}8 - 275{\cdot}1 = 1091{\cdot}7 \text{ kJ/kg}$$

Refrigerating effect

$$= 2 \times 1091{\cdot}7 = 2183{\cdot}4 \text{ kJ/min} \quad \text{Ans. (ii)}$$

Specific volume of refrigerant leaving evaporator and entering compressor = $0{\cdot}96 \times 0{\cdot}6237$ m^3/kg

Volume taken into compressor per minute

$$= 2 \times 0{\cdot}96 \times 0{\cdot}6237 = 1{\cdot}198 \text{ m}^3\text{/min} \quad \text{Ans. (iii)}$$

4. Since there is a change only in the dryness fraction of the refrigerant through the evaporator, the enthalpy of saturated liquid (h_f) and of evaporation (h_{fg}) being the same for h_1 as for h_4 then:

Specific enthalpy gain of the CO_2 through evaporator

$$= h_1 - h_4$$
$$= (h_f + 0{\cdot}92h_{fg}) - (h_f + 0{\cdot}28h_{fg})$$
$$= (0{\cdot}92 - 0{\cdot}28) \times 290{\cdot}7$$
$$= 0{\cdot}64 \times 290{\cdot}7 = 186{\cdot}1 \text{ kJ/kg} \quad \text{Ans. (i)}$$

Heat to be extracted from water to make one kg of ice

$$= 4{\cdot}2 \times 14 + 335 + 2{\cdot}04 \times 5$$
$$= 58{\cdot}8 + 335 + 10{\cdot}2 = 404 \text{ kJ/kg}$$

Let m [kg] = mass of ice made per second
when 0·5 kg/s = mass flow of CO_2

Heat transfer:

from water = to CO_2

$$m \times 404 = 0{\cdot}5 \times 186{\cdot}1$$
$$m = \frac{0{\cdot}5 \times 186{\cdot}1}{404} = 0{\cdot}2303 \text{ kg/s}$$

Mass of ice in tonnes per 24 hours

$$= \frac{0{\cdot}2303 \times 3600 \times 24}{10^3} = 19{\cdot}9 \text{ tonne/day} \quad \text{Ans. (ii)}$$

5. Quantity of heat extracted per kg of water

$$= 4{\cdot}2 \times 18 + 335 + 2{\cdot}04 \times 7$$
$$= 75{\cdot}6 + 335 + 14{\cdot}28 = 424{\cdot}88 \text{ kJ/kg}$$

Heat energy extracted per second (refrigerating effect)

$$= \frac{2 \cdot 5 \times 10^3}{24 \times 3600} \times 424 \cdot 88 \text{ kJ/s}$$

Energy supplied per second, kJ/s = kW = 2·25

$$\text{Coeff. of performance} = \frac{\text{heat energy extracted}}{\text{heat energy supplied}}$$

$$= \frac{2 \cdot 5 \times 10^3 \times 424 \cdot 88}{24 \times 3600 \times 2 \cdot 25}$$

$$= 5 \cdot 464 \quad \text{Ans. (i)}$$

If ice at 0°C was made from water at 0°C, heat extracted would be equal to enthalpy of fusion only = 335 kJ/kg

∴ Capacity from and at 0°C

$$= \frac{2 \cdot 5 \times 424 \cdot 88}{335} = 3 \cdot 171 \text{ tonne/day} \quad \text{Ans. (ii)}$$

6. From tables page 13, Freon-12,
   5·673 bar, sat. temp. = 20°C
   ∴ at 50°C refrigerant is superheated by 30°
   h_2 (compressor discharge) = 216·75
   1·826 bar, sat. temp. = −15°C
   ∴ at 0°C refrigerant is superheated by 15°
   h_1 (compressor suction and evaporator exit) = 190·15
   h_3 (condenser outlet) = h_f at 5·673 bar = 54·87
   h_4 (evaporator inlet) = h_3 = 54·87

$$\text{Coeff. of performance} = \frac{\text{refrigerating effect [kJ/kg]}}{\text{work transfer [kJ/kg]}}$$

$$= \frac{h_1 - h_4}{h_2 - h_1}$$

$$= \frac{190 \cdot 15 - 54 \cdot 87}{216 \cdot 75 - 190 \cdot 15} = \frac{135 \cdot 28}{26 \cdot 6}$$

$$= 5 \cdot 087 \quad \text{Ans.}$$

7. Tables page 12, NH_3:

14·7 bar, sat. temp. = 38°C
∴ vapour is superheated (63 − 38) = 25°

By interpolation,
14·7 bar 50° supht, h = 1620·1

$$
\begin{aligned}
14{\cdot}7 \text{ bar no supht, } h &= 1472{\cdot}6 \\
\text{increase for } 50^\circ &= 147{\cdot}5 \\
\text{increase for } 25^\circ &= \tfrac{1}{2} \times 147{\cdot}5 = 73{\cdot}75 \\
\therefore\ 14{\cdot}7 \text{ bar } 25^\circ \text{ supht, } h &= 1472{\cdot}6 + 73{\cdot}75 = 1546{\cdot}35 = h_2 \\
14{\cdot}7 \text{ bar } 50^\circ\text{C supht, } s &= 5{\cdot}340 \\
14{\cdot}7 \text{ bar no supht, } s &= 4{\cdot}898 \\
\text{increase for } 50^\circ &= 0{\cdot}442 \\
\text{increase for } 25^\circ &= \tfrac{1}{2} \times 0{\cdot}442 = 0{\cdot}221 \\
\therefore\ 14{\cdot}7 \text{ bar } 25^\circ \text{ supht, } s &= 4{\cdot}898 + 0{\cdot}221 = 5{\cdot}119 = s_2 \\
2{\cdot}077 \text{ bar, } h_f &= 98{\cdot}8 \quad h_g = 1422{\cdot}7 \\
h_{fg} &= 1422{\cdot}7 - 98{\cdot}8 = 1323{\cdot}9 \\
s_f &= 0{\cdot}404 \quad s_g = 5{\cdot}593 \\
s_{fg} &= 5{\cdot}593 - 0{\cdot}404 = 5{\cdot}189
\end{aligned}
$$

Isentropic compression in compressor:

$$
\begin{aligned}
s_1 &= s_2 \\
0{\cdot}404 + x_1 \times 5{\cdot}189 &= 5{\cdot}119 \\
x_1 &= 0{\cdot}9086 \\
h_1 &= 98{\cdot}8 + 0{\cdot}9086 \times 1323{\cdot}9 = 1301{\cdot}8 \\
h_4 &= h_3 = h_f \text{ at } 14{\cdot}7 \text{ bar} = 362{\cdot}1 \\
\text{Refrig. effect [kJ/s]} &= (h_1 - h_4) \text{ [kJ/kg]} \times \text{mass flow [kg/s]} \\
&= (1301{\cdot}8 - 362{\cdot}1) \times 0{\cdot}15 \\
&= 140{\cdot}9 \text{ kJ/s} \quad \text{Ans. (i)} \\
\text{Work transfer [kJ/s]} &= (h_2 - h_1) \text{ [kJ/kg]} \times \text{mass flow [kg/s]} \\
&= (1546{\cdot}35 - 1301{\cdot}8) \times 0{\cdot}15 \\
&= 36{\cdot}68 \text{ kJ/s} \quad \text{Ans. (ii)}
\end{aligned}
$$

$$
\text{Coeff. of performance} = \frac{\text{refrigerating effect}}{\text{work transfer}}
$$

$$
= \frac{140{\cdot}9}{36{\cdot}68} = 3{\cdot}842 \quad \text{Ans. (iii)}
$$

8.

$$
\begin{aligned}
\text{At } 1{\cdot}004 \text{ bar} h_1 = h_g &= 174{\cdot}2 \text{ kJ/kg} \\
s_1 = s_g &= 0{\cdot}7170 \text{ kJ/kg K} \\
v_1 = v_g &= 0{\cdot}1594 \text{ m}^3\text{/kg}
\end{aligned}
$$

$$
\text{Mass flow of refrigerant} = \dot{m} = \frac{0{\cdot}15}{0{\cdot}1594}
$$

$$
= 0{\cdot}941 \text{ kg/s}
$$

$$
\text{Isentropic compression } s_1 = s_2 = 0{\cdot}717 \text{ kJ/kg K}
$$

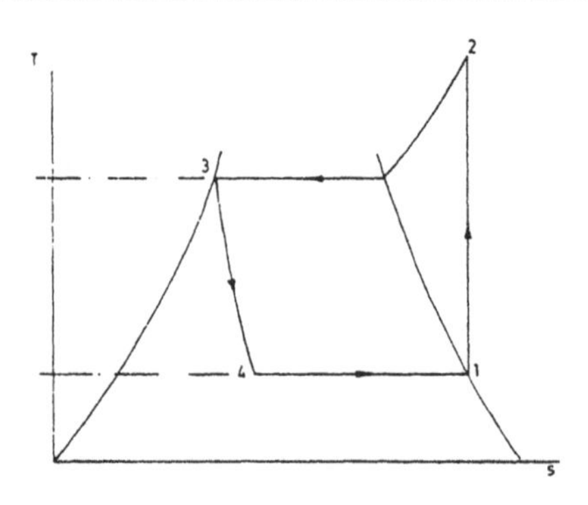

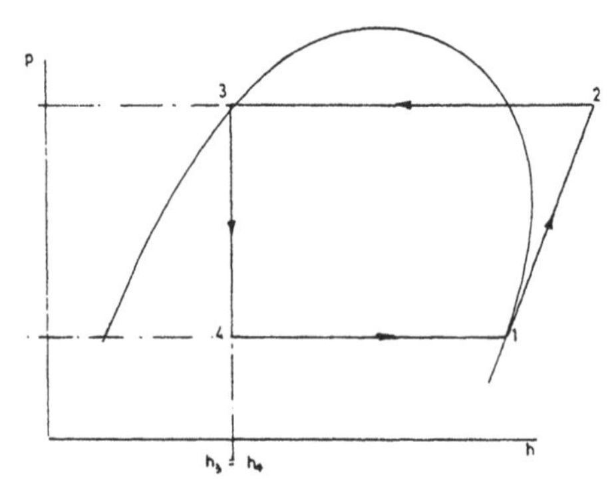

By interpolation

$$0.717 = 0.6901 + \frac{\theta}{15}(0.7251 - 0.6901)$$

$$\theta = 11.53°C$$

$$h_2 = 193.78 + \frac{11.53}{15}(204.1 - 193.78)$$

$$= 201.7 \text{ kJ/kg}$$

At 4·914 bar $h_4 = h_3 = h_f = 50.1$ kJ/kg

$$\text{c.o.p.} = \frac{h_1 - h_4}{h_2 - h_1}$$

$$= \frac{174.2 - 50.1}{201.7 - 174.2}$$

$$= 4.513 \quad \text{Ans. (a)}$$

$$\text{power input} = \dot{m}\,[h_2 - h_1]$$

$$= 0.941 \times [201.7 - 174.2]$$

$$= 25.88 \text{ kW} \quad \text{Ans. (b)}$$

9. At 40°C specific enthalpy

$$\text{leaving compressor} = h_2$$
$$h_2 = h_g = 1473{\cdot}3 \text{ kJ/kg}$$
$$s_2 = 4{\cdot}877 \text{ kJ/kg K}$$
$$\text{With isentropic compression } s_1 = s_2 = 4{\cdot}877 \text{ kJ/kg K}$$

$$\therefore 4{\cdot}877 = 0{\cdot}44 + x\,(5{\cdot}563 - 0{\cdot}44)$$
$$x = 0{\cdot}8661 \text{ dry}$$
$$\text{at } -16^\circ\text{C} \therefore h_1 = h_f + 0{\cdot}8661\,(h_g - h_f)$$
$$h_1 = 107{\cdot}9 + 0{\cdot}8661\,(1425{\cdot}3 - 107{\cdot}9)$$
$$= 1248{\cdot}9 \text{ kJ/kg}$$

If the ammonia leaves the heat exchanger as saturated liquid then at 40°C $h_3 = h_t = 371{\cdot}9$ kJ/kg

$$\text{Work done in compressor} = h_2 - h_1$$
$$= 1473{\cdot}3 - 1248{\cdot}9$$
$$= 224{\cdot}4 \text{ kJ/kg}$$
$$\text{Heat transfer in condenser} = h_2 - h_3$$
$$= 1473{\cdot}3 - 371{\cdot}9$$
$$= 1101{\cdot}4 \text{ kJ/kg}$$
$$\text{c.o.p.} = \frac{h_2 - h_3}{h_2 - h_1}$$
$$= \frac{1101{\cdot}4}{224{\cdot}4} = 4{\cdot}908. \quad \text{Ans.}$$
$$\text{Mass of air changed } \dot{m}_1 = \frac{p\dot{V}}{RT}$$
$$= \frac{1{\cdot}013 \times 1200}{287 \times 293} \times 10^5$$
$$= 1445{\cdot}6 \text{ kg/h}$$
$$\dot{m}_1 \times c_p \times (T_2 - T_1) = (h_2 - h_3) \times \dot{m}$$
$$1445{\cdot}6 \times \frac{1005}{10^3}(293 - 286) = 1101{\cdot}4 \times \dot{m}$$
$$\dot{m} = 9{\cdot}233 \text{ kg/h} \quad \text{Ans.}$$

10. Tables Freon - 12 page 13:

$$25^\circ\text{C}, \quad h_f = 59{\cdot}7 \quad h_g = 197{\cdot}33$$
$$-15^\circ\text{C}, \quad h_f = 22{\cdot}33 \quad h_g = 180{\cdot}97 \quad (h_{fg} = 158{\cdot}64)$$

$$\text{Refrigerating effect} = h_1 - h_4$$
$$= 180{\cdot}97 - 59{\cdot}7$$
$$= 121{\cdot}27 \text{ kJ/kg}$$

$$\text{Cooling load} = \text{refrigerating effect} \times \text{mass flow}$$
$$h_3 = h_4 \text{ as throttling, no undercool}$$
$$73{\cdot}3 = 121{\cdot}27 \times \dot{m}$$
$$\dot{m} = 0{\cdot}6044 \text{ kg/s} \quad \text{Ans. (a)}$$
$$h_2 = 208{\cdot}5$$

i.e. 6·516 bar, 40°C (15° superheat), from tables

$$\text{Compressor work done} = h_2 - h_1$$
$$= 208{\cdot}5 - 180{\cdot}97$$
$$= 27{\cdot}53 \text{ kJ/kg}$$
$$\text{Compressor power} = 27{\cdot}53 \times 0{\cdot}6044$$
$$= 16{\cdot}64 \text{ kW} \quad \text{Ans. (b)}$$
$$h_4 = h_f + x h_{fg}$$
$$59{\cdot}7 = 22{\cdot}33 + 158{\cdot}64 \times x$$
$$\text{Condition after expansion valve} = 0{\cdot}2356 \text{ dry} \quad \text{Ans. (c)}$$

SELECTION OF EXAMINATION QUESTIONS
CLASS TWO

1. The efficiency of an auxiliary boiler is 70% when dry saturated steam at 8 bar is generated from feed water at 43°C. If the calorific value of the fuel is 42·5 MJ/kg and it is burned at the rate of 6 tonne per day, calculate (i) the hourly steam production, (ii) the equivalent evaporation per kg of fuel from and at 100°C.

2. An electric motor was tested by coupling it to a dynamometer and the brake load was 112 N at 40 rev/s. A steady flow of water passed through the brake and was raised in temperature from 15°C to 48°C. If the power absorbed by this brake is given by $Wn/330$ where W is the brake load in newtons and n is the rotational speed in rev/s, and assuming 98% of the heat generated at the brake is carried away by the cooling water, find the quantity of water passing through the dynamometer in litres per minute.

3. (a) A refrigerated hold of 580m^2 surface area is lined with a 100 mm thick layer of insulation of thermal conductivity 0·17 W/m K. The interior and exterior surface temperatures of the insulation are 0°C and 17°C respectively. Calculate the heat flow rate from the hold.

(b) The exterior surface temperature is increased to 50°C. Calculate the additional thickness of insulating material of thermal conductivity 0·07 W/m K required to be placed on the inside surfaces to keep the heat flow rate at the same level.

4. The bore of an I.C. engine is exactly 300 mm at 20°C. The diameters of the piston at 20°C are 298 mm at the crown and 299 mm at mid-depth of the body. Under working conditions the mean temperatures are: piston crown 250°C, piston body 100°C, cylinder 180°C. Take the coefficients of linear expansion of the piston and cylinder materials as $1{\cdot}2 \times 10^{-5}$ and $1{\cdot}1 \times 10^{-5}$/°C respectively and calculate the diametrical clearances at the crown and mid-depth of the piston under working conditions.

5. An indicator card taken off one cylinder of a six-cylinder, two-stroke, single-acting engine, has an area of 2850 mm^2 and length 75 mm when running at 1·75 rev/s. One millimetre height on the card represents 0·4 bar. If the cylinder bore is 550 mm and stroke

850 mm, calculate the indicated power of the engine assuming equal powers are developed in all cylinders.

6. A vessel of volume 0·65 m^3 contains air at 27·6 bar and 18°C. Calculate the final pressure after 3·5 kg of air is added if the final temperature is 20·5°C. Take R for air = 0·287 kJ/kg K.

7. One kilogramme of steam at 7·0 bar, 0·95 dry, is expanded according to the law $pV^{1 \cdot 3}$ = a constant until the pressure is 3·5 bar. Calculate the final dryness fraction of the steam.

8. A turbine plant consisting of H.P. and L.P. units is supplied with steam at 15 bar 300°C. The steam is expanded in the H.P. and leaves at 2·5 bar 0·97 dry. At this point some steam is bled off to the feed heater, the remaining steam passes to the L.P. where it is expanded to 0·14 bar 0·84 dry. If the same quantity of work transfer takes place in each unit, calculate the amount of steam bled off expressed as a percentage of the steam supplied.

9. The mass analysis of a fuel is 87% carbon, 11% hydrogen, and 2% oxygen. Calculate the volume of air in cubic metres at 1·0 bar and 25°C required for stoichiometric combustion per kg of fuel. Take the values: R for air = 0·287 kJ/kg K. Mass analysis of air = 23% oxygen, 77% nitrogen. Atomic weights, hydrogen 1, carbon 12, oxygen 16.

10. Gas at pressure 0·95 bar, volume 0·2 m^3 and temperature 17°C, is compressed until the pressure is 2·75 bar and volume 0·085 m^3, calculate the compression index and the final temperature.

11. A ship's cold room has dimensions of 9 m × 4 m × 2·5 m and is lined on the inside with 15 mm thick boarding. The deck, deckhead and bulkheads are of 10 mm thick steel and a 70 mm thick layer of cork insulation is sandwiched between the steel plate and the boarding. The thermal conductivities of steel, cork and boarding are 45 W/m K, 0·06 W/m K and 0·11 W/m K respectively. The surface heat transfer coefficients at the exposed board inner and steel outer surfaces are 1·62 $W/m^2 K$ and 13 $W/m^2 K$ respectively. Calculate the cooling load required to maintain the cold room at −6°C when the ambient temperature is 27°C.

12. A boiler working at 16 bar produces 9000 kg of steam per hour from feed water at 95°C, the dryness fraction of the steam being

0·98. If the boiler efficiency is 87% and the calorific value of the fuel 42 MJ/kg, calculate the daily fuel consumption.

13. The scavenge ports of a two-stroke diesel engine are just covered when the piston is 800 mm from the top of its stroke. The contents of the cylinder at this instant are at a pressure of 1·21 bar and temperature 40°C. The piston diameter is 700 mm and the clearance is equivalent to 70 mm. Find the mass of air taken in per cycle if the scavenge efficiency is 0·95. R for air = 0·287 kJ/kg K.

14. The diameter of an air compressor cylinder is 130 mm, the stroke is 180 mm, and the clearance volume is 73 cm³. The pressure in the cylinder at the beginning of the stroke is 1·0 bar and the pressure during delivery is constant at 4·6 bar. Taking the law of compression as $pV^{1\cdot2}$ = constant, calculate the distance moved by the piston during the delivery period and express this as a fraction of the stroke.

15. In a single-stage impulse turbine the steam leaves the nozzles at a velocity of 500 m/s at 18° to the plane of rotation of the blades, and the linear velocity of the blades is 230 m/s. Neglecting friction across the blades and assuming the steam leaves the blade wheel in an axial direction, calculate (i) the inlet angle of the blades so that the steam enters without shock, (ii) the outlet blade angle.

16. The clearance volume of a reciprocating compressor of 100 mm bore and 100 mm stroke is 5% of its swept volume. At the end of the suction stroke the air in the cylinder is at 1 bar 25°C.

(a) Show the cycle on a pressure volume diagram.
(b) Calculate the mass of air in the cylinder at the beginning of the delivery stroke.
(c) Explain why the compressor takes in less than the swept volume of air during each suction stroke.

Note: for air R = 287 J/kg K

17. In an NH_3 refrigerating plant the ammonia leaves the condenser as a saturated liquid at 10·34 bar. The evaporator pressure is 2·265 bar and the refrigerant leaves the evaporator as a vapour 0·95 dry. If the circulation of the refrigerant through the plant is 4 kg/min, calculate (i) the dryness fraction at inlet to the evaporator, (ii) the refrigerating effect per minute, and (iii) the volume of refrigerant taken into the compressor per minute.

18. (a) Write down the combustion equations for the complete combustion of carbon to carbon dioxide, hydrogen to water and sulphur to sulphur dioxide.

(b) A fuel oil of mass analysis 86·3% carbon, 12·8% hydrogen and 0·9% sulphur is burned with 25% excess air. The flue gases are at 1·5 bar 370°C. Calculate the mass and volume of flue gases per kg of fuel burned.

Note: air contains 23% oxygen by mass
relative atomic masses: hydrogen 1, carbon 12, oxygen 16, sulphur 32
for flue gases R = 276 J/kg K

19. Water at 100°C flows through a steel pipe of 150 mm inside diameter and 160 mm outside diameter. The surface heat transfer coefficients at the inside and outside surfaces are 240 $W/m^2 K$ and 12 $W/m^2 K$ respectively. The air surrounding the pipe is at 15°C. The thermal resistance of the steel may be neglected and the diameter of the pipe is large compared to its wall thickness.

Calculate the rate of heat loss from the water per metre length of pipe.

20. A volume of 0·8 m^3 of steam at 17 bar 0·95 dry is passed through a reducing valve and throttled to 6 bar. Calculate the dryness fraction and the volume after throttling.

21. At the beginning of a voyage a boiler contains 6 tonne of water having 120 p.p.m. dissolved solids. The feed rate is 1250 kg/h and after 24 hours the boiler water contains 1080 p.p.m. dissolved solids. Calculate the average dissolved solids in the feed water.

22. 1·5 m^3 of wet steam at 2·8 bar are blown into 36 kg of water at 16°C and the resulting temperature of the mixture is 55°C. Calculate the dryness fraction of the steam.

23. A single stage impulse turbine has a mean blade diameter of 600 mm and a blade velocity of 120 m/s. The nozzle angle is 18° and the enthalpy drop across the nozzles is 465 kJ/kg.

Determine:

(a) the turbine rotational speed;
(b) the blade inlet angle;
(c) the axial component of the steam at the blade inlet.

24. The air in a ship's saloon is maintained at 19°C and is changed twice every hour from the outside atmosphere which is at 7°C. The saloon is 27 m by 15 m by 3 m high. Calculate the kilowatt loading to heat this air, taking the saloon to be at atmospheric pressure = 1·013 bar, R for air = 0·287 kJ/kg K, c_p = 1·005 kJ/kg K.

25. A six-cylinder, single-acting, four-stroke oil engine, of 200 mm stroke and 225 mm bore runs at 5 rev/s when the mean effective pressure is 17 bar. If the mechanical efficiency is 85% calculate the indicated and brake powers.

26. A single cylinder single acting compressor of 200 mm stroke and 100 mm bore runs at 5 rev/s and takes in air at 1 bar 17°C. It is used to charge a 3 m^3 capacity receiver from 1 bar 25°C to 7 bar 25°C. The compressor has negligible piston clearance.

(a) Sketch the cycle on a pressure volume diagram.
(b) Calculate the time required to charge the receiver.

27. An ammonia refrigerator has a cooling load of 3·5 kW. The ammonia leaves the condenser as a liquid at 24°C, is throttled to 2·68 bar and leaves the evaporator as a dry saturated vapour.

(a) Sketch component and pressure enthalpy diagrams.
(b) Calculate the mass flow rate of the ammonia.

28. A boiler working at 15 bar generates 7000 kg of steam per hour. The steam leaves the boiler steam drum dry and saturated and then passes through the superheater tubes at constant pressure. The flue gases enter the nests of superheaters at 822°C and leaves at 690°C. The fuel consumption is 750 kg per hour and 24 kg of air are supplied per kg of fuel burned. Find (i) the temperature of the superheated steam, (ii) the mass of injection water to the de-superheater, at 21°C, required to desuperheat each kg of steam. Take the specific heat of the flue gases as 1·007 kJ/kg K.

29. A gas initially at 12 bar, 216°C and volume 9900 cm^3 is expanded in a cylinder. The volumetric expansion is 6 and the index of expansion is 1·33. Calculate the final volume, pressure, and temperature.

30. 1 kg of steam initially occupying a volume of 0·08 m^3 at 40 bar is expanded to 2 bar according to the law $pV^{1 \cdot 3}$ = const.

Calculate:
(a) the work transfer;
(b) the heat transfer.

31. Steam leaves the nozzles and enters the blade wheel of a single-stage impulse turbine at a velocity of 840 m/s and at an angle of 20° to the plane of rotation. The blade velocity is 350 m/s and the exit angle of the blades is 25° 12'. Due to friction, the steam loses 20% of its relative velocity across the blades. Calculate (i) the blade inlet angle, (ii) the magnitude and direction of the absolute velocity of the steam at exit.

32. The mass composition of a fuel oil is 84·6% carbon, 11·4% hydrogen, 0·4% sulphur, 2·4% oxygen, and 1·2% impurities. Calculate the calorific value of the fuel and the theoretical mass of air required for stoichiometric combustion of one kg of fuel. Take the values:

	CALORIFIC VALUE MJ/kg	ATOMIC WEIGHT
Hydrogen	144	1
Carbon	33·7	12
Sulphur	9.3	32
Oxygen	—	16

Mass analysis of air = 23% oxygen, 77% nitrogen.

33. In a Freon-12 refrigerating plant, the refrigerant leaves the condenser with a specific enthalpy of 50 kJ/kg. The pressure in the evaporator is 1·826 bar and the refrigerant leaves the evaporator at this pressure and at a temperature of 0°C. Calculate (i) the dryness fraction of the freon at inlet to the evaporator, and (ii) the refrigerating effect per minute if the flow rate of the refrigerant is 0·4 kg/s.

34. A single stage double-acting air compressor deals with 18·2 m^3 of air per minute measured at conditions of 1·01325 bar 15°C. The condition at the beginning of compression is 0·965 bar 27°C and the discharge pressure is 4·82 bar. The compression is according to the law $pV^{1.32}$ = constant. If the mechanical efficiency of the compressor is 0·9 calculate the input power required to drive the compressor.

35. The mean area of indicator cards taken from a cylinder of a double-acting two-stroke engine is 346 mm^2 and the length is 75 mm. The spring used in the indicator deflects one mm under a force of 60 N and the movement of the stylus is six times that of the indicator piston. The diameter of the indicator piston is 7 mm. Calculate (i) the mean effective pressure. If the diameter of the engine cylinder is 600 mm, stroke 900 mm, and rotational speed 2·1 rev/s, calculate (ii) the indicated power per cylinder.

36. At the entrance of a nozzle of circular cross-section, the velocity of the steam is 457 m/s and the specific volume is 0·2765 m^3/kg. At exit, the velocity is 1524 m/s and the specific volume 7·404 m^3/kg. If the mass flow of steam through the nozzle is 0·315 kg/s, calculate the entrance and exit diameters in millimetres.

37. A quantity of air of volume 0·2 m^3 at 1·1 bar and 15°C is heated at constant pressure until its temperature is 150°C, and then compressed to 7·15 bar according to the law $pV^{1.32}$ = a constant. Calculate (i) the amount of heat energy transferred to the air at constant pressure, and (ii) the temperature at the end of compression. For air, $R = 0.287$ kJ/kg K, $c_p = 1.005$ kJ/kg K.

38. A cold store wall 6 m long and 3 m high is constructed of 120 mm thick brick with an inside layer of 80 mm thick cork insulation. The thermal conductivity of the brick is 1·15 W/m K whilst that of cork is 0·043 W/m K. The inner and outer wall surface temperatures are –4°C and 21°C respectively. Calculate:
(a) the amount of heat flow through the wall in 24 hours;
(b) the interface temperature between the cork and the brick.

39. The clearance volume of a reciprocating compressor of 381 mm stroke is 7% of the swept volume. The piston travels 267 mm from bottom dead centre to the point at which the delivery valves open and raises the air pressure from 1·013 bar to 4 bar. The compression process is polytropic.
(a) Sketch the cycle on a pV diagram.
(b) Calculate the value of the polytropic index n.

40. The mass analysis of a hydrocarbon fuel A is 88·5% carbon and 11·5% hydrogen. Another hydrocarbon fuel B requires 6% more air than fuel A for stoichiometric combustion. Calculate the mass analysis of fuel B taking the following values: Atomic weights, carbon 12, hydrogen 1, oxygen 16, mass content of oxygen in air = 23%.

41. The stroke of an internal combustion engine is 90 mm and the clearance volume is equivalent to a linear clearance of 15 mm. If the clearance is reduced by 2·5 mm, find the pressure at the end of compression before and after the alteration, taking the initial pressure in each case as 1·0 bar and the index of compression as 1·33.

42. The kinetic energy of the steam jet leaving the nozzles of a single-stage impulse turbine is equivalent to 250 kJ/kg. The entrance and exit angles of the blades are equal at 35°, and the steam leaves the blade wheel in an axial direction. Neglecting friction across the blades and assuming shockless flow, calculate (i) the nozzle angle, and (ii) the blade velocity.

43. A four-cylinder, single-acting, two-stroke engine develops 600 kW indicated power when the mean effective pressure is 12·56 bar and the speed is 4·5 rev/s.

(a) If the stroke is 25% greater than the cylinder diameter, calculate the diameter of the cylinders and the stroke to the nearest millimetre.

(b) When burning fuel of calorific value 42 MJ/kg the fuel consumption is 0·225 kg/kWh (indicated), find the indicated thermal efficiency.

44. The refrigerating effect of a plant using ammonia as the refrigerant is 800 kJ/min. At the exit points of the components the conditions of the refrigerant are:

Evaporator, dry saturated vapour at 1·902 bar
Compressor, vapour at 7·529 bar and 66°C
Condenser, saturated liquid at 7·529 bar

Calculate (i) the mass flow of the refrigerant through the plant, in kg/min, (ii) the heat rejected in the condenser, in kJ/min, and (iii) the output power of the compressor in kW.

45. Steam enters a desuperheater at 30 bar and 400°C and leaves at the same pressure as dry saturated steam. The temperature of the water injected into the desuperheater is 38°C. Calculate (i) the mass of injection water used per kg of steam desuperheated, and (ii) the percentage change in volume from that occupied by one kg of superheated steam to the volume occupied by the dry saturated steam resulting from the mixture of one kg of superheated steam and its injected water.

46. A gas is compressed in a cylinder from 1 bar and 35°C at the beginning of the stroke to 37 bar at the end of the stroke. If the clearance volume is 850 cm^3 and the compression index 1·32, calculate the stroke volume and the temperature at the end of compression.

47. When 1 kg of a fuel containing only carbon and hydrogen is burned in air, 15·6 kg of exhaust containing only nitrogen, carbon dioxide and water vapour is produced.

Determine the fuel mass analysis.

Air contains 23% oxygen by mass

Atomic mass relationships: hydrogen 1, carbon 12, oxygen 16

48. (a) A ship's ice making machine produces 27·2 kg/h of ice at 0°C from water at 14·4°C with a coefficient of performance of 7·51. Calculate the power input to the machine.
(b) Using the concept of a reversed heat engine, explain why refrigerators require some power input.

Note: *specific heat capacity of water is 4·186 kJ/kg K*
specific enthalpy of fusion of ice is 332·6 kJ/kg

49. The following data relate to a single stage impulse turbine

Blade:	mean diameter	1·32 m
	inlet angle	34°
	outlet angle	30°
Steam:	inlet axial velocity component	550 m/s
	relative outlet velocity	850 m/s
	flow rate	0·0833 kg/s
Turbine:	power developed	67 kW

Determine the turbine rotational speed.

50. Water is kept at a temperature of 66°C in a closed tank 1 m long × 0·75 m wide × 1·6 m high. The heat loss from the water must not exceed 200 W when the ambient temperature is 18°C. The tank is to be lagged with insulating material of thermal conductivity 0·048 W/m K. The exterior surface heat transfer coefficient is 1 $W/m^2 K$. The thermal resistances of the tank walls and lid and the heat loss through the base are negligible.

Calculate:

(a) the thickness of lagging required;
(b) the additional thickness of lagging required to reduce the heat loss to 100 W.

SOLUTIONS TO EXAMINATION QUESTIONS
CLASS TWO

1. From steam tables

Steam 8 bar, h_g = 2769 kJ/kg

Water 43°C, by interpolation,

$$h \text{ at } 44^\circ\text{C} = 184{\cdot}2$$
$$h \text{ at } 42^\circ\text{C} = 175{\cdot}8$$
$$\text{increase for } 2^\circ = 8{\cdot}4$$
$$\text{increase for } 1^\circ = 4{\cdot}2$$
$$\therefore\ h \text{ at } 43^\circ\text{C} = 175{\cdot}8 + 4{\cdot}2 = 180 \text{ kJ/kg}$$

$$\text{Fuel consumption} = \frac{6 \times 10^3}{24} = 250 \text{ kg/h}$$

$$\text{Boiler efficiency} = \frac{\text{heat energy transferred to steam [kJ/h]}}{\text{heat energy supplied by fuel [kJ/h]}}$$

$$0{\cdot}7 = \frac{\text{mass of steam [kg/h]} \times (2769 - 180)}{250 \times 42{\cdot}5 \times 10^3}$$

$$\text{Mass of steam}/h = \frac{0{\cdot}7 \times 250 \times 42{\cdot}5 \times 10^3}{2589}$$

$$= 2873 \text{ kg/h} \quad \text{Ans. (i)}$$

From tables, h_{fg} at 100°C = 2256·7 kJ/kg

Equivalent evaporation from and at 100°C, per kg of fuel,

$$= \frac{2873 \times (2769 - 180)}{250 \times 2256{\cdot}7}$$

$$= 13{\cdot}18 \text{ kg steam/kg fuel} \quad \text{Ans. (ii)}$$

2. $$\text{Brake power} = \frac{Wn}{330} = \frac{112 \times 40}{330} = 13{\cdot}58 \text{ kW}$$

$$\text{Energy at brake} = 13{\cdot}58 \text{ kJ/s}$$

Energy carried away by water

$$= 0{\cdot}98 \times 13{\cdot}58 = 13{\cdot}3 \text{ kJ/s}$$

From steam tables,

$$\text{Water } 48^\circ\text{C}, h = 200{\cdot}9$$
$$\ldots\ 15^\circ\text{C}, h = 62{\cdot}9$$

Enthalpy gain of cooling water

$$= 200{\cdot}9 - 62{\cdot}9 = 138 \text{ kJ/kg}$$

Heat energy transferred to water [kJ/s]

$$= \text{mass flow [kg/s]} \times \text{spec. enthalpy gain [kJ/kg]}$$

$$\therefore \text{ mass flow} = \frac{13{\cdot}3}{138} = 0{\cdot}09638 \text{ kg/s}$$

$$0{\cdot}09638 \times 60 = 5{\cdot}783 \text{ kg/min}$$

$$= 5{\cdot}783 \text{ litre/min} \quad \text{Ans.}$$

3.

$$Q = \frac{kAt\,(T_1 - T_2)}{S}$$

$$= \frac{0{\cdot}17 \times 580 \times 1\ (0 - 17)}{0{\cdot}1}$$

$= -16762$ W i.e. *into* the hold

$= -16{\cdot}762$ kW Ans. (a)

For the two thicknesses:

$$T_1 - T_3 = \frac{Q}{At}\left\{\frac{S_1}{k_1} + \frac{S_2}{k_2}\right\}$$

$$0 - 50 = \frac{-16762}{580 \times 1}\left\{\frac{S_1}{0{\cdot}07} + \frac{0{\cdot}1}{0{\cdot}17}\right\}$$

$$S_1 = 0{\cdot}07\left\{\frac{50 \times 580 \times 1}{16\,762} - \frac{0{\cdot}1}{0{\cdot}17}\right\}$$

$$= 0{\cdot}07993 \text{ m}$$

$$= 79{\cdot}93 \text{ mm} \quad \text{Ans. (b)}$$

4. Linear (diametrical) expansion $= \alpha \times d \times (\theta_2 - \theta_1)$

Increase in cylinder diameter

$= 1{\cdot}1 \times 10^{-5} \times 300 \times (180 - 20) = 0{\cdot}5279$ mm

Working diameter $= 300 + 0{\cdot}5279 = 300{\cdot}5279$ mm

Increase in piston crown diameter

$= 1{\cdot}2 \times 10^{-5} \times 298 \times (250 - 20) = 0{\cdot}8224$ mm

Working diameter $= 298 + 0{\cdot}8224 = 298{\cdot}8224$ mm

Increase in piston body diameter

$1{\cdot}2 \times 10^{-5} \times 299 \times (100 - 20) = 0{\cdot}2871$mm

Working diameter $= 299 + 0{\cdot}2871 = 299{\cdot}2871$ mm

Diametrical clearances at working temperatures:

$$\left.\begin{array}{l}\text{Piston crown} = 300{\cdot}5279 - 298{\cdot}8224 = 1{\cdot}7055 \text{ mm}\\ \text{Piston body} = 300{\cdot}5279 - 299{\cdot}2871 = 1{\cdot}2408 \text{ mm}\end{array}\right\} \text{Ans.}$$

5. Mean height of card $= \dfrac{2850}{75} = 38$ mm

Mean effective pressure $= 38 \times 0{\cdot}4$ bar

$38 \times 0{\cdot}4 \times 10^2 = 1520$ kN/m²

Indicated power of 6 cylinders

$$= p_m ALn \times 6$$
$$= 1520 \times 0{\cdot}7854 \times 0{\cdot}55^2 \times 0{\cdot}85 \times 1{\cdot}75 \times 6$$
$$= 3224 \quad \text{Ans.}$$

6. $pV = mRT$

Initial mass of air,

$$m = \frac{pV}{RT} = \frac{27{\cdot}6 \times 10^2 \times 0{\cdot}65}{0{\cdot}287 \times 291} = 21{\cdot}48 \text{ kg}$$

Final mass $= 21{\cdot}48 + 3{\cdot}5 = 24{\cdot}98$ kg

$$\text{Final pressure, } p = \frac{mRT}{V}$$
$$= \frac{24{\cdot}98 \times 0{\cdot}287 \times 293{\cdot}5}{0{\cdot}65}$$
$$= 3238 \text{ kN/m}^2 = 32{\cdot}38 \text{ bar} \quad \text{Ans.}$$

7. From steam tables

7 bar, $v_g = 0{\cdot}2728$ 3·5 bar, $v_g = 0{\cdot}5241$

Volume of one kg of steam at 7 bar 0·95 dry

$$= 0{\cdot}95 \times 0{\cdot}2728 = 0{\cdot}2592 \text{ m}^3$$

$$p_1 v_1^{1{\cdot}3} = p_2 v_2^{1{\cdot}3}$$
$$7 \times 0{\cdot}2592^{1{\cdot}3} = 3{\cdot}5 \times V_2^{1{\cdot}3}$$
$$v_2 = 0{\cdot}2592 \times \sqrt[1{\cdot}3]{\frac{7}{3{\cdot}5}}$$
$$= 0{\cdot}4417 \text{ m}^3$$

Specific volume of dry sat. steam at 3·5 bar is 0·5241 m³/kg therefore dryness fraction of expanded steam

$$= \frac{0{\cdot}4417}{0{\cdot}5241} = 0{\cdot}8428 \quad \text{Ans.}$$

8. From steam tables,

15 bar 300°C, $h = 3039$
2·5 bar $h_f = 535$ $h_{fg} = 2182$
0·14 bar, $h_f = 220$ $h_{fg} = 2376$

Specific enthalpy at 2·5 bar 0·97 dry

$$= 535 + 0{\cdot}97 \times 2182 = 2651 \text{ kJ/kg}$$

Specific enthalpy at 0·14 bar 0·84 dry

$$= 220 + 0{\cdot}84 \times 2376 = 2215 \text{ kJ/kg}$$

Specific enthalpy drop through H.P.

$$= 3039 - 2651 = 388 \text{ kJ/kg}$$

Specific enthalpy drop through L.P.

$$= 2651 - 2215 = 436 \text{ kJ/kg}$$

Total enthalpy drop through each unit is to be the same, let 1 kg of steam be supplied and x kg bled off, then 1 kg passes through H.P. and $(1 - x)$ kg passes through L.P.

$$1 \times 388 = (1 - x) \times 436$$
$$436x = 436 - 388$$
$$x = \frac{48}{436} = 0{\cdot}1101$$

Expressed as a percentage
Amount bled off = 11·01 % Ans.

9. Mass of air required per kg of fuel

$$= \frac{100}{23} \times \text{oxygen required}$$
$$= \frac{100}{23}\left\{2\tfrac{2}{3}\text{C} + 8\,(\text{H}_2 - \frac{\text{O}_2}{8})\right\}$$
$$= \frac{100}{23}\{2\tfrac{2}{3}\text{C} + 8\text{H}_2 - \text{O}_2\}$$
$$= \frac{100}{23}\{2\tfrac{2}{3} \times 0{\cdot}87 + 8 \times 0{\cdot}11 - 0{\cdot}02\}$$
$$= \frac{100}{23} \times 3{\cdot}18 = 13{\cdot}82 \text{ kg air/kg fuel}$$

$$pV = mRT$$
$$V = \frac{13{\cdot}82 \times 0{\cdot}287 \times 298}{1 \times 10^2}$$
$$= 11{\cdot}82 \text{ m}^3 \text{ air/kg fuel} \quad \text{Ans.}$$

10.

$$p_1V_1^n = p_2V_2^n$$
$$0{\cdot}95 \times 0{\cdot}2^n = 2{\cdot}75 \times 0{\cdot}085^n$$
$$n = 1{\cdot}243 \quad \text{Ans. (i)}$$

$$\frac{p_1V_1}{T_1} = \frac{p_2V_2}{T_2}$$

$$T_2 = \frac{290 \times 2{\cdot}75 \times 0{\cdot}085}{0{\cdot}95 \times 0{\cdot}2} = 356{\cdot}8 \text{ K}$$

Final temp. = 83·8°C Ans. (ii)

11.

Area of top and bottom = 9 × 4 × 2
Area of sides = 9 × 2·5 × 2
Area of ends = 4 × 2·5 × 2
∴ Area exposed to heat source = 72 + 45 + 20 = 137 m²

Heat passing across inner film:

$$Q = h_i \text{ At } (T_i - T_1) \quad \therefore T_i - T_1 = \frac{Q}{h_i AT}$$

Temperature rise across the three thicknesses:

$$T_1 - T_o = \frac{Q}{At}\left\{\frac{S_1}{k_1} + \frac{S_2}{k_2} + \frac{S_3}{k_3}\right\}$$

Heat passing across outer film:

$$Q = h_o At\,(T_4 - T_o) \quad \therefore T_4 - T_0 = \frac{Q}{h_o At}$$

$$T_i - T_4 = \frac{Q}{At}\left\{\frac{1}{h_i} + \frac{S_1}{k_1} + \frac{S_2}{k_2} + \frac{S_3}{k_3} + \frac{1}{h_o}\right\}$$

$$(-6 - 27) = \frac{Q}{137 \times 1}\left\{\frac{1}{1{\cdot}62} + \frac{0{\cdot}015}{0{\cdot}11} + \frac{0{\cdot}07}{0{\cdot}06} + \frac{0{\cdot}01}{45} + \frac{1}{13}\right\}$$

$$-33 \times 137 = Q\,(0{\cdot}6172 + 0{\cdot}1364 + 1{\cdot}167 + 0{\cdot}0002 + 0{\cdot}077)$$
$$Q = -2263{\cdot}3 \text{ W} \quad \text{i.e. into the room}$$
$$= 2{\cdot}2633 \text{ kW} \quad \text{cooling load} \quad \text{Ans.}$$

Note: $\frac{1}{U}$ equals the term above in brackets i.e. $U = 0{\cdot}5006$

$$Q = UAt(T_i - T_4)$$

12. From steam tables,
Steam 16 bar, $h_f = 859$ $h_{fg} = 1935$
Water 95°C, $h = 398$
Boiler steam $h_1 = 859 + 0{\cdot}98 \times 1935 = 2755$
Feed water $h_w = 398$

Heat energy transferred to water to make steam

$$= h_1 - h_w = 2755 - 398 = 2357 \text{ kJ/kg}$$
$$= 2357 \times 9000 \text{ kJ per hour}$$

Heat energy released by m kg of fuel per hour

$$= 42 \times 10^3 \times m \text{ kJ per hour}$$

$$\text{Boiler efficiency} = \frac{\text{heat energy transferred to steam}}{\text{heat energy supplied by fuel}}$$

$$0{\cdot}87 = \frac{2357 \times 9000}{42 \times 10^3 \times m}$$

$$m = \frac{2357 \times 9000}{0{\cdot}87 \times 42 \times 10^3} = 580{\cdot}6 \text{ kg/h}$$

Tonnes of fuel used per day

$$= 580{\cdot}6 \times 24 \times 10^3 = 13{\cdot}93 \text{ tonne/day} \quad \text{Ans.}$$

13. When scavenge ports are just closed, distance from cylinder cover to piston = 800 + 70 = 870 mm
Volume enclosed $= 0{\cdot}7854 \times 0{\cdot}7^2 \times 0{\cdot}87 \text{ m}^3$
Volume of air taken in (scavenge effic. being 0·95)

$$= 0{\cdot}95 \times 0{\cdot}7854 \times 0{\cdot}7^2 \times 0{\cdot}87$$
$$= 0{\cdot}3181 \text{ m}^3$$
$$pV = mRT$$
$$m = \frac{pV}{RT} = \frac{1{\cdot}21 \times 10^2 \times 0{\cdot}3181}{0{\cdot}287 \times 313}$$
$$= 0{\cdot}4285 \text{ kg} \quad \text{Ans.}$$

14. $$\text{Linear clearance [mm]} = \frac{\text{volumetric clearance [mm}^3]}{\text{area of cylinder [mm}^2]}$$

$$= \frac{73 \times 10^3}{0{\cdot}7854 \times 130^2} = 5{\cdot}5 \text{ mm}$$

Representing volumes by linear dimensions:

$$V_1 = 180 + 5{\cdot}5 = 185{\cdot}5 \text{ mm}$$
$$p_1V_1^{1{\cdot}2} = p_2V_2^{1{\cdot}2}$$
$$1 \times 185{\cdot}5^{1{\cdot}2} = 4{\cdot}6 \times V_2^{1{\cdot}2}$$
$$V_2 = \frac{185{\cdot}5}{\sqrt[1{\cdot}2]{4{\cdot}6}} = 52 \text{ mm}$$

Movement of piston during delivery period

$= 52 - 5{\cdot}5 = 46{\cdot}5$ mm Ans. (i)

as a fraction of the stroke

$= \dfrac{46{\cdot}5}{180} = 0{\cdot}2583$ Ans. (ii)

15.

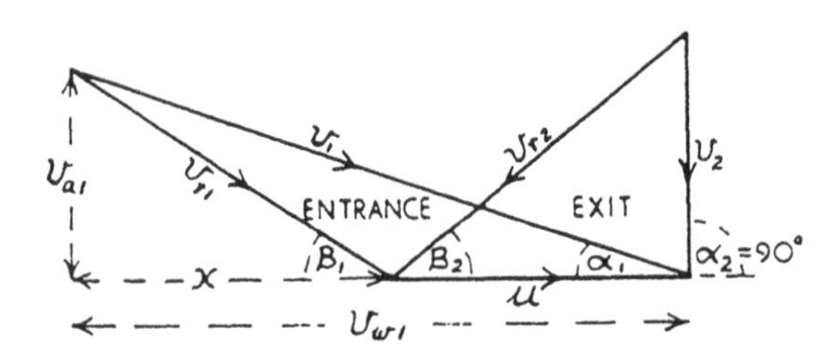

$$v_{a1} = v_1 \sin \alpha_1 = 500 \times \sin 18° = 154{\cdot}5 \text{ m/s}$$
$$v_{w1} = v_1 \cos \alpha_1 = 500 \times \cos 18° = 475{\cdot}5 \text{ m/s}$$
$$x = v_{w1} - u = 475{\cdot}5 - 230 = 245.5$$

$$\tan \beta_1 = \frac{v_{a1}}{x} = \frac{154{\cdot}5}{245{\cdot}5} = 0{\cdot}6292$$

∴ Inlet blade angle = 32° 11′ Ans. (i)

Neglecting friction across the blades

$$v_{r2} = v_{r1} = \frac{v_{a1}}{\sin \beta_1} = \frac{154{\cdot}5}{\sin 32° 11'} = 290 \text{ m/s}$$

$$\cos \beta_2 = \frac{u}{v_{r2}} = \frac{230}{290} = 0{\cdot}7930$$

∴ Outlet blade angle = 37° 32′ Ans. (ii)

16. Refer to Fig. 34 Ans. (a)

Stroke volume (s) $= 0{\cdot}7854 \times 0{\cdot}1^2 \times 0{\cdot}1$

$= 0{\cdot}0007854 \text{ m}^3$

$$\text{Clearance volume } (c) = 0{\cdot}0007854 \times 0{\cdot}05$$
$$= 0{\cdot}00003927 \text{ m}^3$$
$$V_1 = s + c \text{ i.e. } 0{\cdot}0008247$$
$$pV = mRT$$
$$m = \frac{1 \times 100 \times 0{\cdot}0008247}{0{\cdot}287 \times 298}$$
$$= 0{\cdot}0009642 \quad \text{Ans. (b) (i)}$$

The air trapped in the clearance space re-expands and reduces the intake volume on each suction stroke Ans. (b) (ii)

17. From tables for NH_3
10·34 bar, $h_f = 303{\cdot}7$
2·265 bar, $h_f = 107{\cdot}9 \quad h_g = 1425{\cdot}3 \quad v_g = 0{\cdot}5296$
$h_{fg} = 1425{\cdot}3 - 107{\cdot}9 = 1317{\cdot}4$
Ref. Fig. 69:

Throttling process through expansion valve:

$$\text{Enthalpy after } (h_4) = \text{Enthalpy before } (h_3)$$
$$107{\cdot}9 + x_4 \times 1317{\cdot}4 = 303{\cdot}7$$
$$x_4 \times 1317{\cdot}4 = 195{\cdot}8$$
$$x_4 = 0{\cdot}1486 \quad \text{Ans. (i)}$$

Enthalpy gain of refrigerant through evaporator

$$= h_1 - h_4$$
$$= (107{\cdot}9 + 0{\cdot}95 \times 1317{\cdot}4) - (107{\cdot}9 + 0{\cdot}1486 \times 1317{\cdot}4)$$
$$= (0{\cdot}95 - 0{\cdot}1486) \times 1317{\cdot}4$$
$$= 1055 \text{ kJ/kg}$$

Refrigerating effect per minute

$$= 4 \times 1055 = 4220 \text{ kJ/min} \quad \text{Ans. (ii)}$$

Specific volume of refrigerant entering compressor

$$= 0{\cdot}95 \times 0{\cdot}5296 \text{ m}^3\text{/kg}$$

Total volume entering compressor per minute

$$= 4 \times 0{\cdot}95 \times 0{\cdot}5296 = 2{\cdot}013 \text{ m}^3\text{/min} \quad \text{Ans. (iii)}$$

18.

$$\left.\begin{aligned} C + O_2 &= CO_2 \\ 2H_2 + O_2 &= 2H_2O \\ S + O_2 &= SO_2 \end{aligned}\right\} \text{Ans. (a)}$$

$$\text{Stoichiometric air} = \frac{100}{23} \times \text{oxygen required}$$

$= \frac{100}{23}(2\,2/3C + 8H_2 + S)$ kg air/kg fuel

$= \frac{100}{23}(2{\cdot}667 \times 0{\cdot}863 + 8 \times 0{\cdot}128 + 0{\cdot}009)$

$= 14{\cdot}5$ kg

Actual air is $14{\cdot}5 \times 1{\cdot}25$

$= 18{\cdot}125$

Mass of the gases/kg fuel burned = 19·125 kg Ans. (b)(i)

$$pV = mRT$$

$$V = \frac{19{\cdot}125 \times 0{\cdot}276 \times 643}{1{\cdot}5 \times 100}$$

Volume of flue gases/kg fuel burned

$= 22{\cdot}63\ m^3$ Ans. (b)(ii)

19. Heat passing across water film:

$$Q = h_i At\,(T_i - T) \quad \therefore T_i - T = \frac{Q}{h_i At}$$

Heat passing across air film:

$$Q = h_o At\,(T - T_o) \quad \therefore T - T_o = \frac{Q}{h_o At}$$

$$T_i - T_o = \frac{Q}{At}\left\{\frac{1}{h_i} + \frac{1}{h_o}\right\}$$

$$100 - 85 = \frac{Q}{\pi \times 0{\cdot}155 \times 1}\left\{\frac{1}{240} + \frac{1}{12}\right\}$$

$$Q = \frac{85 \times \pi \times 0{\cdot}155 \times 1 \times 240}{21}$$

$= 473{\cdot}1$ W Ans.

Note: area A of large diameter pipe, of small thickness, is mean circumference (per unit length)

20. From steam tables,

17 bar, $h_f = 872$ $h_{fg} = 1923$ $v_g = 0{\cdot}1167$

6 bar, $h_f = 670$ $h_{fg} = 2087$ $v_g = 0{\cdot}3156$

Throttling process:

Enthalpy after = Enthalpy before

$670 + x \times 2087 = 872 + 0{\cdot}95 \times 1923$

$x \times 2087 = 872 + 1827 - 670$

$$x \times 2087 = 2029$$
$$x = 0{\cdot}9721 \quad \text{Ans. (i)}$$

Spec. vol. of steam before throttling

$$= 0{\cdot}95 \times 0{\cdot}1167 \text{ m}^3\text{/kg}$$

Mass of steam passed through

$$= \frac{0{\cdot}8}{0{\cdot}95 \times 0{\cdot}1167} \text{ kg}$$

Spec. vol. of steam after throttling

$$= 0{\cdot}9721 \times 0{\cdot}3156 \text{ m}^3\text{/kg}$$

$$\text{Final volume of steam} = \frac{0{\cdot}8 \times 0{\cdot}9721 \times 0{\cdot}3156}{0{\cdot}95 \times 0{\cdot}1167}$$
$$= 2{\cdot}214 \text{ m}^3 \quad \text{Ans. (ii)}$$

21. Total feed in 24 hours
$= 1250 \times 10^{-3} \times 24 = 30$ tonne

$$\begin{matrix}\text{Solids in boiler} \\ \text{initially}\end{matrix} + \begin{matrix}\text{solids} \\ \text{put in}\end{matrix} = \begin{matrix}\text{solids in} \\ \text{finally}\end{matrix}$$

$$\begin{matrix}\text{water in} \\ \text{boiler}\end{matrix} \times \begin{matrix}\text{initial} \\ \text{p.p.m.}\end{matrix} + \begin{matrix}\text{amount} \\ \text{of feed}\end{matrix} \times \begin{matrix}\text{feed} \\ \text{p.p.m.}\end{matrix} = \begin{matrix}\text{water in} \\ \text{boiler}\end{matrix} \times \begin{matrix}\text{final} \\ \text{p.p.m.}\end{matrix}$$

$$6 \times 120 + 30 \times \text{feed p.p.m.} = 6 \times 1080$$
$$30 \times \text{feed p.p.m.} = 6480 - 720$$
$$\text{feed p.p.m.} = \frac{5760}{30}$$
$$= 192 \text{ p.p.m.} \quad \text{Ans.}$$

22. From steam tables
Steam 2·8 bar, $= h_f = 551 \quad h_{fg} = 2171 \quad v_g = 0{\cdot}6462$
Water 16°C, $= h = 67{\cdot}1$
Water 55°C, $= h = 230{\cdot}2$

Let x = dryness fraction
m = mass of steam [kg]

$$\begin{matrix}\text{Total enthalpy of steam} \\ \text{and water before mixing}\end{matrix} = \begin{matrix}\text{Total enthalpy of} \\ \text{water after mixing}\end{matrix}$$

$$m\,(551 + x \times 2171) + (36 \times 67{\cdot}1) = (m + 36) \times 230{\cdot}2$$
$$551m + 2171mx + 2415 = 230{\cdot}2m + 8287$$
$$320{\cdot}8m + 2171mx = 5872$$

$$320{\cdot}8 + 2171x = \frac{5872}{m} \quad \dots \dots \dots \dots \dots \quad \text{(i)}$$

Also, spec. vol. of steam = $x \times 0{\cdot}6462$ m³/kg

∴ mass of 1·5 m³ of wet steam

$$m = \frac{1{\cdot}5}{x \times 0{\cdot}6462} \text{ kg} \quad \dots \dots \dots \quad \text{(ii)}$$

Substituting for m from (ii) into (i):

$$320{\cdot}8 + 2171x = \frac{5872 \times 0{\cdot}6462x}{1{\cdot}5}$$

$$320{\cdot}8 + 2171x = 2528x$$

$$320{\cdot}8 = 357x$$

$$x = 0{\cdot}8984 \quad \text{Ans.}$$

23.

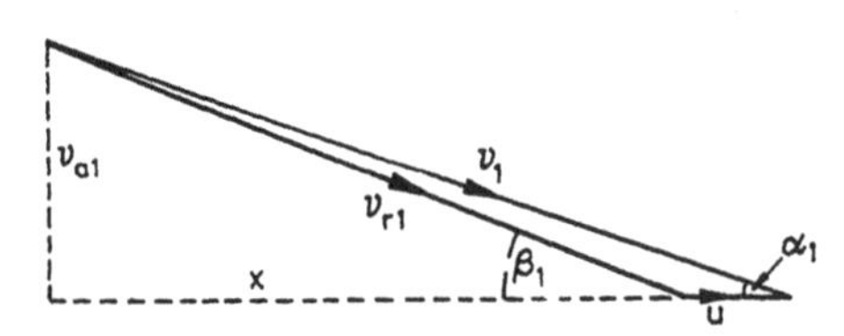

$$u = \pi dn$$

$$120 = \pi \times 0{\cdot}6 \times n$$

$$n = 63{\cdot}65 \text{ rev/s} = 3819{\cdot}0 \text{ rev/min} \quad \text{Ans. (a)}$$

$$v_1 = \sqrt{2 \times 10^3 \times \text{spec. enthalpy drop}}$$

$$= \sqrt{2 \times 10^3 \times 465}$$

$$= 974 \text{ m/s}$$

$$x = v_1 \cos \alpha_1 - u$$

$$= 974 \cos 18° - 120$$

$$= 806{\cdot}4$$

$$v_{a1} = 974 \sin 18°$$

$$= 301 \text{ m/s}$$

$$\tan \beta_1 = \frac{301}{974}$$

$$\beta_1 = 20{\cdot}5°$$

Blade inlet angle = 20·5° Ans. (b)

Axial component of steam at blade inlet = 301 m/s Ans. (c)

24. Volume of air to be heated every hour

$$= 27 \times 15 \times 3 \times 2 = 2430 \text{ m}^3$$

Mass of air to be heated every hour:

$$pV = mRT$$

$$m = \frac{pV}{RT}$$

$$= \frac{1{\cdot}013 \times 10^2 \times 2430}{0{\cdot}287 \times 291} = 2937 \text{ kg/h}$$

Heat energy supplied = mass × spec. heat × temp. rise

Energy supplied per second

$$= \frac{2937}{3600} \times 1{\cdot}005 \times (19 - 7)$$

$$= 9{\cdot}837 \text{ kJ/s}$$

$$\text{Kilowatt loading} = 9{\cdot}837 \text{ kW} \quad \text{Ans.}$$

25. For a 4-stroke single-acting engine

$$n = \text{rev/s} \div 2 = 2{\cdot}5$$

For six cylinders:

$$\text{ip} = p_m A L n \times 6$$

$$= 17 \times 10^2 \times 0{\cdot}7854 \times 0{\cdot}225^2 \times 0{\cdot}2 \times 2{\cdot}5 \times 6$$

$$= 202{\cdot}8 \text{ kW} \quad \text{Ans. (i)}$$

$$\text{bp} = \text{ip} \times \text{mech. efficiency}$$

$$= 202{\cdot}8 \times 0{\cdot}85$$

$$= 172{\cdot}36 \text{ kW} \quad \text{Ans. (ii)}$$

26. Refer to Fig. 35a Ans. (a)

$$\text{Volume of receiver} = 3 \text{ m}^3$$

$$\text{Volume of free air} = 3 \times \frac{7}{1} = 21 \text{ m}^3$$

(at the same temperature)

$$\text{Volume to compress} = 21 - 3 = 18 \text{ m}^3 \text{ at } 25^\circ\text{C}$$

$$\frac{V_1}{T_1} = \frac{V_2}{T_2}$$

$$\frac{V_1}{290} = \frac{18}{298}$$

$$V_1 = 17{\cdot}52 \text{ m}^3$$

$$\text{Volume to compress} = 17{\cdot}52 \text{ m}^3 \text{ at } 17^\circ\text{C}$$

$$\text{suction volume/s} = 0{\cdot}7854 \times 0{\cdot}1^2 \times 0{\cdot}2 \times 5$$

$$= 0{\cdot}007854 \text{ m}^3\text{/s}$$

$$\text{Time to charge receiver} = \frac{17{\cdot}52}{0{\cdot}007854 \times 60}$$

$$= 37{\cdot}18 \text{ min} \quad \text{Ans. (b)}$$

27. Refer to Figs. 69 and 70 Ans. (a)

$$h_3 = h_4 = 294{\cdot}1 \text{ kJ/kg}$$

i.e. from tables, page 12, at 24°C

$$h_4 = h_f + x_4(h_g - h_f)$$
$$294{\cdot}1 = 126{\cdot}2 + x_4\,(1430{\cdot}5 - 126{\cdot}2)$$

i.e. from tables, page 12, at 2·68 bar

$$x_4 = 0{\cdot}1287$$
$$\text{Heat extracted} = h_1 - h_4$$
$$= h_g - x_4\,h_{fg}$$
$$= 1430{\cdot}5 - 0{\cdot}1287\,(1430{\cdot}5 - 126{\cdot}2)$$
$$= 1262{\cdot}2 \text{ kJ/kg}$$

$$3{\cdot}5 = \dot{m} \times 1262{\cdot}6$$
$$\dot{m} = 0{\cdot}00308 \text{ kg/s}$$

$$\text{Mass flows rate of ammonia} = 0{\cdot}00308 \text{ kg/s} \quad \text{Ans. (b)}$$

28. From steam tables, 15 bar, $h_g = 2792$
Mass of flue gases per kg of fuel

$$= 1 \text{ kg fuel} + 24 \text{ kg air} = 25 \text{ kg}$$

Heat energy transferred from flue gases per hour

$$= \text{mass} \times \text{spec. heat} \times \text{temp. change}$$
$$= 25 \times 750 \times 1{\cdot}007 \times (822 - 690)$$
$$= 2{\cdot}492 \times 10^6 \text{ kJ/h}$$

Let h = spec. enthalpy of the superheated steam,

Heat energy transferred to steam per hour to superheat it

$$= 7000 \times (h - 2792)$$

$$\text{Heat gained by steam} = \text{Heat lost by gases}$$
$$7000(h - 2792) = 2{\cdot}492 \times 10^6$$

$$h - 2792 = \frac{2{\cdot}492 \times 10^6}{7000}$$

$$h = 356 + 2792 = 3148 \text{ kJ/kg}$$

From superheated steam tables,
h for 15 bar 350°C reads 3148
∴ Temperature of steam = 350°C Ans. (i)

From steam tables, water 21°C, $h = 88.0$

Enthalpy of 1 kg sup. steam and m kg water entering desuperheater	=	Enthalpy of $(1 + m)$ kg of dry sat. steam leaving desuperheater

$$(1 \times 3148) + m \times 88 = (1 + m) \times 2792$$
$$3148 + 88m = 2792 + 2792\,m$$
$$356 = 2704\,m$$
$$m = 0{\cdot}1316 \text{ kg} \quad \text{Ans. (ii)}$$

29. $$\text{Ratio of expansion} = \frac{\text{final volume}}{\text{initial volume}}$$

$$\therefore \text{ final volume} = 6 \times 9900 = 59400 \text{ cm}^3$$

or, 59·4 litre, or, 0·0594 m³ Ans. (i)

$$p_1 V_1^{1{\cdot}33} = p_2 V_2^{1{\cdot}33}$$

where V_1 and V_2 may be represented by 1 and 6 respectively.

$$12 \times 1^{1{\cdot}33} = p_2 \times 6^{1{\cdot}33}$$

$$p_2 = \frac{12}{6^{1{\cdot}33}} = 1{\cdot}108 \text{ bar} \quad \text{Ans. (ii)}$$

$$\frac{p_1 V_1}{T_1} = \frac{P_2 V_2}{T_2}$$

$$\frac{12 \times 1}{489} = \frac{1{\cdot}108 \times 6}{T_2}$$

$$T_2 = \frac{489 \times 1{\cdot}1018 \times 6}{12} = 270{\cdot}9 \text{ K}$$

$$= -2{\cdot}1°\text{C} \quad \text{Ans. (iii)}$$

30.
$$p_1 v_1^n = p_2 v_2^n$$
$$40 \times 0{\cdot}08^{1{\cdot}3} = 2 \times v_2^{1{\cdot}3}$$
$$v_2 = 0{\cdot}8013$$
$$0{\cdot}8856x = 0{\cdot}8013$$
$$x = 0{\cdot}9048$$

$$\text{work transfer} = \frac{p_1 v_1 - p_2 v_2}{n-1}$$

$$= \frac{40 \times 10^2 \times 0{\cdot}08 - 2 \times 10^2 \times 0{\cdot}8013}{0{\cdot}3}$$

$$= 532{\cdot}5 \text{ kJ} \quad \text{Ans. (a)}$$

$$u_1 + q_{in} = u_2 + w_{out}$$

At 40 bar, 0·08 m³/kg, steam is 450°C (tables)

$$3010 + q_{in} = (505 + 0{\cdot}9048 \times 2025) + 532{\cdot}5$$

$$q_{in} = -3010 + 2337{\cdot}2 + 532{\cdot}5$$

$$\text{Heat transfer} = -140{\cdot}3 \text{ kJ i.e. loss} \quad \text{Ans. (b)}$$

Note: this is non-flow work, u not h is used.

$$u_1 = h - pv_g$$
$$= 3330 - 40 \times 10^2 \times 0{\cdot}08$$
$$= 2337{\cdot}2 \text{ (as above)}$$

Similarly for u_2, by calculation

31. Referring to Fig. 57:

$$v_{a1} = v_1 \sin \alpha_1 = 840 \times \sin 20° = 287{\cdot}4 \text{ m/s}$$
$$v_{w1} = v_1 \cos \alpha_1 = 840 \times \cos 20° = 789{\cdot}4 \text{ m/s}$$
$$x = v_{w1} - u = 789{\cdot}4 - 350 = 439{\cdot}4 \text{ m/s}$$

$$\tan \beta_1 = \frac{V_{a1}}{x} = \frac{287{\cdot}4}{439{\cdot}4} = 0{\cdot}6539$$

$\therefore$ Blade inlet angle = 33°11′ Ans. (i)

$$v_{r1} = \frac{v_{a1}}{\sin \beta_1} = \frac{287{\cdot}4}{\sin 33°\ 11'} = 525 \text{ m/s}$$

$$v_{r2} = 0{\cdot}8 \times 525 = 420 \text{ m/s}$$
$$v_{a2} = v_{r2} \times \sin \beta_2 = 420 \times \sin 25°\ 12' = 178{\cdot}8 \text{ m/s}$$
$$v_{w2} + u = v_{r2} \times \cos \beta_2 = 420 \times \cos 25°\ 12' = 380 \text{ m/s}$$
$$v_{w2} = 380 - 350 = 30 \text{ m/s}$$

$$\tan \varphi = \frac{v_{w2}}{v_{a2}} = \frac{30}{178{\cdot}8} = 0{\cdot}1678$$

$$\varphi = 9°\ 32'$$

$$v_2 = \frac{v_{a2}}{\cos \varphi} = \frac{178{\cdot}8}{\cos 9°\ 32'} = 181{\cdot}3 \text{ m/s}$$

Absolute velocity of steam at exit:

Magnitude = 181·3 m/s
Direction to axis = 9° 32′
or, direction to plane of wheel = 80 28′ } Ans. (ii)

32. Available hydrogen $= H_2 - \frac{O_2}{8}$

$= 0{\cdot}114 - \frac{0{\cdot}024}{8} = 0{\cdot}111$ kg

$$\text{c.v.} = 33{\cdot}7\,C + 144\left[H_2 - \frac{O_2}{8}\right] + 9{\cdot}3\,S$$

$$= 33{\cdot}7 \times 0{\cdot}846 + 144 \times 0{\cdot}111 + 9{\cdot}3 \times 0{\cdot}004$$
$$= 28{\cdot}51 + 15{\cdot}99 + 0{\cdot}0372$$
$$= 44{\cdot}5372 \text{ MJ/kg} \quad \text{Ans. (i)}$$

$$\text{Air required} = \frac{100}{23} \times \text{oxygen required}$$

$$= \frac{100}{23}\left\{2\tfrac{2}{3}\,C + 8\left[H_2 - \frac{O_2}{8}\right] + S\right\}$$

$$= \frac{100}{23}\{2\tfrac{2}{3} \times 0{\cdot}846 + 8 \times 0{\cdot}111 + 0{\cdot}004\}$$

$$= \frac{100}{23} \times 3{\cdot}148$$

$$= 13{\cdot}69 \text{ kg air/kg fuel} \quad \text{Ans. (ii)}$$

33. From tables, Freon-12, 1·826 bar,

$h_f = 22{\cdot}33 \qquad h_g = 180{\cdot}97$
$h_{fg} = 180{\cdot}97 - 22{\cdot}33 = 158{\cdot}64$

Sat. Temp. at 1·826 bar = – 15°C therefore freon at evaporator outlet at 0°C is superheated by 15°.

1·826 bar 15° superheat, $h = 190{\cdot}15$

Referring to Fig. 69:

Throttling effect through expansion valve between condenser outlet and evaporator inlet:

$$\text{Enthalpy after throttling } (h_4) = \text{Enthalpy before } (h_3)$$
$$22{\cdot}33 + x_4 \times 158{\cdot}64 = 50$$
$$x_4 \times 158{\cdot}64 = 27{\cdot}67$$
$$x_4 = 0{\cdot}1744 \quad \text{Ans. (i)}$$

$$\text{Refrig. effect/kg} = h_1 - h_4 \text{ (note } h_4 = h_3)$$
$$= 190{\cdot}15 - 50$$
$$= 140{\cdot}15 \text{ kJ/kg}$$

Refrigerating effect per minute

$$= 0{\cdot}4 \times 60 \times 140{\cdot}15$$
$$= 3364 \text{ kJ/min} \quad \text{Ans. (ii)}$$

34.
$$\frac{p_1V_1}{T_1} = \frac{p_2V_2}{T_2}$$

$$\frac{1{\cdot}01325 \times 18{\cdot}2}{288} = \frac{0{\cdot}965 \times V_2}{300}$$

$$V_2 = 19{\cdot}91 \text{ m}^3/\text{min}$$
$$p_2V_2^n = p_3V_3^n$$
$$0{\cdot}965 \times 19{\cdot}91^{1{\cdot}32} = 4{\cdot}82 \times V_3^{1{\cdot}32}$$
$$V_3 = 5{\cdot}891 \text{ m}^3/\text{min}$$
$$\text{Work/cycle} = \frac{n}{n-1}(p_3V_3 - p_2V_2)$$
$$\text{Work/s} = \frac{1{\cdot}32 \times 10^2}{0{\cdot}32}\left(\frac{4{\cdot}82 \times 5{\cdot}891}{60 \times 2} - \frac{0{\cdot}965 \times 19{\cdot}91}{60 \times 2}\right)$$
$$= 31{\cdot}56 \text{ kW}$$

$$\text{Input power} = \frac{31{\cdot}56}{0{\cdot}9}$$
$$= 35{\cdot}06 \text{ kW} \quad \text{Ans.}$$

35. For one mm movement of the stylus (one mm on height of card) indicator piston deflection = ⅙ mm and this would be under a force of 60 ÷ 6 = 10 N in the indicator cylinder.

force [N] = pressure [N/m²] × area [m²]

10 = pressure [N/m²] × 0·7854 × $7^2 \times 10^{-4}$

∴ pressure scale on card per mm of height

$$= \frac{10}{0{\cdot}7854 \times 7^2 \times 10^{-6}}$$

$$= 2{\cdot}598 \times 10^5 \text{ N/m}^2 = 2{\cdot}598 \text{ bar}$$

$$\text{Mean height of diagram [mm]} = \frac{\text{area [mm}^2]}{\text{length [mm]}}$$
$$= \frac{346}{75} \text{ mm}$$
$$\therefore \text{Mean effective press.} = \frac{346}{75} \times 2{\cdot}598$$
$$= 11{\cdot}99 \text{ bar} \quad \text{Ans. (i)}$$

$$\text{ip} = p_m ALn$$
where n = 2 × rev/s for a double-acting 2-stroke
$$\text{ip} = 11{\cdot}99 \times 10^2 \times 0{\cdot}7854 \times 0{\cdot}6^2 \times 0{\cdot}9 \times 2{\cdot}1 \times 2$$
$$= 1281 \text{ kW} \quad \text{Ans. (ii)}$$

36. $$\text{Cross–sect. area } [\text{m}^2] = \frac{\text{volume flow } [\text{m}^3/\text{s}]}{\text{velocity } [\text{m/s}]}$$

$$= \frac{\text{mass flow } [\text{kg/s}] \times \text{spec. vol. } [\text{m}^3/\text{kg}]}{\text{velocity } [\text{m/s}]}$$

$$\text{Diameter} = \sqrt{\frac{\text{area}}{0{\cdot}7854}}$$

$$\text{Entrance diameter} = \sqrt{\frac{0{\cdot}315 \times 0{\cdot}2765}{457 \times 0{\cdot}7854}}$$

$$= 0{\cdot}01558 \text{ m} = 15{\cdot}58 \text{ mm} \quad \text{Ans. (ii)}$$

37. $$pV = mRT$$

$$m = \frac{pV}{RT} = \frac{1{\cdot}1 \times 10^2 \times 0{\cdot}2}{0{\cdot}287 \times 288} = 0{\cdot}2662 \text{ kg}$$

Heat energy supplied [kJ]

$$= \text{mass [kg]} \times \text{spec. heat [kJ/kg K]} \times \text{temp. rise [K]}$$
$$= 0{\cdot}2662 \times 1{\cdot}005 \times (423 - 288)$$
$$= 36{\cdot}1 \text{ kJ} \quad \text{Ans. (i)}$$

$$p_1 V_1^{1{\cdot}32} = p_2 V_2^{1{\cdot}32}$$
$$1{\cdot}1 \times 0{\cdot}2^{1{\cdot}32} = 7{\cdot}15 \times V_2^{1{\cdot}32}$$

$$V_2 = 0{\cdot}2 \times \sqrt[1{\cdot}32]{\frac{1{\cdot}1}{7{\cdot}15}}$$

$$= 0{\cdot}04844 \text{ m}^3$$

$$\frac{p_1 V_1}{T_1} = \frac{p_2 V_2}{T_2}$$

$$\frac{1{\cdot}1 \times 0{\cdot}2}{288} = \frac{7{\cdot}15 \times 0{\cdot}04844}{T_2}$$

$$T_2 = \frac{288 \times 7{\cdot}15 \times 0{\cdot}04844}{1{\cdot}1 \times 0{\cdot}2} = 453{\cdot}4 \text{ K}$$

$$= 180{\cdot}4 ^\circ\text{C} \quad \text{Ans. (ii)}$$

38. For the two thicknesses:

$$T_1 - T_3 = \frac{Q}{At}\left\{\frac{S_1}{k_1} + \frac{S_2}{k_2}\right\}$$

$$-4-21 = \frac{Q}{6\times3\times1}\left\{\frac{0{\cdot}12}{1{\cdot}15}+\frac{0{\cdot}08}{0{\cdot}043}\right\}$$

$$Q = \frac{-25\times18}{0{\cdot}1043+1{\cdot}860}$$

$$= -229{\cdot}09 \text{ W i.e. } \textit{into} \text{ the store}$$

$$Q = 229{\cdot}09\times3600\times24 \text{ J/day}$$

$$= 19{\cdot}793 \text{ MJ/day} \quad \text{Ans. (a)}$$

$$T_1-T_2 = \frac{-229{\cdot}09}{6\times3\times1}\left\{\frac{0{\cdot}08}{0{\cdot}043}\right\}$$

$$-4-T_2 = -23{\cdot}67$$

$$T_2 = 19{\cdot}67^\circ\text{C} \quad \text{Ans. (b)}$$

39. Refer to Fig. 33 Ans. (a)

$$\text{Clearance volume} = 0{\cdot}07\times381$$

$$= 26{\cdot}67 \text{ mm}$$

$$V_1 = 381+26{\cdot}67$$

$$= 407{\cdot}67 \text{ mm}$$

$$V_2 = 407{\cdot}67-267$$

$$= 140{\cdot}67 \text{ mm}$$

$$p_1V_1^n = p_2V_2^n$$

$$1{\cdot}013\times407{\cdot}67^n = 4\times0{\cdot}1407^n$$

$$2{\cdot}898^n = 3{\cdot}949$$

$$n = 1{\cdot}291 \quad \text{Ans. (b)}$$

40. Stoichiometric air required per kg of fuel A

$$= \frac{100}{23}\{2\tfrac{2}{3}\times0{\cdot}885+8\times0{\cdot}115\}$$

$$= \frac{100}{23}\times3{\cdot}28 \text{ kg}$$

Stoichiometric air required per kg of fuel B

$$= \frac{100}{23}\{2\tfrac{2}{3}\,C+8\,H_2\}$$

and this is 6% more than for fuel A, therefore,

$$\frac{100}{23}\{2\tfrac{2}{3}C+8H_2\} = 1{\cdot}06\times\frac{100}{23}\times3{\cdot}28$$

Cancelling $\frac{100}{23}$ and multiplying throughout by $\tfrac{3}{8}$:

$$C + 3H_2 = 1{\cdot}304 \quad \ldots \quad (i)$$

Also, fractional analysis of fuel B:

$$C + H_2 = 1$$
$$\therefore C = 1 - H_2 \quad \ldots \quad (ii)$$

Substituting value of C from (ii) into (i):

$$C + 3H_2 = 1{\cdot}304$$
$$1 - H_2 + 3H_2 = 1{\cdot}304$$
$$2H_2 = 0{\cdot}304$$
$$H_2 = 0{\cdot}152$$
$$\text{and, } C = 1 - 0{\cdot}152 = 0{\cdot}848$$

∴ Mass analysis of fuel B =

84·8% carbon, 15·2% hydrogen Ans.

41. Before alteration:

$$V_1 = 90 + 15 = 105 \qquad V_2 = 15$$
$$p_1 V_1^{1\cdot33} = p_2 V_2^{1\cdot33}$$
$$1 \times 105^{1\cdot33} = p_2 \times 15^{1\cdot33}$$
$$p_2 = \left\{\frac{105}{15}\right\}^{1\cdot33} = 13{\cdot}3 \text{ bar} \quad \text{Ans. (i)}$$

After alteration:

$$V_1 = 90 + 12{\cdot}5 = 102{\cdot}5 \qquad V_2 = 12{\cdot}5$$
$$1 \times 102{\cdot}5^{1\cdot33} = p_2 \times 12{\cdot}5^{1\cdot33}$$
$$p_2 = \left\{\frac{102{\cdot}5}{12{\cdot}5}\right\}^{1\cdot33} = 16{\cdot}43 \text{ bar} \quad \text{Ans. (ii)}$$

42.

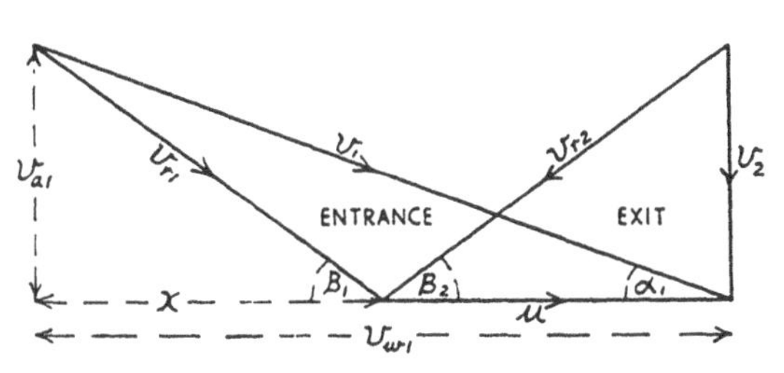

Kinetic energy = ½mv^2

Kinetic energy being in joules when m is the mass in kg and v is the velocity in m/s, hence,

$$250 \times 10^3 = \tfrac{1}{2} \times 1 \times v^2$$
$$v = \sqrt{2 \times 250 \times 10^3} = 707{\cdot}1 \text{ m/s}$$

With no friction, $v_{r2} = v_{r1}$
and, since $\beta_2 = \beta_1$
then $x = u = \frac{1}{2}v_{w1}$

$$\frac{v_{a1}}{x} = \tan\beta_1 \qquad \therefore \frac{v_1 \sin\alpha_1}{\frac{1}{2} v_1 \cos\alpha_1} = \tan 35°$$

v_1 cancels, $\sin\alpha_1 \div \cos\alpha_1 = \tan\alpha_1$, therefore

$$2\tan\alpha_1 = \tan 35°$$

$$\tan\alpha_1 = \frac{0{\cdot}7002}{2} = 0{\cdot}3501$$

$\therefore$ Nozzle angle $\alpha_1 = 19°\ 18'$ Ans. (i)

Blade velocity $u = \frac{1}{2}v_{w1} = \frac{1}{2}v_1 \cos\alpha_1$

$= \frac{1}{2} \times 707{\cdot}1 \times \cos 19°\ 18'$

$= 333{\cdot}7$ m/s Ans. (ii)

43. Let d = diameter of cylinder
then $1{\cdot}25d$ = stroke

For 4 cylinders:

$$\text{ip} = p_m ALn \times 4$$

$$600 = 12{\cdot}56 \times 10^2 \times 0{\cdot}7854d^2 \times 1{\cdot}25d \times 4{\cdot}5 \times 4$$

$$d = \sqrt[3]{\frac{600}{1256 \times 0{\cdot}7854 \times 1{\cdot}25 \times 4{\cdot}5 \times 4}}$$

$= 0{\cdot}3$ m $= 300$ mm

Cylinder diameter = 300 mm
Stroke = $1{\cdot}25 \times 300$ = 375 mm
Ans. (a)

Indicated thermal efficiency

$$= \frac{3{\cdot}6\ [\text{MJ/kW h}]}{\text{kg fuel/ind. kW h} \times \text{c.v. [MJ/kg]}}$$

$$= \frac{3{\cdot}6}{0{\cdot}225 \times 42}$$

$= 0{\cdot}381$ or $38{\cdot}1\%$ Ans. (b)

44. From NH_3 tables,

1·902 bar, $h_g = 1420$

7·529 bar, $h_f = 256$, sat. temp. = 16°C

$\therefore$ at 66°C vapour is superheated 50°

7·529 bar 50° superheat, $h = 1591{\cdot}7$

Referring to Fig. 69

Enthalpy gain per kg through evaporator

$$= h_1 - h_4 \text{ (note } h_4 = h_3\text{)}$$
$$= 1420 - 256 = 1164 \text{ kJ/kg}$$

Refrigerating effect [kJ/min]

$$= \text{mass flow [kg/min]} \times \text{enthalpy gain [kJ/kg]}$$
$$\therefore\ m = \frac{800}{1164} = 0{\cdot}6873 \text{ kg/min} \quad \text{Ans.}$$

Enthalpy drop per kg through condenser

$$= h_2 - h_3$$
$$= 1591{\cdot}7 - 256 = 1335{\cdot}7 \text{ kJ/kg}$$

Heat rejected in condenser

$$= 0{\cdot}6873 \times 1335{\cdot}7 = 918{\cdot}1 \text{ kJ/min} \quad \text{Ans. (ii)}$$

Enthalpy gain per kg in compressor

$$= h_2 - h_1$$
$$= 1591{\cdot}7 - 1420 = 171{\cdot}7 \text{ kJ/kg}$$

Energy given to refrigerant in compressor

$$= 0{\cdot}6873 \times 171{\cdot}7 \text{ kJ/min}$$
$$\text{Power [kW} = \text{kJ/s]}$$
$$= \frac{0{\cdot}6873 \times 171{\cdot}7}{60}$$
$$= 1{\cdot}967 \text{ kW} \quad \text{Ans. (iii)}$$

45. From saturated steam tables:

Steam 30 bar, $h_g = 2803$ $v_g = 0{\cdot}06665$
Water 38°C, $h = 159{\cdot}1$

From superheated steam tables

Steam 30 bar 400°C, $h = 3231$ $v = 0{\cdot}0993$

Let m[kg] = mass of injection water per kg of superheated steam.

Mixing in desuperheater:

$$\text{Enthalpy before mixing} = \text{Enthalpy after}$$
$$1 \times 3231 + m \times 159{\cdot}1 = (1 + m) \times 2803$$
$$3231 + 159{\cdot}1\,m = 2803 + 2803m$$
$$428 = 2643{\cdot}9m$$
$$m = 0{\cdot}1618 \text{ kg} \quad \text{Ans. (i)}$$

Volume of one kg superheated steam = $0{\cdot}0993$ m^3

Volume of $(1 + m)$ kg of dry saturated steam

$= 1{\cdot}1618 \times 0{\cdot}06665 = 0{\cdot}07743 \text{ m}^3$

Percentage change in volume

$$= \frac{0{\cdot}0993 - 0{\cdot}07743}{0{\cdot}0993} \times 100$$

$= 22{\cdot}03\%$ Ans. (ii)

46. Let V_1 = volume of gas in cylinder at beginning of stroke, this is stroke volume + clearance volume.

V_2 = volume of gas in cylinder at end of stroke, this is the clearance volume.

$$p_1 V_1^{1{\cdot}32} = p_2 V_2^{1{\cdot}32}$$

$$1 \times V_1^{1{\cdot}32} = 37 \times 850^{1{\cdot}32}$$

$$V_1 = 850 \times \sqrt[1{\cdot}32]{37} = 13\,100 \text{cm}^3$$

Stroke volume $= 13100 - 850 = 12250 \text{ cm}^3$

$= 12{\cdot}25$ litre or $0{\cdot}01225 \text{ m}^3$ Ans. (i)

$$\frac{p_1 V_1}{T_1} = \frac{p_2 V_2}{T_2}$$

$$\frac{1 \times 13100}{308} = \frac{37 \times 850}{T_2}$$

$$T_2 = \frac{308 \times 37 \times 850}{13100} = 739{\cdot}4 \text{ K}$$

$= 466{\cdot}4°\text{C}$ Ans.

47. Let x be kg of C/kg fuel

$1 - x$ is kg of H_2/kg fuel

Oxygen required for C $= 2{\cdot}667x$

Oxygen required for H_2 $= 8(1 - x)$

$$\text{Stoichiometric air} = \frac{100}{23}(2{\cdot}667x + 8 - 8x)$$

Exhaust gas includes 1 kg of fuel burned

$$15{\cdot}6 - 1 = \frac{100}{23}(2{\cdot}667x + 8 - 8x)$$

$$3{\cdot}358 - 8 = -5{\cdot}333x$$

$$x = 0{\cdot}8704$$

$$1 - x = 0{\cdot}1296$$

Mass of carbon $= 0{\cdot}8704$ kg
Mass of hydrogen $= 0{\cdot}1296$ kg
Ans.

48.

$$\text{Heat extracted} = 27{\cdot}2(4{\cdot}186 \times 14{\cdot}4 + 332{\cdot}6)\ \text{kJ/h}$$
$$= \frac{10686{\cdot}3 \times 10^3}{3600}\ \text{J/s}$$
$$= 2968{\cdot}4$$
$$\text{c.o.p.} = \frac{\text{heat extracted by refrigerant}}{\text{work done on refrigerant}}$$
$$\text{Work done on refrigerant} = \frac{2968{\cdot}4}{7{\cdot}51}$$
$$= 395\ \text{W}$$

Power input to machine = 395W Ans. (a)

For a heat engine; heat is taken in and heat is given out (at a lower temperature) and overall there is a net work output. Theoretically the cycle can be operated in reverse (reversibility), i.e. heat taken in at a lower temperature and given out at a high temperature but it does require a net work input (heat pump or refrigerator) Ans.

49.

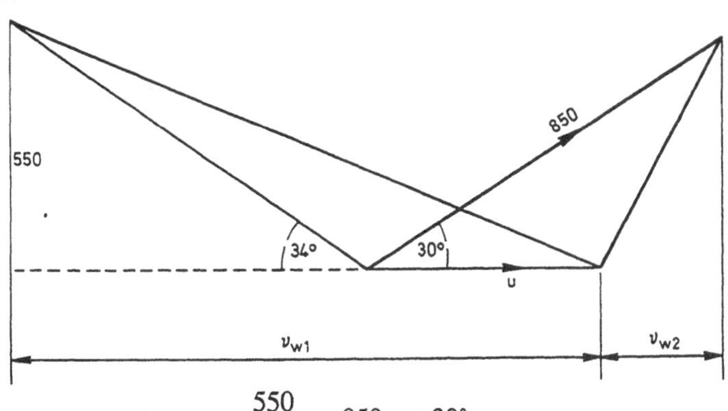

$$v_{w1} + v_{w2} = \frac{550}{\tan 34^\circ} + 850 \cos 30^\circ$$
$$= 815{\cdot}4 + 736{\cdot}1$$
$$v_w = 1551{\cdot}5\ \text{m/s}$$
$$\text{Power} = \text{Force} \times \text{velocity}$$
$$= \dot{m} v_w u$$
$$1000 \times 67 = 0{\cdot}0833 \times 1551{\cdot}5 \times \pi \times 1{\cdot}32 \times n$$
$$n = 124{\cdot}9\ \text{rev/s}$$

Turbine rotational speed = 124·9 rev/s Ans.

50.

$$\text{Area of top} = 1 \times 0{\cdot}75$$
$$\text{Area of sides} = 1 \times 1{\cdot}6 \times 2$$
$$\text{Area of ends} = 1{\cdot}6 \times 0{\cdot}75 \times 2$$
$$\text{Area exposed to heat source} = 0{\cdot}75 + 3{\cdot}2 + 2{\cdot}4 = 6{\cdot}35 \text{ m}^2$$

Temperature fall across the lagging:

$$T_i - T = \frac{QS}{Atk}$$

Heat passing across air film:

$$Q = h_o At\,(T - T_o) \quad \therefore\ T - T_o = \frac{Q}{h_o At}$$

$$T_i - T_o = \frac{Q}{AT}\left\{\frac{S}{0{\cdot}048} + \frac{1}{1}\right\}$$

$$66 - 18 = \frac{200}{6{\cdot}35 \times 1}\left\{\frac{S}{0{\cdot}048} + 1\right\}$$

$$S = 0{\cdot}048\left\{\frac{48 \times 6{\cdot}35}{200} - 1\right\}$$

$$= 0{\cdot}02515 \text{ m}$$
$$= 25{\cdot}15 \text{ mm} \quad \text{Ans. (a)}$$

$$S' = 0{\cdot}048\left\{\frac{48 \times 6{\cdot}35}{100} - 1\right\}$$

$$= 0{\cdot}0983 \text{ m}$$
$$= 98{\cdot}3 \text{ mm}$$
$$S' - S = 73{\cdot}15 \text{ mm} \quad \text{Ans. (b)}$$

A SELECTION OF EXAMINATION QUESTIONS
CLASS ONE

1. A simple open-cycle gas turbine plant operates at a pressure ratio of 5:1 and at an air/fuel ratio of 50:1. The power output of the plant is 5 MW with maximum and minimum temperatures in the cycle of 925 K and 300 K respectively. The isentropic efficiencies of the compressor and turbine are 0·82 and 0·86 respectively. Calculate:
(a) the temperature of the exhaust gases leaving the turbine;
(b) the mass of fuel used per second

For air: $c_p = 1005$ J/kg K, $\frac{c_p}{c_v} = 1{\cdot}4$

For the turbine gases: $c_p = 1150$ J/kg K, $\frac{c_p}{c_v} = 1{\cdot}3$

2. A vessel of volume 15 m^3 contains air and dry saturated steam at a total pressure of 0·09 bar and temperature 39°C. Taking R for air = 0·287 kJ/kg K calculate the masses of steam and air in the vessel, and give the mass ratio of air to steam.

3. An air compressor is to compress 0·4 kg/s of air from 100 kN/m^2 and 20°C to 1500 kN/m^2 and 40°C. The air inlet velocity is negligible and the exit velocity is 100m/s. The cooling water for the compressor has a mass flow rate of 0·1 kg/s and an inlet temperature of 20°C. Power input to the compressor is 13 kW. Calculate the cooling water outlet temperature.
Note: c_p air = 1·005 kJ/kg K, c_p water = 4·18 kJ/kg K

4. The mass analysis of the fuel burned in a boiler is 87% carbon, 11% hydrogen, and 2% oxygen, and the fuel is burned at the rate of 1·8 tonne/h. Calculate the mass flow rate [kg/s] of each of the constituents of the flue gases if combustion is stoichiometric. Express the composition of the flue gases as percentages by mass.
Atomic weights: hydrogen 1, carbon 12, nitrogen 14, oxygen 16. Mass composition of air: 23% oxygen, 77% nitrogen.

5. Dry saturated steam enters a convergent-divergent nozzle at 9 bar, and the pressures at the throat and exit are 5 bar and 0·14 bar respectively. The specific enthalpy drop of the steam from entrance to throat is 107 kJ/kg, and from entrance to exit it is 633

kJ/kg. Assuming 8% of the enthalpy drop is lost to friction in the divergent part of the nozzle, calculate the areas in mm^2 of the nozzle at the throat and exit to pass 23 kg of steam per minute.

6. In a Freon-12 refrigerating plant the compressor takes the refrigerant in at 0·8071 bar and discharges it at 12·19 bar and 65°C. At condenser outlet the Freon is saturated liquid at 12·19 bar. If compression is isentropic and the flow of the refrigerant is 15 kg/min, calculate the refrigerating effect and the coefficient of performance.

7(a) A quantity of a perfect gas is compressed at constant temperature from initial pressure p_1 to final pressure p_2. Show that the area under the p–V curve for this process can be written as:

$$RT \ln\left(\frac{p_1}{p_2}\right)$$

Note: Area under pV curve is given by $\int p dV$.

(b) An isothermal compression from 1 bar to 8 bar takes place on a perfect gas. The initial volume of the gas is 0·25 m^3. Calculate the heat energy transfer during the process.

8. The compressor of an open cycle gas turbine unit receives air at 1 bar 18°C and delivers it at 4 bar 200°C to the combustion chamber where the temperature is raised at constant pressure to 650°C. The products of combustion pass through the turbine, which has an isentropic efficiency of 0·85, and exhaust at 1 bar. The power required by the compressor is provided by the turbine. Calculate:
(a) the isentropic efficiency of the compressor;
(b) the net work output of the turbine per kilogram of air.

Note: For air $c_p = 1{\cdot}005$ kJ/kg K, $\gamma = 1{\cdot}4$
For combustion gases $c_p = 1{\cdot}15$ kJ/kg K, $\gamma = 1{\cdot}333$

9. In a steam turbine plant steam expands in the high pressure turbine from 50 bar 500°C to 6 bar with an isentropic efficiency of 0·9. The steam is then reheated at constant pressure to 500°C before entering the low pressure turbine. In the low pressure turbine steam is expanded to 0·05 bar with an isentropic efficiency of 0·85. Neglecting feed pump work,
(a) sketch the expansion and reheat processes on a temperature–entropy diagram;
(b) calculate using chart and tables the thermal efficiency of the plant.

10. The walls of a cold room consist of an outer layer of wood of thickness 30 mm and thermal conductivity 0·18 W/m K, and a cork lining of thickness 70 mm and thermal conductivity 0·05 W/m K. If the surface heat transfer coefficient from and to each exposed surface is 10 W/m^2 K and the heat flow through the wall is 24 W/m^2, calculate, (i) the temperature differences across the thicknesses of the wood and cork, (ii) the total temperature difference between the outside atmosphere and inside of room, and (iii) the temperature of the room when the external ambient temperature is 20°C.

11. A surface type feed heater is supplied with steam at 2 bar and 0·95 dry. The temperature of the feed water entering and leaving the heater are 55°C and 105°C respectively. The mass flow rate of feed water through the heater is 20 000 kg/hour and the overall heat transfer coefficient is 4540 W/m^2 K. Calculate, stating any assumptions made:

(a) the mass flow rate of heating steam in kg/hour;

(b) the effective heating surface area of the feed heater.

For the feed water: Specific heat capacity = 4·18 kJ/kg K.

Note: Logarithmic mean temperature difference

$$\theta_m = \frac{\theta_1 - \theta_2}{\ln\left(\dfrac{\theta_1}{\theta_2}\right)}$$

where θ_1 = temperature difference between hot and cold fluid in inlet.

θ_2 = temperature difference between hot and cold fluid at outlet.

12. In a test on a single-cylinder, two-stroke, diesel engine, the mean effective pressure was 8·9 bar, running speed 2·3 rev/s, brake load 8 kN, brake radius 1·25 m, specific fuel consumption 0·251 kg/ kWh (brake). The diameter of the cylinder is 360 mm, stroke 780 mm, calorific value of the fuel 41·5 MJ/kg. Calculate (i) the indicated power, (ii) brake power, (iii) indicated thermal efficiency, and (iv) the total heat energy loss per second.

13. The volumetric analysis of a mixture of gases shows it to contain 80% hydrogen and 20% oxygen. A vessel holds 0·7 m^3 of this mixture at 38°C and 3·5 bar.

Calculate the masses of hydrogen and oxygen in the vessel.

Universal Gas Constant = 8·3143 kJ/mol K.

Atomic mass relationships: hydrogen = 1, oxygen = 16

14. In a single-stage impulse turbine the steam enters the nozzle at 7 bar 300°C and is discharged at 1·2 bar dry saturated, directed at 20° to the plane of rotation. The blade velocity is 40% of the steam jet velocity and the relative velocity of the steam at exit is 80% of the relative velocity at entrance. The outlet angle of the blades is 35° and the steam flow 0·5 kg/s. Calculate (i) the axial thrust, (ii) the power developed.

15. Air expands isentropically through a convergent nozzle from 6 bar 260°C to 4 bar. The velocity of the air at nozzle inlet is 900 m/s and the nozzle cross-sectional area at exit is 0·025 m^2. Calculate:
(a) the air velocity at nozzle exit;
(b) the mass flow rate of air;
(c) the nozzle cross-sectional area at inlet.

For air: c_p = 1005 J/kg K , $\frac{c_p}{c_v}$ = 1·4

16. The pressure, volume and temperature of a gas mixture sample in a closed vessel is 1·01 bar, 500 cm^3 and 20°C, and is composed of 14% carbon dioxide and 86% nitrogen by volume. Taking R for CO_2 = 0·189 kJ/kg K, calculate (i) the partial pressure of each gas, (ii) the mass of carbon dioxide in the sample.

17. A perfect gas is compressed, polytropically from 1 bar, 22°C, 0·037 m^3 to 35 bar, 420°C. Determine:
(a) the index of compression;
(b) the work done;
(c) the change of internal energy;
(d) the heat transfer.
For the gas c_v = 718 J/kg K, R = 282 J/kg K.

18. In a Freon-12 refrigeration plant the refrigerant leaves the condenser as a liquid at 25°C. The refrigerant leaves the evaporator as dry saturated vapour at –15°C and leaves the compressor at 6·516 bar and 40°C. The cooling load is 73·3 kW. Calculate:
(a) the mass flow rate of refrigerant;
(b) the compressor power;
(c) the coefficient of performance of the plant.

19. In a compression ignition engine working on the ideal dual-combustion cycle, the volumetric compression ratio is 12·5:1.

The cycle consists of (a) adiabatic compression from 1·013 bar, 35°C, (b) heat received at constant volume to a maximum pressure of 40 bar, (c) heat received at constant pressure to a maximum temperature of 1425°C, (d) adiabatic expansion to the initial volume, (e) heat rejected at constant volume. Make a sketch of the pV diagram and calculate the mean effective pressure. Take $\gamma = 1{\cdot}4$ for air and products of combustion.

20. A rigid vessel contains a mixture of superheated steam and air at a total pressure of 0·02 bar and at 30°C. The steam/air mass ratio is 20:1. Assume the superheated steam has the properties of a perfect gas. Calculate:
(a) the partial pressures of steam and air;
(b) the density of the mixture
Note: For air $R = 287$ J/kg K. For superheated steam $R = 462$ J/kg K.

21. The mass analysis of a fuel burned in a boiler is 85·5% carbon, 13·5% hydrogen, and 1% oxygen, and the air supply is 25% in excess of the minimum required for stoichiometric combustion. Calculate (i) the percentage mass analysis of the wet flue gases, (ii) the percentage volumetric analysis of the dry flue gases. Take mass composition of air = 23% oxygen, 77% nitrogen. Atomic weights: hydrogen 1, carbon 12, nitrogen 14, oxygen 16.

22. The wall of a cold room is composed of two materials, an inner material of thickness 100 mm, having a thermal conductivity of 0·115 W/m K and a layer of cork of thermal conductivity 0·06 W/m K. The external air temperature is 24°C and the room air temperature is –23°C. The surface heat transfer coefficient of exposed surfaces is 12 W/m^2 K. The heat transfer through the wall is 30 W/m^2. Calculate:
(a) the temperature of the exposed surfaces;
(b) the temperature of the interface;
(c) the thickness of the cork.

23. In a refrigerating plant using Freon as the refrigerating agent, the refrigerant leaves the condenser as saturated liquid at 15°C, leaves the evaporator and enters the compressor at 1·509 bar and –5°C, and is delivered from the compressor into the condenser at 4·914 bar 45°C. Calculate the coefficient of performance.

24. The gravimetric analysis of a liquid fuel is: carbon 82%,

hydrogen 18%. Assume stoichiometric combustion. Calculate:
(a) the air/fuel ratio by mass;
(b) the volumetric analysis of the wet products of combustion.
Air contains 23% oxygen by mass.
Atomic mass relationships: oxygen = 16, nitrogen = 14, carbon = 12, hydrogen = 1.

25. Steam expands in a turbine from 25 bar 320°C to 0·04 bar with an isentropic efficiency of 0·73. The power output of the turbine is 3 MW. The system is to be modified by fitting a new boiler, generating steam at 60 bar 370°C, which supplies a new higher pressure turbine exhausting to the original turbine at 25 bar. The high pressure turbine has an isentropic efficiency of 0·76. Between the turbines the steam is reheated to 320°C at constant pressure. Using tables and chart as required calculate:
(a) the enthalpy change of the steam during reheating;
(b) the condition of the steam at condenser inlet;
(c) the percentage reduction in steam flow due to the plant modification when total power output is unchanged.

26. The power absorbed by a single-acting, single-stage reciprocating air compressor is 13·58 kW when the mean piston speed is 2·8 m/s and rotational speed 3·5 rev/s. The air is compressed from 1·0 bar and delivered at 10 bar, the index of the law of compression being 1·32. Neglecting clearance, calculate (i) the stroke of the compressor piston, (ii) the cylinder diameter, and (iii) the mean effective pressure.

27. In an engine working on the ideal diesel cycle, the temperature of the air at the beginning of compression is 37°C, compression takes place according to the law $pV^{1\cdot4}$ = a constant, and the volumetric compression ratio is 13:1. At the end of compression the air receives heat energy at constant pressure and is then expanded to the original volume. If 1 kg of fuel is burned per 35 kg of air compressed, the calorific value of the fuel being 42 MJ/kg, calculate (i) the temperature at the end of compression, (ii) the temperature at the end of heat reception, (iii) the volumetric expansion ratio. Take c_p = 1·02 kJ/kg K

28. At a certain stage of a reaction turbine, the steam leaves the guide blades at a velocity of 135 m/s, the exit angle being 20°. The linear velocity of the moving blades is 87 m/s. Assuming the channel section of fixed and moving blades to be identical, and

assuming ideal conditions, calculate (i) the entrance angle of the moving blades, (ii) the stage power per kg/s steam flow.

29. In an open cycle gas turbine plant, a heat exchanger is included to heat the air before entering the combustion chamber by the exhaust gases from the engine. The gases enter the heat exchanger at 300°C and 140 m/s and leave at 240°C and 10 m/s. The air enters the exchanger at 200°C and the air/fuel ratio is 84. Calculate the temperature of the air at the exchanger exit, taking c_P as 1·1 kJ/kg K for the gases and 1·005 kJ/kg K for air.

30. A CO_2 refrigerating machine produces 250 kg of ice per hour at –10°C from water at 15°C. The refrigerant enters the evaporator 0·2 dry and leaves 0·95 dry. The compressor is single-acting, runs at 4·15 rev/s, and the stroke/bore ratio is 2:1. Calculate (i) the mass flow of the refrigerant through the circuit, and (ii) the diameter and stroke of the compressor piston. Take the following values:

Specific heat of ice = 2·04 kJ/kg K
Latent heat of fusion = 335 kJ/kg
Specific heat of water = 4·2 kJ/kg K
CO_2 vapour at evaporator pressure,
h_{fg} = 290·2 kJ/kg, v_g = 0·02168 m^3/kg

31. Air is taken into a single-stage air compressor at 1 bar and delivered at 5 bar. The piston swept volume is 1440 cm^3 and the clearance volume is 40 cm^3. Taking the index of compression and expansion as 1·3, calculate (i) the fraction of the stroke when the delivery valves open, (ii) the fraction of the stroke when the suction valves open, (iii) the mean indicated pressure.

32. Steam is supplied to a turbine at 30 bar 350°C and the condenser pressure is 0·045 bar. The power developed is 5 MW when the steam consumption is 22·5 Mg/h. Calculate (i) the ideal efficiency of the Rankine cycle, (ii) the actual efficiency of the engine, (iii) the efficiency ratio.

33. A sample of steam at 10 bar is tested by a combined separating and throttling calorimeter, the data obtained were:

Mass of water collected in separator = 0·113 kg
Mass of condensed water after throttling = 3·03 kg
Pressure of steam in throttling calorimeter = 1·2 bar
Temperature of steam in throttling calorimeter = 109·8°C

Take specific heat of superheated steam at calorimetric pressure as 2·02 kJ/kg K and calculate the dryness fraction of the sample.

34. The piston swept volume of an engine working on the ideal dual combustion cycle is 0·1068 m^3 and the clearance volume is 8900 cm^3. At the beginning of compression the pressure is 1 bar and temperature 42°C. The maximum pressure in the cycle is 45 bar and maximum temperature 1500°C. Taking $\gamma = 1{\cdot}4$ and $c_V = 0{\cdot}715$ kJ/kg K for air and products of combustion, calculate the proportion of heat received at constant volume to that received at constant pressure.

35. A fuel has an analysis by mass of 85% C, 11% H_2 and 3% O_2. This fuel is burnt in a combustion chamber with an air/fuel ratio by mass of 11:1. For every kg of fuel burnt calculate:
(a) the mass of carbon burnt to produce carbon monoxide;
(b) the mass of carbon burnt to produce carbon dioxide.
Atomic mass relationships are: C = 12, O = 16, H = 1. Air contains 23% O_2 by mass.

36. At one stage of a steam turbine the inlet velocity of the steam to the rotating blades is 590 m/s and the relative velocity at outlet is 620 m/s. The blade speed is 320 m/s and the inlet and outlet angles are 37° and 26° respectively. Calculate the force on the blades and the power developed at this stage for a steam flow of 0·075 kg/s.

37. A single-acting air compressor takes in air at 1 bar and delivers it at 4 bar, the cylinder is 300 mm diameter, stroke 450 mm, and it runs at 5 rev/s. Initially the index of compression was 1·15 and after running for some time the index of compression was found to be 1·35. Neglecting clearance, calculate the power absorbed in each case and the percentage increase in power.

38. A steam pipe 140 mm outside diameter and 23 m long is lagged with insulating material of thermal conductivity 0·13 W/m K. Steam passes along the pipe at the rate of 1200 kg/h, entering at 18 bar dry saturated and leaving at the same pressure 0·985 dry. The outside surface temperature of the lagging is 35°C and the inside surface may be taken as equal to the steam temperature. Calculate the thickness of the lagging taking the rate of heat transfer through the insulating material, in J/s per unit length of pipe, as:

$$\frac{2\pi k\,(T_1 - T_2)}{\ln\,(r_2/r_1)}$$

39. Steam is supplied to a turbine at 20 bar 400°C and exhausts at 0·04 bar and 0·85 dry. At the stage in the turbine where the pressure is 1·4 bar, 13·4% of the steam is withdrawn and passed to the feed heater and this heats the feed water to the saturation temperature of the tapped off steam. Compare the thermal efficiencies with and without feed heating.

40. Air is compressed in a cylinder according to the law pV^n = a constant. The initial condition of the air is 0·125 m³, 1·01 bar and 19°C, and the final condition is 36 bar and 508°C. Taking c_p = 1·005 kJ/kg K and c_V = 0·718 kJ/kg K, calculate (i) the index of compression, (ii) the mass of air compressed, (iii) the work done during compression, (iv) the change of internal energy, (v) the transfer of heat to or from the air.

41. The following data were taken during a test on a four-cylinder, four-stroke, compression ignition engine of cylinder diameter 320 mm and stroke 480 mm while running at 4 rev/s. Mean effective pressure 14·9 bar, brake load 12 kN on a radius of 960 mm, fuel consumption 99 kg/h, calorific value of fuel 44·5 MJ/kg, mass flow of engine cooling water 154 kg/min, water inlet and outlet temperatures 14°C and 47°C. Calculate the indicated and brake thermal efficiencies, percentage of heat carried away in the cooling water, and draw up a heat balance. Specific heat of cooling water = 4·2 kJ/kg K.

42. In a simple gas turbine working on the ideal cycle, the pressure ratio of both the compressor and turbine is 4·3:1. The temperatures at inlet to the compressor and at inlet to the turbine are 16°C and 600°C respectively. Calculate (i) the temperature at the outlet from the compressor, (ii) the temperature at the outlet from the turbine, (iii) the heat supplied per kilogramme of working fluid, (iv) the thermal efficiency. Take γ = 1·4 and c_p = 1·005 kJ/kg K.

43. A simple Freon-12 refrigerator operates with evaporator at 1·509 bar –20°C and condenser at 8·477 bar 35°C. The compression between these two states is isentropic and the refrigerant leaving the compressor is dry saturated vapour. Calculate:

(a) using thermodynamic tables:
   (i) the refrigerating effect per kg of refrigerant;
   (ii) the coefficient of performance.

(b) the coefficient of performance of a reversed Carnot cycle operating between the same temperatures.

44. A two pass oil cooler consists of 350 tubes and is required to cool 4 kg/s of oil from 50°C to 20°C. The overall heat transfer coefficient of the tubes is 70 W/m² K, cooling water temperature 15°C. Determine:
(a) Logarithmic mean temperature difference of the oil in the cooler.
(b) Surface area of the tubes.
(c) Length of the thin tubes if their diameter is 19 mm.

$$\text{Mean temperature difference } \theta_m = \frac{T_1 - T_2}{\ln\left(\frac{T_1 - T_C}{T_2 - T_C}\right)}$$

where T_1 = oil inlet temperature
T_2 = oil outlet temperature
T_C = cooling water temperature.

The specific heat capacity of oil is 1395·6 J/kg K

45. Dry saturated steam at 7 bar is throttled to 0·5 bar and then passed through two oil heaters in series. The steam leaves the first heater 0·98 dry and is throttled to 0·16 bar before entering the second heater. If there is no pressure drop in the heaters determine using the enthalpy/entropy chart:
(a) final steam condition if there has been no overall change in entropy;
(b) mass flow of steam if 0·72 kg/s of oil, specific heat capacity 2·1 kJ/kg K is raised in temperature through 72°C.

46. A marine boiler installation is fired with methane (CH_4). For stoichiometric combustion calculate:
(a) the correct air to fuel mass ratio;
(b) the percentage composition of the dry flue gases by volume.

Atomic mass relationships: hydrogen 1, oxygen 16, carbon 12, nitrogen 14. Air contains 23% oxygen and 77% nitrogen by mass.

47. Gas enters a rotary compressor at 15°C with a velocity of 75 m/s and a specific enthalpy of 80 kJ/kg. It is discharged at 200°C with a velocity of 175 m/s and a specific enthalpy of 300 kJ/kg. If the compressor loses 10 kJ/kg of gas flowing through the compressor, calculate:
(a) the external work transfer per kilogram of gas;
(b) the datum temperature on which the specific enthalpies given are based.

48. Saturated steam at 50 bar is supplied through a pipe which has two layers of insulation. Outside diameter of the pipe is 200 mm, inner layer of insulation 100 mm thick thermal conductivity 0·05 W/m K, outer layer 50 mm thick thermal conductivity 0·15 W/m K which has a surface heat transfer coefficient of 8 W/m^2 K. If the ambient temperature is 20°C, calculate the condensation rate per metre of pipe length.

49. Steam at 65 bar and 500°C is supplied to a three stage turbine. It leaves the first stage at 15 bar, 330°C then passes through a reheater, which it leaves at 500°C with specific enthalpy of 3475 kJ/kg. After leaving the second stage at 330°C and 3·5 bar the steam passes through the third stage and exhausts at 0·05 bar, 0·94 dry.

Determine using the enthalpy/entropy chart:

(a) the pressure drop in the reheater;
(b) the isentropic efficiency of each stage;
(c) the ratio of powers developed in each stage.

50. The following successive processes are performed on one kilogram of air from a complete cycle.

(i) Isentropic compression from 120°C and 1 bar to 10 bar.
(ii) Heating at constant volume to 800°C.
(iii) Isentropic expansion.
(iv) Heat rejection at constant pressure.

Sketch the cycle on *p–V* and *T–s* axes and determine (a) cycle efficiency, (b) mean effective pressure.

SOLUTIONS TO EXAMINATION QUESTIONS
CLASS ONE

1. Refer to Fig. 67

$$\frac{T_3}{T_4} = \left(\frac{p_3}{p_4}\right)^{\frac{\gamma-1}{\gamma}}$$

$$\frac{925}{T_4} = 5^{\frac{0\cdot4}{1\cdot4}}$$

$$T_4 = 637{\cdot}8 \text{ K}$$

$$\frac{T_3 - T_4{}^1}{T_3 - T_4} = 0{\cdot}86$$

$$925 - T_4{}^1 = (925 - 637{\cdot}8) \times 0{\cdot}86$$

$$T_4{}^1 = 678 \text{ K}$$

Temperature of exhaust gases leaving turbine = 403°C Ans. (a)

$$\frac{T_2}{T_1} = \left(\frac{p_2}{p_1}\right)^{\frac{\gamma-1}{\gamma}}$$

$$T_2 = 300 \times 5^{\frac{0\cdot4}{1\cdot4}}$$

$$= 475 \text{ K}$$

$$\frac{T_2 - T_1}{T_2{}^1 - T_1} = 0{\cdot}82$$

$$475 - 300 = (T_2{}^1 - 300) \times 0{\cdot}82$$

$$T_2{}^1 = 513{\cdot}7 \text{ K}$$

$$\text{Net work} = \dot{m}_g\, c_{pg}\,(925 - 678) - \dot{m}_a\, c_{pa}\,(513{\cdot}7 - 300)$$

$$5000 = \frac{51}{50}\dot{m}_a \times 1{\cdot}150 \times 247 - \dot{m}_a \times 1{\cdot}005 \times 213{\cdot}7$$

$$100 = \dot{m}_f\,(289{\cdot}73 - 214{\cdot}8) \text{ i.e. } \dot{m}_f = \frac{\dot{m}_a}{50}$$

$$\dot{m}_f = 1{\cdot}334$$

$$\text{Fuel mass} = 1{\cdot}334 \text{ kg/s} \quad \text{Ans. (b)}$$

2. From steam tables, for a saturation temperature of 39°C, pressure is 0·07 bar, and spec. volume 20·53 m^3/kg. Therefore mass of steam in volume of 15 m^3

$$= \frac{15}{20{\cdot}53} = 0{\cdot}7307 \text{ kg} \quad \text{Ans. (i)}$$

Partial pressure due to air

$$= \text{total pressure} - \text{steam pressure}$$
$$= 0{\cdot}09 - 0{\cdot}07 = 0{\cdot}02 \text{ bar} = 2 \text{ kN/m}^2$$
$$pV = mRT$$
$$\text{mass of air } m = \frac{pV}{RT}$$
$$= \frac{2 \times 15}{0{\cdot}287 \times 312} = 0{\cdot}335 \text{ kg} \quad \text{Ans. (ii)}$$

Ratio of air to steam

$$= 0{\cdot}335 : 0{\cdot}7307$$
$$= \frac{0{\cdot}335}{0{\cdot}7307} : \frac{0{\cdot}7307}{0{\cdot}7307}$$
$$= 0{\cdot}4584 : 1 \quad \text{Ans. (iii)}$$

3. $h_1 + ½\, c_1{}^2 + q = h_2 + ½\, c_2{}^2 + w$

is the Steady Flow Energy Equation.

$$H_1 + ½\, mc_1{}^2 + Q = H_2 + ½\, mc_2{}^2 + W$$

$$Q = H_2 - H_1 + ½\text{m}\,(c_2{}^2 - c_1{}^2) + W$$
$$= mc_p\,(T_2 - T_1) + ½\text{m}\,(c_2{}^2 - c_1{}^2) + W$$
$$= 0{\cdot}4\,[1{\cdot}005 \times 10^3(40 - 20) + 0{\cdot}5(100^2 - 0^2)] - 13000$$
$$= 0{\cdot}4\,(20100 + 5000) - 13000$$
$$= 10040 - 13000$$
$$Q = -2960\ W$$

Q negative (heat out) and W negative (work on air)

$$Q = mc_p\,(T_o - T_i)$$
$$2960 = 0{\cdot}1 \times 10^3 \times 4{\cdot}18\,(T_o - 20)$$
$$T_o = 27{\cdot}08°\text{C}$$

Cooling water outlet temperature = 27·08°C Ans.

Note: the pressures given are not required

4. Stoichiometric oxygen required per kg of fuel

$$= 2⅔\text{C} + 8\left\{\text{H}_2 - \frac{\text{O}_2}{8}\right\}$$
$$= 2⅔\text{C} + 8\text{H}_2 - \text{O}_2$$
$$= 2⅔ \times 0{\cdot}87 + 8 \times 0{\cdot}11 - 0{\cdot}02$$
$$= 3{\cdot}18 \text{ kg}$$

$$\text{Stoichiometric air} = \frac{100}{23} \times 3{\cdot}18 = 13{\cdot}82 \text{ kg}$$

Mass of nitrogen in 13·82 kg of air

$$= 0{\cdot}77 \times 13{\cdot}82 = 10{\cdot}64 \text{ kg}$$

$$(\text{or, } 13{\cdot}82 \text{ kg air} - 3{\cdot}18 \text{ kg } O_2 = 10{\cdot}64 \text{ kg } N_2)$$

$$CO_2 \text{ formed} = 3\tfrac{2}{3} \times 0{\cdot}87 = 3{\cdot}19 \text{ kg}$$

$$H_2O \text{ formed} = 9 \times 0{\cdot}11 = 0{\cdot}99 \text{ kg}$$

Total gases/kg fuel

$$= 10{\cdot}64 + 3{\cdot}19 + 0{\cdot}99 = 14{\cdot}82 \text{ kg}$$

$$\text{also, } 13{\cdot}82 \text{ kg air} + 1 \text{ kg fuel} = 14{\cdot}82 \text{ kg}$$

At 1·8 tonne of fuel per hour, fuel rate

$$= \frac{1{\cdot}8 \times 10^3}{3600} = 0{\cdot}5 \text{ kg/s}$$

Mass flow rate of each of the gases, Ans. (i):

$$\text{Nitrogen} = 0{\cdot}5 \times 10{\cdot}64 = 5{\cdot}32 \text{ kg/s}$$

$$CO_2 = 0{\cdot}5 \times 3{\cdot}19 = 1{\cdot}595 \text{ kg/s}$$

$$H_2O = 0{\cdot}5 \times 0{\cdot}99 = 0{\cdot}495 \text{ kg/s}$$

As a percentage analysis, Ans. (ii):

$$\text{Nitrogen} = \frac{10{\cdot}64}{14{\cdot}82} \times 100 = 71{\cdot}79\%$$

$$CO_2 = \frac{3{\cdot}19}{14{\cdot}82} \times 100 = 21{\cdot}53\%$$

$$H_2O = \frac{0{\cdot}99}{14{\cdot}82} \times 100 = 6{\cdot}68\%$$

5. From steam tables,

9 bar, $h_g = 2774$

5 bar, $h_f = 640$ $h_{fg} = 2109$ $v_g = 0{\cdot}3748$

0·14 bar, $h_f = 220$ $h_{fg} = 2376$ $v_g = 10{\cdot}69$

Enthalpy drop from 9 bar to 5 bar =

$$2774 - (640 + x \times 2109) = 107$$

$$2774 - 640 - 107 = x \times 2109$$

$$2027 = x \times 2109$$

$$\text{dryness at throat, } x = 0{\cdot}9614$$

Specific volume of steam at throat

$$= 0{\cdot}9614 \times 0{\cdot}3748 = 0{\cdot}3603 \text{ m}^3/\text{kg}$$

$$\text{Velocity [m/s]} = \sqrt{2 \times \text{spec. enthalpy drop [J/kg]}}$$

Velocity through throat

$$= \sqrt{2 \times 107 \times 10^3} = 462{\cdot}6 \text{ m/s}$$

area [m²] × velocity [m/s] = mass flow [kg/s] × spec. vol. [m³/kg]

∴ Area at throat [mm²]

$$= \frac{23 \times 0{\cdot}3603}{60 \times 462{\cdot}6} \times 10^6 = 298{\cdot}5 \text{ mm}^2 \quad \text{Ans. (i)}$$

Effective enthalpy drop from entrance to exit

$$= 0{\cdot}92 \times 633 = 582{\cdot}4 \text{ kJ/kg}$$

Enthalpy drop from 9 bar to 0·14 bar =

$$2774 - (220 + x \times 2376) = 582{\cdot}4$$
$$2774 - 220 - 582{\cdot}4 = x \times 2376$$
$$1971{\cdot}6 = x \times 2376$$
$$\text{dryness at exit, } x = 0{\cdot}8299$$

Specific volume of steam at exit

$$= 0{\cdot}8299 \times 10{\cdot}69 = 8{\cdot}869 \text{ m}^3/\text{kg}$$

Velocity at exit

$$= \sqrt{2 \times 582{\cdot}4 \times 10^3} = 1079 \text{ m/s}$$

Area at exit [mm²]

$$= \frac{\text{mass flow [kg/s]} \times \text{spec. vol. [m}^3\text{/kg]}}{\text{velocity [m/s]}} \times 10^6$$

$$= \frac{23 \times 8{\cdot}869}{60 \times 1079} \times 10^6 = 3150 \text{ mm}^2 \quad \text{Ans. (ii)}$$

6. From Freon-12 tables,

0·8071 bar, $h_f = 4{\cdot}42 \qquad h_g = 171{\cdot}9$

$\therefore h_{fg} = 171{\cdot}9 - 4{\cdot}42 = 167{\cdot}48$

$s_f = 0{\cdot}0187 \qquad s_g = 0{\cdot}7219$

$\therefore s_{fg} = 0{\cdot}7219 - 0{\cdot}0187 = 0{\cdot}7032$

12·19 bar, $h_f = 84{\cdot}94$, sat. temp. = 50°C

∴ at 65°C Freon is superheated by 15°

$$h = 218{\cdot}64 \qquad s = 0{\cdot}7166$$

Ref. Fig. 69,
Isentropic compression: $s_1 = s_2$

$$0{\cdot}0187 + x_1 \times 0{\cdot}7032 = 0{\cdot}7166$$

$$x_1 \times 0{\cdot}7032 = 0{\cdot}6979$$
$$x_1 = 0{\cdot}9924$$

$h_1 = h$ leaving evaporator $= h$ entering compressor
$= 4{\cdot}42 + 0{\cdot}9924 \times 167{\cdot}48$
$= 4{\cdot}42 + 166{\cdot}3 = 170{\cdot}72$

$h_4 = h$ entering evaporator $= h$ leaving condenser (h_3)
Refrigerating effect/kg $= h_1 - h_4$
$= 170{\cdot}72 - 84{\cdot}94 = 85{\cdot}78$ kJ/kg

Refrigerating effect for flow of 15 kg/min
$= 15 \times 85{\cdot}78 = 1286$ kJ/min Ans. (i)
Work transfer in compressor/kg $= h_2 - h_1$
$= 218{\cdot}64 - 170{\cdot}72 = 47{\cdot}92$ kJ/kg

$$\text{Coeff. of performance} = \frac{\text{refrigerating effect}}{\text{work transfer}}$$

$$= \frac{85{\cdot}78}{47{\cdot}92} = 1{\cdot}79 \quad \text{Ans. (ii)}$$

7. Area = Work done $= \int p\,\mathrm{d}V$
Constant temperature (isothermal) $pV = C$

$$p = \frac{C}{V}$$

$$\text{Area} = C\int_{V_1}^{V_2} \frac{\mathrm{d}V}{V}$$

$$= C \ln\left(\frac{V_2}{V_1}\right)$$

$$= C \ln\left(\frac{p_1}{p_2}\right) \quad \text{as } p_1V_1 = p_2V_2$$

$$= pV \ln\left(\frac{p_1}{p_2}\right) \quad \text{as } pV = C$$

$$= RT \ln\left(\frac{p_1}{p_2}\right) \quad \text{as } pV = RT$$

$\therefore$ Area under p–V curve $= RT \ln\left(\frac{p_1}{p_2}\right)$ Ans. (a)

Isothermal process, no change of internal energy

Heat extracted = Work done on the gas

$$\text{Work done} = \int p\,dV$$

$$= p_1 V_1 \ln\left(\frac{p_1}{p_2}\right)$$

$$= 1 \times 100 \times 0{\cdot}25 \ln\left(\frac{1}{8}\right)$$

$$= -51{\cdot}99 \text{ kJ}$$

Negative result confirming work done on the gas

Heat energy transfer = –51·99 kJ Ans. (b)

8. Refer to Figs. 65 and 67:

$$\frac{T_2}{T_1} = \left(\frac{p_2}{p_1}\right)^{\frac{\gamma-1}{\gamma}}$$

$$T_2 = 291 \times 4^{\frac{0{\cdot}333}{1{\cdot}333}}$$

$$= 432{\cdot}4 \text{ K}$$

$$\frac{T_3}{T_4} = \left(\frac{p_3}{p_4}\right)^{\frac{\gamma-1}{\gamma}}$$

$$\frac{923}{T_4} = 4^{\frac{0{\cdot}4}{1{\cdot}4}}$$

$$T_4 = 652{\cdot}8 \text{ K}$$

$$\eta_c = \frac{T_2 - T_1}{T_2' - T_1} \times 100$$

$$= \frac{432{\cdot}4 - 291}{473 - 291} \times 100$$

Compressor isentropic efficiency = 77·69% Ans. (a)

$$\eta_T = \frac{T_3 - T_4'}{T_3 - T_4}$$

$$0{\cdot}85 = \frac{923 - T_4'}{923 - 652{\cdot}8}$$

$$T_4' = 693{\cdot}3 \text{ K}$$

$$\text{Turbine work} = c_P\,(T_3 - T_4')$$
$$= 1{\cdot}15\,(923 - 693{\cdot}3)$$
$$= 264{\cdot}2 \text{ kJ/kg}$$

$$\text{Compressor work} = c_P\,(T_2' - T_1)$$
$$= 1{\cdot}005\,(473 - 291)$$
$$= 182{\cdot}9 \text{ kJ/kg}$$

Net turbine work $= 81{\cdot}3$ kJ/kg Ans. (b)

9.

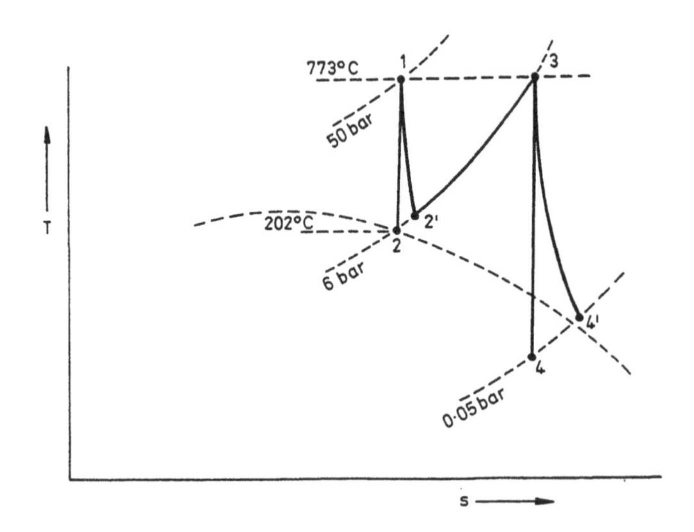

See the T–s diagram Ans. (a)

$s_1 = 6{\cdot}975$ $h_1 = 3433$ (tables or $h - s$ chart)
$s_2 = 6{\cdot}975$ $h_2 = 2854$ (202°C, just superheated)

$$\Delta h = 3433 - 2854$$
$$= 579$$
$$0{\cdot}9\,\Delta h = 521$$
$$h_2' = 2912 \quad \text{i.e., } 3433 - 521$$

$s_3 = 8{\cdot}001$ $h_3 = 3483$ (tables or $h - s$ chart)
$s_4 = 8{\cdot}001$ (0·05 bar, steam wet)

$$8{\cdot}001 = 0{\cdot}476 + 7{\cdot}918\,x$$
$$x = 0{\cdot}9504 \text{ (dryness fraction)}$$

$$h_4 = 138 + 0{\cdot}9504 \times 2423$$
$$= 2441 \qquad \text{(or from } h\text{–}s \text{ chart directly)}$$

$$\Delta h = 3483 - 2441$$
$$= 1042$$
$$0{\cdot}85\,\Delta h = 885{\cdot}7$$
$$h_4' = 2598{\cdot}7 \text{ i.e., } 3483 - 885{\cdot}7$$

$$\text{Heat supplied in boiler} = h_{500} - h_{f\,0{\cdot}05}$$
$$= 3433 - 138 = 3295$$
$$\text{Heat supplied in reheat} = h_3 - h_2'$$
$$= 3483 - 2912 = 571$$
$$\text{Work output} = 521 + 885{\cdot}7 = 1406{\cdot}7$$
$$\text{Thermal efficiency} = \frac{1406{\cdot}7}{3866} \times 100 = 36{\cdot}39\% \quad \text{Ans. (b)}$$

Note: The h–s chart is more used; the T–s sketch was specified here.

10. Temperature difference across thickness of wood

$$= \frac{QS_W}{k_W A t}$$

$$= \frac{24 \times 0{\cdot}03}{0{\cdot}18 \times 1 \times 1} = 4\text{ K} \quad \text{Ans. (i) (a)}$$

Temperature difference across thickness of cork

$$= \frac{QS_C}{k_C A t}$$

$$= \frac{24 \times 0{\cdot}07}{0{\cdot}05 \times 1 \times 1} = 33{\cdot}6\text{ K} \quad \text{Ans. (i) (b)}$$

Temperature difference between inside and outside atmospheres and their respective exposed surfaces

$$= \frac{Q}{hAt}$$

$$= \frac{24}{10 \times 1 \times 1} = 2{\cdot}4\text{ K}$$

Total temperature difference between outside atmosphere and inside atmosphere

$$= 2{\cdot}4 + 4 + 33{\cdot}6 + 2{\cdot}4 = 42{\cdot}4\text{ K or } 42{\cdot}4^\circ\text{C} \quad \text{Ans. (ii)}$$

$$\text{Room temperature} = 20 - 42{\cdot}4 = -22{\cdot}4^\circ\text{C} \quad \text{Ans. (iii)}$$

11.
$$\text{Heat lost by steam} = \text{Heat gained by feed water}$$
$$m_s \times x \times h_{fg} = m_w \times c_p \times \text{temp. rise}$$
$$m_s = \frac{20\,000 \times 4{\cdot}18\,(105 - 55)}{3600 \times 0{\cdot}95 \times 2202}$$
$$= 0{\cdot}555 \text{ kg/s}$$
$$= 1998 \text{ kg/h} \quad \text{Ans. (a)}$$

Assumption: no undercooling of condensed steam

$$\theta_m = \frac{(120{\cdot}2 - 55) - (120{\cdot}2 - 105)}{\ln \dfrac{(120{\cdot}2 - 55)}{(120{\cdot}2 - 105)}}$$
$$= \frac{50}{1{\cdot}456}$$
$$= 34{\cdot}34°\text{C}$$
$$Q = m_w \times c_p \times \text{temp. rise}$$
$$= \frac{20000 \times 4{\cdot}18 \times 50}{3600}$$
$$= 1161{\cdot}1 \text{ kW}$$
$$Q = UAt\,\theta_m$$
$$1161100 = 4540 \times A \times 1 \times 34{\cdot}34$$
$$A = 7{\cdot}448 \text{ m}^2 \quad \text{Ans. (b)}$$

12.
$$\text{ip} = p_m ALn$$
$$= 8{\cdot}9 \times 10^2 \times 0{\cdot}7854 \times 0{\cdot}36^2 \times 0{\cdot}78 \times 2{\cdot}3$$
$$= 162{\cdot}5 \text{ kW} \quad \text{Ans. (i)}$$
$$\text{bp} = T\omega$$
$$= 8 \times 1{\cdot}25 \times 2\pi \times 2{\cdot}3$$
$$= 144{\cdot}5 \text{ kW} \quad \text{Ans. (ii)}$$

Heat energy supplied per second

$$= \frac{0{\cdot}251 \times 144{\cdot}5 \times 41{\cdot}5 \times 10^3}{3600} = 418{\cdot}1 \text{ kJ/s}$$

Heat energy converted into work in cylinder per second

$$= 162{\cdot}5 \text{ kJ/s}$$

Indicated thermal efficiency

$$= \frac{162{\cdot}5}{418{\cdot}1} = 0{\cdot}3887 \text{ or } 38{\cdot}87\% \quad \text{Ans. (iii)}$$

Total heat energy loss per second

$$= \text{heat supplied} - \text{heat converted into work}$$
$$= 418{\cdot}1 - 162{\cdot}5 = 255{\cdot}6 \text{ kJ/s} \quad \text{Ans. (iv)}$$

13.

$$\text{Hydrogen } R_1 = \frac{8{\cdot}3143}{2} = 4{\cdot}1572$$

$$\text{Oxygen } R_2 = \frac{8{\cdot}3143}{32} = 0{\cdot}2598$$

Ratio of partial pressures = Ratio of volumes

Hydrogen, partial pressure = $0{\cdot}8 \times 3{\cdot}5 = 2{\cdot}8$ bar
Oxygen, partial pressure = $0{\cdot}2 \times 3{\cdot}5 = 0{\cdot}7$ bar

$$pV = mRT$$

$$\text{Mass of hydrogen} = \frac{2{\cdot}8 \times 100 \times 0{\cdot}7}{4{\cdot}1572 \times 311}$$
$$= 0{\cdot}1516 \text{ kg} \quad \text{Ans.}$$

$$\text{Mass of oxygen} = \frac{0{\cdot}7 \times 100 \times 0{\cdot}7}{0{\cdot}2598 \times 311}$$
$$= 0{\cdot}6064 \text{ kg} \quad \text{Ans.}$$

14. From steam tables:

1·2 bar, $h_g = 2683$
7 bar 300°C, $h = 3060$

Enthalpy drop through nozzle

$$= 3060 - 2683 = 377 \text{ kJ/kg}$$

Velocity of steam at nozzle exit [m/s]

$$= \sqrt{2 \times \text{spec. enthalpy drop [J/kg]}}$$
$$= \sqrt{2 \times 377 \times 10^3} = 868{\cdot}4 \text{ m/s}$$

Ref. Fig. 58:

$$v_{w1} = v_1 \cos \alpha_1 = 868{\cdot}4 \times \cos 20° = 816 \text{ m/s}$$
$$u = 40\% \text{ of } v_1 = 0{\cdot}4 \times 868{\cdot}4 = 347{\cdot}4 \text{ m/s}$$
$$x = v_{w1} - u = 816 - 347{\cdot}4 = 468{\cdot}6 \text{ m/s}$$
$$v_{a1} = v_1 \sin \alpha_1 = 868{\cdot}4 \times \sin 20° = 297 \text{ m/s}$$
$$v_{r1} = \sqrt{297^2 + 468{\cdot}6^2} = 554{\cdot}8 \text{ m/s}$$
$$v_{r2} = 0{\cdot}8 \times 554{\cdot}8 = 443{\cdot}8 \text{ m/s}$$
$$v_{a2} = v_{r2} \times \sin \beta_2 = 443{\cdot}8 \times \sin 35° = 254{\cdot}6 \text{ m/s}$$

Axial force on blades [N]

= mass flow [kg/s] × change of axial velocity [m/s]
= $0{\cdot}5 \times (297 - 254{\cdot}6)$
= $21{\cdot}2$ N Ans. (i)

$$v_{w2} = v_{r2} \cos \beta_2 - u$$
$$= 443{\cdot}8 \times \cos 35° - 347{\cdot}4 = 16{\cdot}2 \text{ m/s}$$

Effective change of velocity,

$$v_w = v_{w1} + v_{w2}$$
$$= 816 + 16{\cdot}2 = 832{\cdot}2 \text{ m/s}$$

Tangential force on blades

$$= \dot{m} v_w$$
$$= 0{\cdot}5 \times 832{\cdot}2 = 416{\cdot}1 \text{ N}$$

Power [W] = force [N] × blade velocity [m/s]

$$= 416{\cdot}1 \times 347{\cdot}4$$
$$= 1{\cdot}445 \times 10^5 \text{ W} = 144{\cdot}5 \text{ kW} \quad \text{Ans. (ii)}$$

15.
$$\frac{T_1}{T_2} = \left(\frac{p_1}{p_2}\right)^{\frac{\gamma-1}{\gamma}}$$

$$\frac{533}{T_2} = \left(\frac{6}{4}\right)^{\frac{0{\cdot}4}{1{\cdot}4}}$$

$$T_2 = 474{\cdot}6 \text{ K}$$

$$h_1 - h_2 = c_P (T_1 - T_2) = \tfrac{1}{2} (c_2^2 - c_1^2)$$
$$1005\,(533 - 474{\cdot}6) = \tfrac{1}{2} (c_2^2 - 90^2)$$
$$c_2 = 354{\cdot}2 \text{ m/s}$$

Air velocity at nozzle exit = $354{\cdot}2$ m/s Ans. (a)

$$p v_s = RT$$

$$4 \times 10^2 \times v_s = 0{\cdot}2871 \times 474{\cdot}6$$
$$v_s = 0{\cdot}3406 \text{ m}^3/\text{kg}$$
$$\dot{m} = \frac{\text{Area} \times \text{velocity}}{\text{Specific volume}}$$
$$= \frac{0{\cdot}025 \times 354{\cdot}2}{0{\cdot}3406}$$

Mass flow rate of air = 26 kg/s Ans. (b)

$$p v_s = RT$$

$$4 \times 10^2 \times v_s = 0{\cdot}2871 \times 533$$
$$v_s = 0{\cdot}255$$
$$26 = \frac{\text{Area} \times 90}{0{\cdot}255}$$
$$\text{Nozzle inlet area} = 0{\cdot}0737 \text{ m}^2 \quad \text{Ans. (c)}$$

16. The ratio of partial pressures is the same as the ratio of partial volumes, therefore,

$$\left.\begin{aligned} \text{partial pressure of } CO_2 &= 0{\cdot}14 \times 1{\cdot}01 = 0{\cdot}1414 \text{ bar} \\ \text{partial pressure of } N_2 &= 0{\cdot}86 \times 1{\cdot}01 = 0{\cdot}8686 \text{ bar} \end{aligned}\right\} \text{Ans. (i)}$$

For mass of CO_2, $pV = mRT$

$$\text{where } p = 0{\cdot}1414 \times 10^2 \text{ kN/m}^2$$
$$V = 500 \times 10^{-6} \text{ m}^3$$
$$m = \frac{pV}{RT} = \frac{0{\cdot}1414 \times 10^2 \times 500 \times 10^{-6}}{0{\cdot}189 \times 293}$$
$$= 1{\cdot}276 \times 10^{-4} \text{ kg or } 0{\cdot}1276 \text{ g} \quad \text{Ans. (ii)}$$

17.
$$\frac{T_2}{T_1} = \left(\frac{p_2}{p_1}\right)^{\frac{n-1}{n}}$$
$$\frac{693}{295} = \left(\frac{35}{1}\right)^{\frac{n-1}{n}}$$
$$n = 1{\cdot}316 \quad \text{Ans. (a)}$$
$$\frac{p_1V_1}{T_1} = \frac{p_2V_2}{T_2}$$
$$\frac{1 \times 0{\cdot}037}{295} = \frac{35 \times V_2}{693}$$
$$V_2 = 0{\cdot}02483$$
$$\text{Work done} = \frac{p_1V_1 - p_2V_2}{n-1}$$
$$= \frac{1 \times 100 \times 0{\cdot}037 - 35 \times 100 \times 0{\cdot}02483}{0{\cdot}316}$$
$$= -15{\cdot}794 \text{ kJ} \quad \text{Ans. (b)}$$

Negative (compression) work done on the gas

$$pV = mRT$$
$$1 \times 100 \times 0{\cdot}037 = m \times 282 \times 295$$
$$m = 0{\cdot}0000444 \text{ kg}$$

$$\text{Change of internal energy} = mc_V (T_2 - T_1)$$
$$= 0{\cdot}0000444 \times 718\,(693 - 295)$$
$$= 12{\cdot}69 \text{ kJ} \quad \text{Ans. (c)}$$

$$\text{Heat transfer} = 12{\cdot}69 - 15{\cdot}794$$
$$= -3{\cdot}104 \text{ kJ} \quad \text{Ans. (d)}$$

18. Refer to Fig. 69

$$h_3 = h_4 = 59{\cdot}7 \text{ kJ/kg}$$

i.e. from tables, page 13, at 25°C

$$h_4 = h_f + x_4\,(h_g - h_f)$$
$$59{\cdot}7 = 22{\cdot}33 + x_4\,(180{\cdot}97 - 22{\cdot}33)$$

i.e. from tables, page 13, at –15°C

$$x_4 = 0{\cdot}2356$$

$$\text{Heat extracted} = \dot{m}\,(h_1 - h_4)$$
$$= \dot{m}\,\{h_g - h_f - x\,(h_g - h_f)\}$$
$$73{\cdot}3 = \dot{m}\,\{180{\cdot}97 - 23{\cdot}33 - 0{\cdot}2356(180{\cdot}97 - 23{\cdot}33)\}$$
$$= \dot{m}\,(180{\cdot}97 - 59{\cdot}71)$$
$$\dot{m} = 0{\cdot}605 \text{ kg/s}$$

$$\text{Refrigerant mass flow rate} = 0{\cdot}605 \text{ kg/s} \quad \text{Ans. (a)}$$
$$h_2 = 208{\cdot}5 \text{ kJ/kg}$$

i.e. from tables, page 13, at 6·516 bar and 40°C

$$\text{Compressor power} = \dot{m}\,(h_2 - h_1)$$
$$= 0{\cdot}605\,(208{\cdot}5 - 180{\cdot}97)$$
$$= 16{\cdot}66 \text{ kW} \quad \text{Ans. (b)}$$

$$\text{c.o.p.} = \frac{h_1 - h_4}{h_2 - h_1}$$
$$= \frac{180{\cdot}97 - 59{\cdot}71}{208{\cdot}5 - 180{\cdot}97}$$
$$= 4{\cdot}405 \quad \text{Ans. (c)}$$

19.

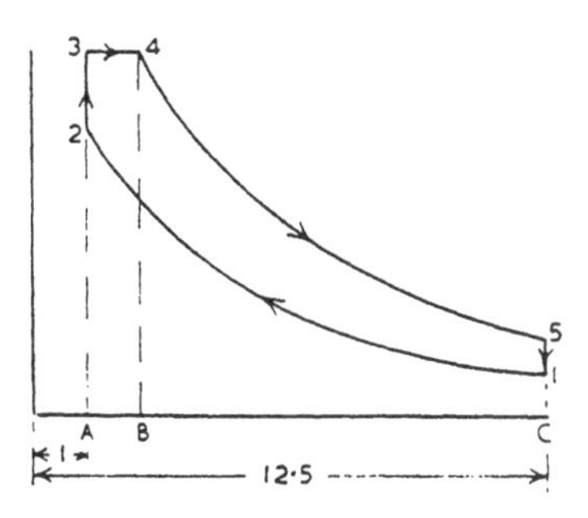

Representing volumes by ratio:

$$V_1 \text{ and } V_5 = 12{\cdot}5 \qquad V_2 \text{ and } V_3 = 1$$

$$p_1 V_1^{\gamma} = p_2 V_2^{\gamma}$$

$$1{\cdot}013 \times 12{\cdot}5^{1{\cdot}4} = p_2 \times 1^{1{\cdot}4} \qquad p_2 = 34{\cdot}77 \text{ bar}$$

$$\frac{p_1 V_1}{T_1} = \frac{p_2 V_2}{T_2}$$

$$T_2 = \frac{308 \times 34{\cdot}77 \times 1}{1{\cdot}013 \times 12{\cdot}5} = 845{\cdot}9 \text{ K}$$

$$\frac{T_3}{T_2} = \frac{p_3}{p_2}$$

$$T_3 = \frac{845{\cdot}9 \times 40}{34{\cdot}77} = 973{\cdot}4 \text{ K}$$

$$\frac{V_4}{V_3} = \frac{T_4}{T_3}$$

$$V_4 = \frac{1 \times 1698}{973{\cdot}4} = 1{\cdot}745$$

$$p_4 V_4^{\gamma} = p_5 V_5^{\gamma}$$

$$40 \times 1{\cdot}745^{1{\cdot}4} = p_5 \times 12{\cdot}5^{1{\cdot}4}$$

$$p_5 = 40 \times \left\{\frac{1{\cdot}745}{12{\cdot}5}\right\}^{1{\cdot}4} = 2{\cdot}541 \text{ bar}$$

Areas representing positive work done:

Area during combustion = A 3 4 B

$= 40 \times (1{\cdot}745 - 1) = 29{\cdot}8$

Area during expansion = B 4 5 C

$$= \frac{p_4V_4 - p_5V_5}{\gamma - 1}$$

$$= \frac{40 \times 1{\cdot}745 - 2{\cdot}541 \times 12{\cdot}5}{1{\cdot}4 - 1} = 95{\cdot}07$$

Gross area = A 3 4 5 C
= 29·8 + 95·07 = 124·87

Area representing negative work done:
Area during compression = C 1 2 A

$$= \frac{p_1V_1 - p_2V_2}{\gamma - 1}$$

$$= \frac{1{\cdot}013 \times 12{\cdot}5 - 34{\cdot}77 \times 1}{1{\cdot}4 - 1} = -55{\cdot}27$$

Net area representing useful work

= 124·87 − 55·27 = 69·6

Mean effective pressure = mean height of net area

$$= \frac{\text{area}}{\text{length}} = \frac{69{\cdot}6}{12{\cdot}5 - 1}$$

= 6·052 bar Ans.

Note: m.e.p. and maximum pressure are low by modern standards.

20.

$$p = p_s + p_A$$
$$mR = m_sR_s + m_AR_A$$
$$21 \times R = 20 \times 0{\cdot}462 + 1 \times 0{\cdot}287$$
$$R = 0{\cdot}4537 \text{ kJ/kg K}$$
$$p_s = 0{\cdot}02 \times \frac{20 \times 0{\cdot}462}{21 \times 0{\cdot}4537}$$
$$p_s = 0{\cdot}0194 \text{ bar}$$
$$p_A = 0{\cdot}02 - 0{\cdot}0194 = 0{\cdot}0006 \text{ bar}$$

partial pressure, steam = 0·0194 bar
partial pressure, air = 0·0006 bar } Ans. (a)

$$pv = RT$$
$$100 \times 0{\cdot}02 \times v = 0{\cdot}4537 \times 303$$
$$v = 68{\cdot}735 \text{ m}^3\text{/kg}$$
$$\rho = 0{\cdot}0145 \text{ kg/m}^3$$

Density of mixture = 0·0145 kg/m³ Ans. (b)

21. Working on the basis of 1 kg fuel:

$$\text{Stoichiometric air} = \frac{100}{23}\left\{2⅔C + 8\left(H_2 - \frac{O_2}{8}\right)\right\}$$
$$= \frac{100}{23}\{2⅔C + 8\,H_2 - O_2\}$$
$$= \frac{100}{23}\{2⅔ \times 0{\cdot}855 + 8 \times 0{\cdot}135 - 0{\cdot}1\}$$
$$= \frac{100}{23} \times 3{\cdot}35 = 14{\cdot}56\text{ kg}$$

Excess air = $0{\cdot}25 \times 14{\cdot}56 = 3{\cdot}64$ kg
Actual air = $14{\cdot}56 + 3{\cdot}64 = 18{\cdot}2$ kg

Mass products of combustion per kg fuel:

CO_2 = $3⅔ \times 0{\cdot}855$ = 3·135 kg
H_2O = $9\,H_2 = 9 \times 0{\cdot}135$ = 1·215 kg
O_2 = 23% of excess air = $0{\cdot}23 \times 3{\cdot}64$ = 0·8372 kg
N_2 = 77% of all air = $0{\cdot}77 \times 18{\cdot}2$ = 14·014 kg

Total products = 1 kg fuel + 18·2 kg air = 19·2 kg
%mass analysis, Ans. (i):

$$CO_2 = \frac{3{\cdot}135}{19{\cdot}2} \times 100 = 16{\cdot}32\%$$
$$H_2O = \frac{1{\cdot}215}{19{\cdot}2} \times 100 = 6{\cdot}33\%$$
$$O_2 = \frac{0{\cdot}8372}{19{\cdot}2} \times 100 = 4{\cdot}36\%$$
$$N_2 = \frac{14{\cdot}014}{19{\cdot}2} \times 100 = 72{\cdot}99\%$$

Dry flue gases = total gases – H_2O
= $19{\cdot}2 - 1{\cdot}215$ = 17·985 kg

∴ %Volumetric analysis of dry flue gases, Ans. (ii):

DFG	*m*%	M	*N*	*N*%
CO_2	16·32	44	0·371	11·91%
O_2	4·36	32	0·1362	4·38%
N_2	72·99	28	2·606	83·71%
		Total = 3·1132		

22. Heat passing across inner surface:

$$Q = h_i At\,(T_i - T_1) \;\therefore\; T_i - T_1 = \frac{Q}{h_i At}$$

$$-23 - T_1 = \frac{-30}{12 \times 1 \times 1}$$

Temperature, inner surface (T_1) = −20·5°C Ans. (a)
i.e. heat flow *into* room is $-Q$

Heat passing across outer surface:

$$Q = h_o At\,(T_3 - T_o) \;\therefore\; T_3 - T_o = \frac{Q}{h_o At}$$

$$T_3 - 24 = \frac{-30}{12 \times 1 \times 1}$$

Temperature, outer surface (T_3) = 21·5°C Ans. (a)

Heat conducted through inner material:

$$Q = \frac{kAt\,(T_1 - T_2)}{S}$$

$$T_1 - T_2 = \frac{QS}{kAt}$$

$$-20{\cdot}5 - T_2 = \frac{-30 \times 0{\cdot}1}{0{\cdot}115 \times 1 \times 1}$$

$$-20{\cdot}5 - T_2 = -26{\cdot}09$$

$$T_2 = 5{\cdot}59^\circ\text{C}$$

Temperature of the interface (T_2) = 5·59°C Ans. (b)

Heat conducted throught cork:

$$T_2 - T_3 = \frac{QS}{kAt}$$

$$5{\cdot}59 - 21{\cdot}5 = \frac{-30 \times S}{0{\cdot}06 \times 1 \times 1}$$

$$S = \frac{15{\cdot}91 \times 0{\cdot}06}{30}$$

$$S = 0{\cdot}0318 \text{ m}$$

Thickness of the cork = 31·8 mm Ans. (c)

23. Reference to Freon-12 tables, and Fig. 69,

Compressor suction and evaporator exit:
1·509 bar, sat. temp. = –20°C,
∴ at –5°C refrigerant is superheated by 15°
h_1 = h at 1·509 bar, supht. 15° = 187·75
Compressor discharge:
4·914 bar, sat. temp. = 15°C
∴ at 45°C refrigerant is superheated by 30°
h_2 = h at 4·914 bar, supht. 30° = 214·35
Condenser outlet: h_3 = h_f at 15°C = 50·1
Evaporator inlet: h_4 = h_3 = 50·1

Coeff. of performance =

$$\frac{\text{refrigerating effect in evaporator [kJ/kg]}}{\text{work transfer in compressor [kJ/kg]}}$$

$$= \frac{h_1 - h_4}{h_2 - h_1}$$

$$= \frac{187{\cdot}75 - 50{\cdot}1}{214{\cdot}35 - 187{\cdot}75} = 5{\cdot}175 \quad \text{Ans.}$$

24. Stoichiometric air $= \frac{100}{23}(2\tfrac{2}{3}C + 8H)$

$$= 4{\cdot}348(2{\cdot}667 \times 0{\cdot}82 + 8 \times 0{\cdot}18)$$
$$= 15{\cdot}77 \text{ kg/kg fuel}$$

Air fuel ratio by mass = 15·77 Ans. (a)

Mass products of combustion per kg fuel:

CO_2 = $3\tfrac{2}{3} \times 0{\cdot}8$ = 3·007 kg
H_2O = $9 \times 0{\cdot}18$ = 1·62
N_2 = $0{\cdot}77 \times 15{\cdot}77$ = 12·143
Total = 16·77 kg

% mass analysis of the wet flue gases:

$$CO_2 = \frac{3{\cdot}007}{16{\cdot}77} \times 100 = 17{\cdot}93$$

$$H_2O = \frac{1{\cdot}62}{16{\cdot}77} \times 100 = 9{\cdot}66 \quad \text{Ans. (a)}$$

$$N_2 = \frac{12{\cdot}14}{16{\cdot}77} \times 100 = 72{\cdot}39$$

% volume analysis of the wet flue gases:

DFG	$m\%$	M	N	$N\%$
CO_2	17·93	44	0·4075	11·54
H_2O	9·66	18	0·5367	15·20
N_2	72·39	28	2·5854	73·26
			3·5296	

Ans. (b)

25.

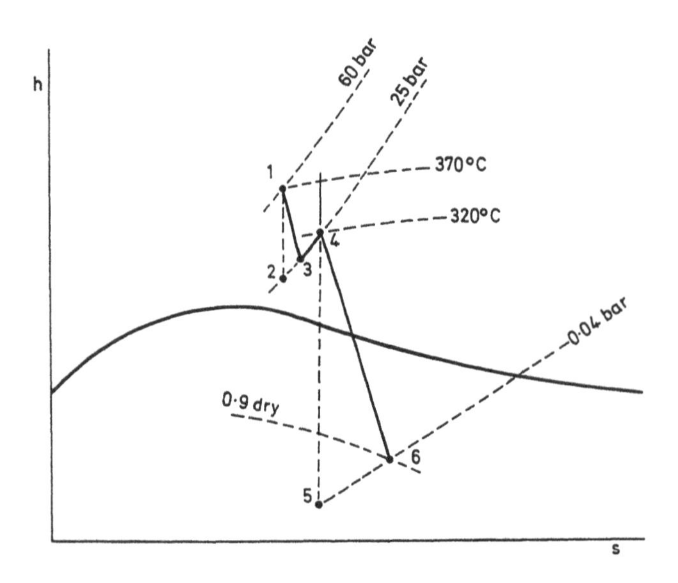

From the h–s chart; in kJ/kg

$$h_4 = 3057$$
$$h_5 = 2030$$
$$h_4 - h_5 = 1027$$
$$h_4 - h_6 = 1027 \times 0{\cdot}73$$
$$= 750$$
$$h_6 = 2307$$
$$P = \dot{m}_A(h_4 - h_6)$$
$$3 \times 10^3 = \dot{m}_A \times 750$$
$$\dot{m}_A = 4 \text{ kg/s}$$
$$h_1 = 3097$$

$$
\begin{aligned}
h_2 &= 2900 \\
h_1 - h_2 &= 197 \\
h_1 - h_3 &= 197 \times 0{\cdot}76 \\
&= 150 \\
h_3 &= 2947 \\
h_4 - h_3 &= 3057 - 2947 \\
&= 110
\end{aligned}
$$

Enthalpy change during reheating is 110 kJ/kg Ans. (a)
Condition of steam at condenser inlet (from chart) 0·9 dry Ans. (b)

$$
\begin{aligned}
P &= \dot{m}_B\{(h_1 - h_3) + (h_4 - h_6)\} \\
3 \times 10^3 &= \dot{m}_B\,(150 + 750) \\
\dot{m}_B &= 3{\cdot}33 \text{ kg/s}
\end{aligned}
$$

$$
\frac{100\,(4 - 3{\cdot}33)}{4} = 16{\cdot}8
$$

Percentage reduction in steam flow is 16·8 Ans. (c)

26.

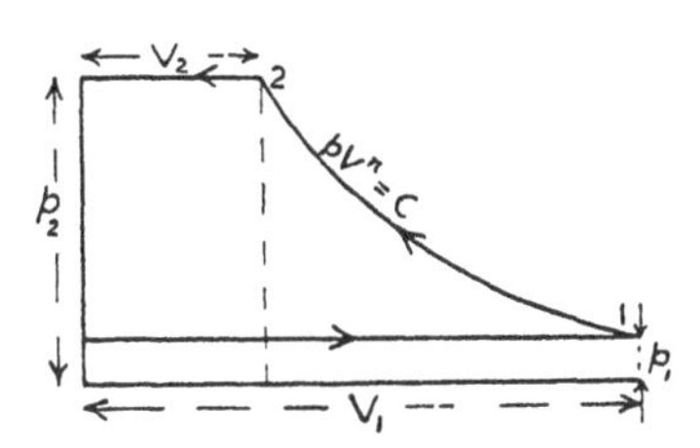

$$
\begin{aligned}
\text{mean piston speed [m/s]} &= \text{distance [m] moved by piston/second} \\
&= 2 \times \text{stroke} \times \text{rev/s} \\
\therefore\ \text{stroke} &= \frac{2{\cdot}8}{2 \times 3{\cdot}5} \\
&= 0{\cdot}4 \text{ m} = 400 \text{ mm} \quad \text{Ans. (i)}
\end{aligned}
$$

$$
\begin{aligned}
\text{work per cycle [kJ]} &= \frac{\text{work per second [kJ/s = kW]}}{\text{cycles per second}} \\
&= \frac{13{\cdot}38}{3{\cdot}5} = 3{\cdot}88 \text{ kJ}
\end{aligned}
$$

Referring to sketch, neglecting clearance:

$$
\text{Area under compression curve} = \frac{p_1V_1 - p_2V_2}{n - 1}
$$

this is the work done *by* the air,

$$\text{work done } on \text{ the air} = \frac{p_2V_2 - p_1V_1}{n-1}$$

work done on the air per cycle

= net area of diagram

= compression area + delivery area – suction area

$$\frac{p_2V_2 - p_1V_1}{n-1} + p_2V_2 - p_1V_1$$

$$\frac{n}{n-1}(p_2V_2 - p_1V_1)$$

$$= \frac{n}{n-1} mR\,(T_2 - T_1)$$

$$= \frac{n}{n-1} mR\,T_1\left[\frac{T_2}{T_1} - 1\right]$$

$$= \frac{n}{n-1} p_1V_1\left[\left\{\frac{p_2}{p_1}\right\}^{\frac{n-1}{n}} - 1\right]$$

$$\therefore\ 3{\cdot}88 = \frac{1{\cdot}32}{0{\cdot}32} \times 10^2 \times V_1 \times \left[\left\{\frac{10}{1}\right\}^{\frac{0{\cdot}32}{1{\cdot}32}} - 1\right]$$

$$3{\cdot}88 \times 0{\cdot}32 = 1{\cdot}32 \times 10^2 \times V_1 \times 0{\cdot}748$$

$$V_1 = \frac{3{\cdot}88 \times 0{\cdot}32}{1{\cdot}32 \times 10^2 \times 0{\cdot}748} = 0{\cdot}01257 \text{ m}^3$$

$$\text{Diameter} = \sqrt{\frac{\text{volume}}{0{\cdot}7854 \times \text{stroke}}} = \sqrt{\frac{0{\cdot}01257}{0{\cdot}7854 \times 0{\cdot}4}}$$

$$= 0{\cdot}2 \text{ m} = 200 \text{ mm} \quad \text{Ans. (ii)}$$

$$\text{Mean eff. press.} = \frac{\text{net area of diagram}}{\text{length of diagram}}$$

$$= \frac{3{\cdot}88}{0{\cdot}01257} = 308{\cdot}6 \text{ kN/m}^2$$

$$= 3{\cdot}086 \text{ bar} \quad \text{Ans. (iii)}$$

27. Referring to Fig. 28:

$$\frac{T_2}{T_1} = \left\{\frac{V_1}{V_2}\right\}^{n-1}$$

$$T_2 = 310 \times 13^{0\cdot4} = 865 \text{ K}$$

Temp. at end of compression = 592°C Ans. (i)

Assume 35 kg of air is compressed and 1 kg of fuel is burned, mass of gases formed = 35 + 1 = 36 kg

$$\begin{matrix}\text{Heat energy given} \\ \text{up by 1 kg fuel}\end{matrix} = \begin{matrix}\text{Heat energy received} \\ \text{by 36 kg gases}\end{matrix}$$

$$\text{c.v.} = m \times c_p \times (T_3 - T_2)$$

$$T_3 - T_2 = \frac{42 \times 10^3}{36 \times 1\cdot02} = 1144 \text{ K}$$

$$T_3 = 1144 + 865 = 2009 \text{ K}$$

Temp. at end of combustion = 1736°C Ans. (ii)

$$\frac{V_3}{V_2} = \frac{T_3}{T_2}$$

$$V_3 = 1 \times \frac{2009}{865} = 2\cdot322$$

$$\text{Ratio of expansion} = \frac{V_4}{V_3} = \frac{13}{2\cdot322} = 5\cdot598 \quad \text{Ans. (iii)}$$

28. Referring to Fig 60:

$$v_{a1} = v_1 \sin \alpha_1 = 135 \times 135 \times \sin 20° = 46\cdot17 \text{ m/s}$$
$$v_{w1} = v_1 \cos \alpha_1 = 135 \times \cos 20° = 126\cdot9 \text{ m/s}$$
$$x = v_{w1} - u = 126\cdot9 - 87 = 39\cdot9 \text{ m/s}$$

$$\tan \beta_1 = \frac{46\cdot17}{39\cdot9} = 1\cdot157$$

∴ Entrance angle β_1 = 49° 10′ Ans. (i)

$$\beta_2 = \alpha_1 \qquad v_{r2} = v_1 \qquad v_{w2} = x$$

Effective change of velocity, $v_w = v_{w1} + v_{w2}$

$$= 126\cdot9 + 39\cdot9 = 166\cdot8 \text{ m/s}$$

$$\begin{aligned}
\text{Force on blades} &= \text{change of momentum per second} \\
\text{Force [N]} &= \text{mass flow [kg/s]} \times v_w \text{ [m/s]} \\
\text{For 1 kg/s, force} &= 1 \times 166{\cdot}8 = 166{\cdot}8 \text{ N} \\
\text{Power [W = J/s = N m/s]} &= \text{force [N]} \times \text{blade velocity [m/s]} \\
&= 166{\cdot}8 \times 87 \\
&= 1{\cdot}451 \times 10^4 \text{ W} \\
&= 14{\cdot}51 \text{ kW} \quad \text{Ans. (ii)}
\end{aligned}$$

29. Kinetic energy = $\frac{1}{2} mv^2$

Change in kinetic energy per kg of exhaust gases through exchanger due to change of velocity

$= \frac{1}{2}(v_1^2 - v_2^2) = \frac{1}{2}(140^2 - 10^2) = 9750$ J/kg

which is converted into heat energy $= 9{\cdot}75$ kJ/kg

Mass of exhaust gases per kg of fuel

$= 84$ kg air + 1 kg fuel = 85 kg gases

Heat energy transferred = mass × c_P × temp. change

∴ Transfer of heat in exchanger, from 85 kg of gases, to 84 kg of air:

$$85 \times 1{\cdot}1 \times (300 - 240) + 85 \times 9{\cdot}75 = 84 \times 1{\cdot}005 \times \text{temp. change}$$

$$\text{Air temp. rise} = \frac{85\,(1{\cdot}1 \times 60 + 9{\cdot}75)}{84 \times 1{\cdot}005} = 76{\cdot}28^\circ$$

Air temp. outlet $= 200 + 76{\cdot}28 = 276{\cdot}28$°C Ans.

30. Heat to be taken from water to make ice [kJ/s]

$$= \frac{250}{3600}(4{\cdot}2 \times 15 + 335 + 2{\cdot}04 \times 10)$$

$$= 29{\cdot}05 \text{ kJ/s}$$

Let $\dot{m}$ [kg/s] = mass flow of refrigerant

Heat absorbed by refrigerant in evaporator [kJ/s]

$$= \dot{m} \times (0{\cdot}95 - 0{\cdot}2) \times 290{\cdot}2$$

$$= \dot{m} \times 217{\cdot}7 \text{ kJ/s}$$

Assuming perfect heat transfer,

$$\dot{m} \times 217{\cdot}7 = 29{\cdot}05$$

$$\dot{m} = 0{\cdot}1335 \text{ kg/s} \quad \text{Ans. (i)}$$

Volume flow of refrigerant leaving evaporator and entering compressor [m³/s] $= 0{\cdot}1335 \times 0{\cdot}02168 \times 0{\cdot}95$

$$= 2{\cdot}749 \times 10^{-3} \text{ m}^3\text{/s}$$

Let d = diameter, stroke = $2d$, assuming 100% volumetric efficiency, volume taken into compressor per second [m³/s] = $0.7854 \times d^2 \times 4.15 = 2.749 \times 10^{-3}$

$$d = \sqrt[3]{\frac{2.749 \times 10^{-3}}{0.7854 \times 2 \times 4.15}}$$

$$= 0.07499 \text{ m say } 75 \text{ mm}$$

$$\text{Stroke} = 2 \times 75 = 150 \text{ mm} \quad \text{Ans. (ii)}$$

31. Referring to Fig. 38:

$$V_1 = \text{stroke volume} + \text{clearance volume}$$
$$= 1440 + 40 = 1480 \text{ cm}^3$$
$$V_3 = \text{clearance volume} = 40 \text{ cm}^3$$
$$p_1 V_1^n = p_2 V_2^n$$
$$1 \times 1480^{1.3} = 5 \times V_2^{1.3}$$
$$V_2 = \frac{1480}{\sqrt[1.3]{5}} = 429.1 \text{ cm}^3$$

Volume swept by piston from beginning of compression stroke to point where delivery valves open

$$= V_1 - V_2$$
$$= 1480 - 429.1 = 1050.9 \text{ cm}^3$$

As a fraction of the stroke when delivery valves open

$$= \frac{1050.9}{1440} = 0.7298 \quad \text{Ans. (i)}$$

$$p_3 V_3^n = p_4 V_4^n$$
$$5 \times 40^{1.3} = 1 \times V_4^{1.3}$$
$$V_4 = 40 \times \sqrt[1.3]{5} = 138 \text{ cm}^3$$

Volume swept by piston from beginning of suction stroke to point when suction valves open

$$= V_4 - V_3$$
$$= 138 - 40 = 98 \text{ cm}^3$$

As a fraction of the stroke when suction valves open

$$= \frac{98}{1440} = 0.06805 \quad \text{Ans. (ii)}$$

$$\text{Area under compression curve} = \frac{p_2 V_2 - p_1 V_1}{n - 1}$$

By using $p_2V_2 - p_1V_1$ instead of $p_1V_1 - p_2V_2$ the area represents work done *on* the air instead of work done *by* the air.

Note that the mean indicated pressure will be obtained from net area ÷ length, therefore, since the actual work is not required, it is not necessary to convert pressures into kN/m^2 and volumes into m^3.

$$\text{Area under compression curve} = \frac{p_2V_2 - p_1V_1}{n-1}$$

$$= \frac{5 \times 429{\cdot}1 - 1 \times 1480}{1{\cdot}3 - 1} = 2218{\cdot}3$$

$$\text{Area under delivery line} = p_2 \times (V_2 - V_3)$$
$$= 5 \times (429{\cdot}1 - 40) = 1945{\cdot}5$$

$$\text{Area under expansion curve} = \frac{p_3V_3 - p_4V_4}{n-1}$$

$$= \frac{5 \times 40 - 1 \times 138}{1{\cdot}3 - 1} = 206{\cdot}7$$

$$\text{Area under suction line} = p_1 \times (V_1 - V_4)$$
$$= 1 \times (1480 - 138) = 1342$$

$$\text{Net area of diagram} = 2218{\cdot}3 + 1945{\cdot}5 - 206{\cdot}7 - 1342$$
$$= 4163{\cdot}8 - 1548{\cdot}7 = 2615{\cdot}1$$

Mean indicated press. = mean height of diagram

$$= \frac{\text{net area}}{\text{length}} = \frac{2615{\cdot}1}{1440} = 1{\cdot}816 \text{ bar} \quad \text{Ans. (iii)}$$

Alternatively, the net area of the diagram, representing work per cycle could be obtained from:

$$\frac{n}{n-1} p_1 (V_1 - V_4) \left[\left\{ \frac{p_2}{p_1} \right\}^{\frac{n-1}{n}} - 1 \right]$$

32. From steam tables,

30 bar 350°C, $h = 3117$ $\quad s = 6{\cdot}744$
0·045 bar, $h_f = 130$ $\quad s_f = 0{\cdot}451$
$h_{fg} = 2428$ $\quad s_{fg} = 7{\cdot}980$

For the Rankine cycle, isentropic expansion from 30 bar 350°C to 0·045 bar:

Entropy after expansion = Entropy before

$$0{\cdot}451 + x \times 7{\cdot}98 = 6{\cdot}744$$
$$x = 0{\cdot}7886$$
$$h_2 = h \text{ at } 0{\cdot}045 \text{ bar, } 0{\cdot}7886 \text{ dry}$$
$$= 130 + 0{\cdot}7886 \times 2428 = 2044$$

$$\text{Rankine efficiency} = \frac{h_1 - h_2}{h_1 - h_{f2}} = \frac{3117 - 2044}{3117 - 130}$$

$$= \frac{1073}{2987} = 0{\cdot}3592 \text{ or } 35{\cdot}92\% \quad \text{Ans. (i)}$$

$$\text{Actual efficiency} = \frac{\text{heat energy converted into work}}{\text{heat energy supplied by steam}}$$

Heat energy into work [kJ/s = kW] = 5×10^3 kJ/s

Heat energy supplied [kJ/s]

= steam consumption [kg/s] × $(h_1 - h_{f2})$ [kJ/kg]

$$= \frac{22{\cdot}5 \times 10^3}{3600} \times 2987$$

$$\therefore \text{ Engine effic.} = \frac{5 \times 10^3 \times 3600}{22{\cdot}5 \times 10^3 \times 2987}$$

$$= 0{\cdot}2679 \text{ or } 26{\cdot}79\% \quad \text{Ans. (ii)}$$

$$\text{Efficiency ratio} = \frac{0{\cdot}2679}{0{\cdot}3592}$$

$$= 0{\cdot}7457 \text{ or } 74{\cdot}57\% \quad \text{Ans. (iii)}$$

33. From steam tables,

10 bar, $h_f = 763$ $\quad h_{fg} = 2015$

1·2 bar, sat. temp. = 104·8°C $\quad h_g = 2683$

∴ at 1·2 bar 109·8°C, steam is superheated 5°

Dryness fraction by separating calorimeter:

$$x_1 = \frac{3{\cdot}03}{3{\cdot}03 + 0{\cdot}113} = 0{\cdot}964$$

Dryness fraction by throttling calorimeter:

Enthalpy before throttling = Enthalpy after

$$763 + x_2 \times 2015 = 2683 + 2{\cdot}02 \times 5$$
$$x_2 \times 2015 = 1930{\cdot}1$$
$$x_2 = 0{\cdot}9579$$

Dryness fraction of sample

$$= 0{\cdot}964 \times 0{\cdot}9579 = 0{\cdot}9234 \text{ Ans.}$$

34. Referring to Fig. 29,

$$T_1 = 315 \text{ K}$$
$$T_4 = 1773 \text{ K}$$
$$V_1 = V_5 = \text{stroke volume} + \text{clearance volume}$$
$$= 0{\cdot}1068 + 0{\cdot}0089 = 0{\cdot}1157 \text{ m}^3$$
$$V_2 = V_3 = \text{clearance volume} = 0{\cdot}0089 \text{ m}^3$$
$$p_1 V_1^{y} = p_2 V_2^{y}$$
$$1 \times 0{\cdot}1157^{1{\cdot}4} = p_2 \times 0{\cdot}0089^{1{\cdot}4}$$
$$p_2 = \left\{\frac{0{\cdot}1157}{0{\cdot}0089}\right\}^{1{\cdot}4} = 36{\cdot}26 \text{ bar}$$
$$\frac{p_1 V_1}{T_1} = \frac{p_2 V_2}{T_2}$$
$$T_2 = \frac{315 \times 36{\cdot}26 \times 0{\cdot}0089}{1 \times 0{\cdot}1157} = 878{\cdot}8 \text{ K}$$
$$\frac{T_3}{T_2} = \frac{p_3}{p_2}$$
$$T_3 = \frac{878{\cdot}8 \times 45}{36{\cdot}26} = 1090 \text{ K}$$

Heat received at constant volume

$$= m \times c_V \times (T_3 - T_2)$$

(per kg of air): $= 1 \times 0{\cdot}715 \times (1090 - 878{\cdot}8)$

$$= 151 \text{ kJ/kg}$$

Spec. heat at constant press. $c_P = c_V \times \gamma$

$$= 0{\cdot}715 \times 1{\cdot}4 = 1{\cdot}001$$

Heat received at constant pressure

$$= m \times c_P \times (T_4 - T_3)$$
$$= 1 \times 1{\cdot}001 \times (1773 - 1090)$$
$$= 683{\cdot}7 \text{ kJ/kg}$$

Ratio: $\dfrac{\text{heat received at const. vol}}{\text{heat received at const. press}}$

$$= \frac{151}{683{\cdot}7} = 0{\cdot}2209{:}1 \text{ Ans.}$$

35. Per kg of fuel:

$$\begin{aligned}
\text{Air supplied} &= 11 \text{ kg} \\
\text{Gases} &= 11 + 0{\cdot}99 \text{ kg} \qquad (1) \\
\text{Let } x &= \text{C to CO}_2 \\
\text{then } (0{\cdot}85 - x) &= \text{C to CO} \\
\text{Air supplied} &= \frac{100}{23}\{2\tfrac{2}{3}x + 1\tfrac{1}{3}(0{\cdot}85 - x) + 8 \times 0{\cdot}11 - 0{\cdot}03\} \\
&= 5{\cdot}796x + 8{\cdot}622 \text{ kg} \\
\text{Gases} &= 5{\cdot}796x + 8{\cdot}622 + 0{\cdot}99 \text{ kg} \qquad (2)
\end{aligned}$$

From (1) and (2):

$$\begin{aligned}
11 + 0{\cdot}99 &= 5{\cdot}796x + 8{\cdot}622 + 0{\cdot}99 \\
x &= 0{\cdot}41 \\
(0{\cdot}85 - x) &= 0{\cdot}44
\end{aligned}$$

Carbon to carbon monoxide = 0·44 kg Ans. (a)
Carbon to carbon dioxide = 0·41 kg Ans. (b)

36.

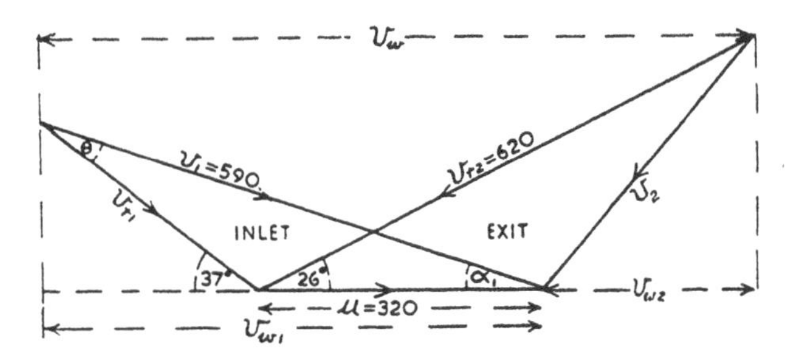

By sine rule referring to inlet triangle:

$$\frac{320}{\sin \theta} = \frac{590}{\sin(180^\circ - 37^\circ)}$$

$$\sin \theta = \frac{320 \times 0{\cdot}6018}{590} = 0{\cdot}3263$$

$$\begin{aligned}
\theta &= 19^\circ\, 3' \quad \alpha_1 = 37^\circ - 19^\circ\, 3' = 17^\circ\, 57' \\
v_{w1} &= 590 \times \cos 17^\circ\, 57' = 561{\cdot}4 \text{ m/s} \\
v_{w2} &= 620 \times \cos 26^\circ - 320 = 237{\cdot}3 \text{ m/s}
\end{aligned}$$

Effective change of velocity $v_w = v_{w1} + v_{w2}$
$= 561{\cdot}4 + 237{\cdot}3 = 798{\cdot}7$ m/s

Force on blades [N] = mass flow [kg/s] × change of velocity [m/s]

$$= 0{\cdot}075 \times 798{\cdot}7 = 59{\cdot}91 \text{ N} \quad \text{Ans. (i)}$$

$$\begin{aligned} \text{Power [W} &= \text{J/s} = \text{N m/s]} = \text{force [N]} \times \text{blade velocity [m/s]} \\ &= 59{\cdot}91 \times 320 \\ &= 1{\cdot}917 \times 10^4 \text{ W} \\ &= 19{\cdot}17 \text{ kW} \quad \text{Ans. (ii)} \end{aligned}$$

37. $V_1 = 0{\cdot}7854 \times 0{\cdot}3^2 \times 0{\cdot}45 = 0{\cdot}03181 \text{ m}^3$

$$\text{Work per cycle} = \frac{n}{n-1} p_1 V_1 \left[\left\{ \frac{p_2}{p_1} \right\}^{\frac{n-1}{n}} - 1 \right]$$

When $n = 1{\cdot}15$:

$$\text{Work/cycle} = \frac{1{\cdot}15}{0{\cdot}15} \times 1 \times 10^2 \times 0{\cdot}03181 \, (4^{\frac{0{\cdot}15}{1{\cdot}15}} - 1)$$

$$= 4{\cdot}828 \text{ kJ/cycle}$$

At 5 cycles per second:

$$\text{Power [kW} = \text{kJ/s]} = 4{\cdot}828 \times 5 = 24{\cdot}14 \text{ kW} \quad \text{Ans. (i)}$$

When $n = 1{\cdot}35$:

$$\text{Work/cycle} = \frac{1{\cdot}35}{0{\cdot}35} \times 1 \times 10^2 \times 0{\cdot}03181 \, (4^{\frac{0{\cdot}35}{1{\cdot}35}} - 1)$$

$$= 5{\cdot}299 \text{ kJ/cycle}$$

$$\text{Power} = 5{\cdot}299 \times 5 = 26{\cdot}5 \text{ kW} \quad \text{Ans. (ii)}$$

$$\% \text{ increase} = \frac{26{\cdot}5 - 24{\cdot}14}{24{\cdot}14} \times 100 = 9{\cdot}777\% \quad \text{Ans. (iii)}$$

An alternative solution to the above is to find the value of V_2 in each case, from $p_1 V_1{}^n = p_2 V_2{}^n$ then calculate the work per cycle from:

$$\frac{n}{n-1} (p_2 V_2 - p_1 V_1)$$

38. From steam tables,

18 bar, $t_s = 207{\cdot}1\text{°C}$ $h_{fg} = 1912$

Enthalpy drop per kg of steam

$$= (1 - 0{\cdot}985) \times 1912 = 28{\cdot}68 \text{ kJ/kg}$$

Rate of heat loss [J/s] per unit length

$$= \frac{1200 \times 28{\cdot}68 \times 10^3}{3600 \times 25} = 382{\cdot}4 \text{ J/s}$$

and this is to be equal to $\dfrac{2\pi k\,(T_1 - T_2)}{\ln\,(r_2/r_1)}$

$$\frac{2\pi \times 0{\cdot}13 \times (207{\cdot}1 - 35)}{\ln\,(r_2/r_1)} = 382{\cdot}4$$

$$\ln\left(\frac{r_2}{r_1}\right) = \frac{2\pi \times 0{\cdot}13 \times 172{\cdot}1}{382{\cdot}4} = 0{\cdot}3676$$

$$\therefore\ \frac{r_2}{r_1} = 1{\cdot}444 \qquad r_2 = 1{\cdot}444 \times 70 = 101{\cdot}1 \text{ mm}$$

Thickness $= r_2 - r_1 = 101{\cdot}1 - 70 = 31{\cdot}1$ mm Ans.

39. From steam tables,

$$\begin{array}{llll} & \text{20 bar 400°C, } h = 3248 & & \\ 1{\cdot}4 \text{ bar, } & t_s = 109{\cdot}3 & h_f = 458 & h_{fg} = 2232 \\ 0{\cdot}04 \text{ bar, } & h_f = 121 & h_{fg} = 2433 & \end{array}$$

Without feed heating:
h at 0·04 bar 0·85 dry

$$= 121 + 0{\cdot}85 \times 2433 = 2188$$

Enthalpy drop through turbine/kg

$$= 3248 - 2188 = 1060$$

Heat supplied to steam in boiler

$$= 3248 - 121 = 3127$$

$$\text{Thermal eff.} = \frac{\text{heat energy converted into work}}{\text{heat energy supplied}}$$

$$= \frac{1060}{2127} = 0{\cdot}3389 \text{ or } 33{\cdot}89\%$$

With feed heating:
0·134 kg of steam at 1·4 bar is tapped off per kg of supply steam, leaving (1 – 0·134) = 0·866 kg to complete its path through the engine.

Considering heater and hotwell as one system,

$$\text{Enthalpy at entry} = \text{Enthalpy at exit}$$

$$0{\cdot}134\,(458 + x \times 2232) + 0{\cdot}866 \times 121 = 1 \times 458$$
$$61{\cdot}38 + 299x + 104{\cdot}8 = 458$$
$$299x = 291{\cdot}82$$
$$\text{dryness fraction at } 1{\cdot}4 \text{ bar}, x = 0{\cdot}976$$

h at 1·4 bar 0·976 dry

$$= 458 + 0{\cdot}976 \times 2232 = 2636$$

Enthalpy drop through turbine/kg

$$= 1 \times (3248 - 2646) + 0{\cdot}866\,(2636 - 2188)$$
$$= 612 + 388 = 1000$$

Heat supplied to steam in boiler

$$= 3248 - 458 = 2790$$

$$\text{Thermal eff.} = \frac{1000}{2790} = 0{\cdot}3584 \text{ or } 35{\cdot}84\%$$

$$\left\{\begin{array}{ll}\text{Comparison:} & \text{With feed heating, } \eta = 35{\cdot}84\% \\ & \text{Without feed heating, } \eta = 33{\cdot}89\%\end{array}\right\} \text{Ans.}$$

40.

$$T_1 = 292 \text{ K}$$
$$T_2 = 781 \text{ K}$$

$$\frac{p_1V_1}{T_1} = \frac{p_2V_2}{T_2}$$

$$V_2 = \frac{1{\cdot}01 \times 0{\cdot}125 \times 781}{36 \times 292} = 0{\cdot}009\,38 \text{ m}^3$$

$$p_1V_1^n = p_2V_2^n$$
$$1{\cdot}01 \times 0{\cdot}125^n = 36 \times 0{\cdot}00938^n$$
$$n = 1{\cdot}38 \quad \text{Ans. (i)}$$

Alternatively, n could be obtained from

$$\frac{T_1}{T_2} = \left\{\frac{p_1}{p_2}\right\}^{\frac{n-1}{n}}$$

$$R = c_p - c_V = 1{\cdot}005 - 0{\cdot}718 = 0{\cdot}287 \text{ kJ/kg K}$$
$$p_1V_1 = mRT_1$$

$$m = \frac{1\cdot01 \times 10^2 \times 0\cdot125}{0\cdot287 \times 292}$$
$$= 0\cdot1506 \text{ kg} \quad \text{Ans. (ii)}$$

$$\text{Work done by the air} = \frac{p_1V_1 - p_2V_2}{n-1}$$
$$= \frac{mR\,(T_1 - T_2)}{n-1}$$
$$= \frac{0\cdot1506 \times 0\cdot287 \times (292 - 781)}{1\cdot38 - 1}$$
$$= -55\cdot62 \text{ kJ} \quad \text{Ans. (iii)}$$

The minus sign indicates that the work is done *on* the air.
Increase in internal energy

$$U_2 - U_1 = m \times c_V \times (T_2 - T_1)$$
$$= 0\cdot1506 \times 0\cdot718 \times (781 - 292)$$
$$= 52\cdot86 \text{ kJ} \quad \text{Ans. (iv)}$$

$$\text{Heat supplied} = \begin{matrix}\text{increase in}\\ \text{internal energy}\end{matrix} + \begin{matrix}\text{external}\\ \text{work done}\end{matrix}$$
$$= 52\cdot86 + (-55\cdot62) = -2\cdot76 \text{ kJ} \quad \text{Ans. (v)}$$

The minus sign indicates that the transfer of heat is from the air to its surrounds.

41. $\text{ip of 4 cylinders} = p_m ALn \times 4$

$$\text{where } n = \text{rev/s} \div 2 \text{ for a four-stroke engine}$$
$$\text{ip} = 14\cdot9 \times 10^2 \times 0\cdot7854 \times 0\cdot32^2 \times 0\cdot48 \times 2 \times 4$$
$$= 460 \text{ kW}$$
$$\text{bp [kW]} = T\text{ [kN m]} \times \omega\text{[rad/s]}$$
$$= 12 \times 0\cdot96 \times 2\pi \times 4 = 289\cdot6 \text{ kW}$$

$$\text{Ind. thermal effic.} = \frac{\text{Heat into work in cylinders [kJ/h]}}{\text{Heat energy supplied [kJ/h]}}$$
$$= \frac{460 \times 3600}{99 \times 44\cdot5 \times 10^3}$$
$$= 0\cdot3758 \text{ or } 37\cdot58\% \quad \text{Ans. (i)}$$

$$\text{Brake thermal effic.} = \frac{\text{Heat into work at brake [kJ/h]}}{\text{Heat energy supplied [kJ/h]}}$$
$$= \frac{289\cdot6 \times 3600}{99 \times 44\cdot5 \times 10^3}$$

$= 0{\cdot}2367$ or $23{\cdot}67\%$ Ans. (ii)

Heat energy carried away by cooling water [kJ/h]

$= \text{mass} \times \text{spec. ht.} \times \text{temp. rise}$
$= 154 \times 60 \times 4{\cdot}2 \times (47 - 14)$

As a percentage of the heat supplied

$$= \frac{154 \times 60 \times 4{\cdot}2 \times 33}{99 \times 44{\cdot}5 \times 10^3} \times 100$$

$= 29{\cdot}07\%$ Ans. (iii)

The remainder of the heat losses may be attributed to the heat carried away in the exhaust gases

$= 100 - (37{\cdot}58 + 29{\cdot}07) = 33{\cdot}35\%$

Friction and pumping losses

$= 37{\cdot}58 - 23{\cdot}67 = 13{\cdot}91\%$

Heat balance diagram:

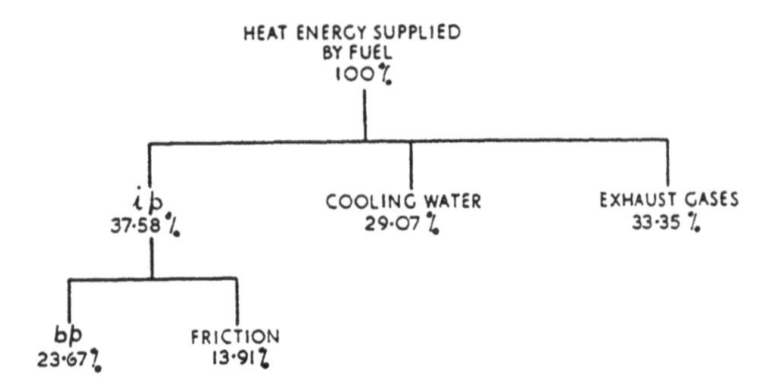

42. Referring to Fig. 65,

$$\frac{T_2}{T_1} = \left\{\frac{p_2}{p_1}\right\}^{\frac{\gamma-1}{\gamma}} \quad \text{where } \frac{\gamma-1}{\gamma} = \frac{0{\cdot}4}{1{\cdot}4} = \frac{2}{7}$$

$$T_2 = 4{\cdot}3^{\frac{2}{7}} \times 289 = 438{\cdot}4 \text{ K}$$

Temperature at compressor outlet

$= 165{\cdot}4°\text{C}$ Ans. (i)

$$\frac{T_4}{T_3} = \frac{T_1}{T_2} \text{ because pressure ratios are equal}$$

$$T_4 = \frac{873 \times 289}{438{\cdot}4} = 575{\cdot}4 \text{ K}$$

Alternatively T_4 could be obtained from

$$\frac{T_4}{T_3} = \left\{\frac{p_4}{p_3}\right\}^{\frac{\gamma-1}{\gamma}}$$

Temperature at turbine outlet

$$= 302{\cdot}4^{\circ}\text{C} \quad \text{Ans. (ii)}$$

Heat supplied per kg of working fluid

$$= \text{mass} \times \text{spec. heat} \times \text{temp. rise}$$
$$= 1 \times 1{\cdot}005 \times (873 - 483{\cdot}4)$$
$$= 436{\cdot}7 \text{ kJ/kg} \quad \text{Ans. (iii)}$$

$$\text{Thermal effic.} = 1 - \frac{T_4 - T_1}{T_3 - T_2} \text{ or } 1 - \frac{T_1}{T_2} \text{ or } 1 - \frac{T_4}{T_3}$$

$$\text{or } 1 - \frac{1}{r_p^{(\gamma-1)/\gamma}}$$

$$1 - \frac{T_1}{T_2} = 1 - \frac{289}{438{\cdot}4} = 1 - 0{\cdot}6592$$

$$= 0{\cdot}3408 \text{ or } 34{\cdot}08\% \quad \text{Ans. (iv)}$$

43. Refer to Fig. 69:
$h_3 = h_4 = 69{\cdot}55$ kJ/kg
i.e. from tables, page 13, at 8·477 bar and 35°C

$$h_4 = h_f + x_4 (h_g - h_f)$$
$$69{\cdot}55 = 17{\cdot}82 + x_4 (178{\cdot}73 - 17{\cdot}82)$$

i.e. from tables, page 13, at 1·509 bar and –20°C

$$x_4 = 0{\cdot}3215$$

$$s_1 = s_2$$

i.e. isentropic compression

$$0{\cdot}0731 + x_1 (0{\cdot}7087 - 0{\cdot}0731) = 0{\cdot}6839$$
$$x_1 = 0{\cdot}961$$

$$
\begin{aligned}
h_1 &= h_f + x_1 (h_g - h_f) \\
&= 17{\cdot}82 + 0{\cdot}961\ (178{\cdot}73 - 17{\cdot}82) \\
&= 172{\cdot}45 \text{ kJ/kg} \\
\text{Refrigerating effect} &= h_1 - h_4 \\
&= 172{\cdot}45 - 69{\cdot}55 \\
&= 102{\cdot}9 \text{ kJ/kg} \quad \text{Ans. (a)(i)}
\end{aligned}
$$

$$
\begin{aligned}
\text{c.o.p.} &= \frac{h_1 - h_4}{h_2 - h_1} \\
&= \frac{102{\cdot}9}{201{\cdot}45 - 172{\cdot}45} \\
&= 3{\cdot}548 \quad \text{Ans. (a)(ii)}
\end{aligned}
$$

$$
\begin{aligned}
\text{Reversed Carnot c.o.p.} &= \frac{T_L}{T_H - T_L} \\
&= \frac{253}{308 - 253} \\
&= 4{\cdot}6 \quad \text{Ans. (b)}
\end{aligned}
$$

44.

$$
\begin{aligned}
\theta_m &= \frac{50 - 20}{\ln \dfrac{50 - 15}{20 - 15}} \\
&= 15{\cdot}42^\circ\text{C} \quad \text{Ans. (a)}
\end{aligned}
$$

$$
Q = UAt\theta_m \quad \text{also } Q = 4(50 - 20) \times 1395{\cdot}6/10^3
$$

$$
\begin{aligned}
\therefore\ UA\theta_m &= 167{\cdot}5 \\
A &= \frac{167{\cdot}5 \times 10^3}{70 \times 15{\cdot}42} \\
&= 155 \text{ m}^2 \quad \text{Ans. (b)} \\
\pi dLn &= A \qquad n = 350 \text{ tubes} \\
L &= \frac{A}{\pi dn} \\
&= \frac{155 \times 10^3}{\pi \times 19 \times 350} \\
&= 7{\cdot}42 \text{ m.} \quad \text{Ans. (c)}
\end{aligned}
$$

45. From h–s chart

$$
\begin{aligned}
h_1 &= 2760 \text{ kJ/kg} \\
h_2 &= 2760 \text{ kJ/kg}
\end{aligned}
$$

$$h_3 = 2595 \text{ kJ/kg}$$
$$h_4 = 2595 \text{ kJ/kg}$$
$$h_5 = 2180 \text{ kJ/kg}$$
Final condition of steam = 0·823 dry. Ans.

Enthalpy lost by steam = Enthalpy gained by the oil

$$\dot{m}\,[(h_2 - h_3) + (h_4 - h_5)] = 0{\cdot}72 \times 2{\cdot}1 \times 72$$
$$\dot{m} = 0{\cdot}186 \text{ kg/s} \quad \text{Ans.}$$

46. Mass of 1 mol of fuel = $12 \times 1 + 1 \times 4$
= 16 kg

$$H_2 \text{ by mass} = \frac{4}{16}$$
= 0·25 kg

$$\text{C by mass} = \frac{12}{16}$$
= 0·75 kg

$$\text{Stoichiometric air required} = \frac{100}{23}(2\tfrac{2}{3}\text{C} + 8\text{H})$$
$$= 4{\cdot}348\,(2\tfrac{2}{3} \times 0{\cdot}75 + 8 \times 0{\cdot}25)$$
= 17·39 kg/kg fuel

Correct air-fuel mass ratio = 17·39 Ans. (a)

Mass of dry products of combustion per kg fuel burnt:

$CO_2 = 3\tfrac{2}{3} \times 0{\cdot}75 = 2{\cdot}75$ kg

$N_2 = 0{\cdot}77 \times 17{\cdot}39 = 13{\cdot}39$

Total = 16·14 kg

% mass analysis of the dry flue gases:

$$CO_2 = \frac{275}{16{\cdot}14} = 17{\cdot}04$$
$$N_2 = \frac{1339}{16{\cdot}14} = 82{\cdot}96$$

% volume analysis of the dry flue gases:

DFG	m%	M	N	N%
CO_2	17·04	44	0·3873	11·56
N_2	82·96	28	2·9629	88·44
			3·3502	

Ans. (b)

47. Steady flow energy equation

$$h_1 + \tfrac{1}{2}c_1^2 + q = h_2 + \tfrac{1}{2}c_2^2 + w$$

$$80 + \tfrac{1}{2} \times \frac{75^2}{10^3} - 10 = 300 + \tfrac{1}{2} \times \frac{175^2}{10^3} + w$$

from which $w = -242{\cdot}5$ kJ/kg. Ans. (a)

If t_h = datum temperature on which the specific enthalpies are based

$$\left\{\begin{array}{l} \text{then } h_1 = c_P\,(t_1 - t_h) \\ \text{and } h_2 = c_P\,(t_2 - t_h) \end{array}\right\} \text{divide}$$

$$\therefore \quad \frac{h_1}{h_2} = \frac{c_P\,(t_1 - t_h)}{c_P\,(t_2 - t_h)}$$

$$\text{i.e.} \quad \frac{80}{300} = \frac{15 - t_h}{200 - t_h}$$

$$\text{i.e.} \quad 80(200 - t_h) = 300\,(15 - t_h)$$

$$16\,000 - 80\,t_h = 4500 - 300\,t_h$$

$$16\,000 - 4500 = 80\,t_h - 300\,t_h$$

$$\frac{11\,500}{220} = -t_h$$

$$\therefore \quad t_h = -52{\cdot}27^\circ\text{C or } 220{\cdot}7\text{ K} \quad \text{Ans. (b)}$$

48.

$$\text{For insulation (1)} \quad Q = \frac{2\pi k_1 (T_1 - T_2)}{\ln\left(\dfrac{D_2}{D_1}\right)} \text{ W/m}$$

$$\text{For insulation (2)} \quad Q = \frac{2\pi k_2 (T_2 - T_3)}{\ln\left(\dfrac{D_3}{D_2}\right)} \text{ W/m}$$

$$\text{For surface film} \quad Q = \frac{hA\,(T_3 - T_4)}{L} \text{ W/m}$$

Total temperature drop $T_1 - T_4$

$$= (T_1 - T_2) + (T_2 - T_3) + (T_3 - T_4)$$

$$= \frac{Q \ln (D_2/D_1)}{2\pi k_1} + \frac{Q \ln (D_3/D_2)}{2\pi k_2} + \frac{Q}{h\pi D_3}$$

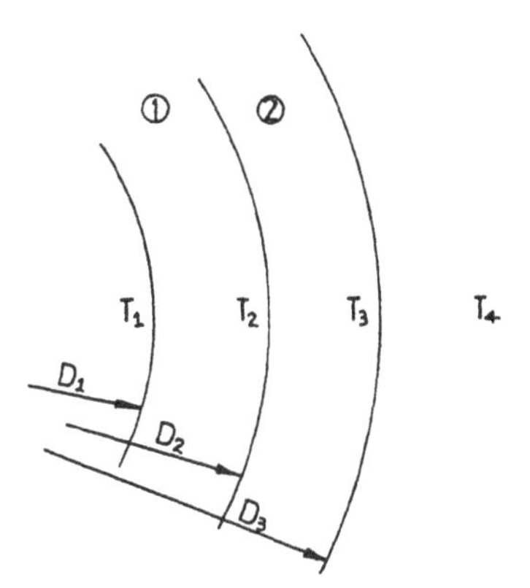

From steam tables $T_1 = 263{\cdot}9°C$

$$\therefore \quad 263{\cdot}9 - 20 = Q\left[\frac{\ln\left(\frac{400}{200}\right)}{2\pi \times 0{\cdot}5} + \frac{\ln\left(\frac{500}{400}\right)}{2\pi \times 0{\cdot}1} + \frac{10^3}{8\pi \times 500}\right]$$

$$243{\cdot}9 = Q\,[2{\cdot}206 + 0{\cdot}355 + 0{\cdot}0796]$$

$$Q = 92{\cdot}35 \text{ W/m.}$$

Q = mass of steam condensed/metre length of pipe/s × latent heat of steam.

$$Q = \dot{m} \times h_{fg}$$

$$\dot{m} = \frac{92{\cdot}35}{1639}$$

$$= 0{\cdot}0563 \text{ kg/s} \quad \text{Ans.}$$

49.

Using the h–s chart h = 3410 kJ/kg

h_2^1 = 3100 kJ/kg

h_2 = 2990 kJ/kg

h_3 = 3475 kJ/kg

h_4^1 = 3130 kJ/kg

h_4 = 3085 kJ/kg

h_5^1 = 2415 kJ/kg

h_5 = 2370 kJ/kg

Pressure drop in reheater = 15 – 13

= 2 bar. Ans. (a)

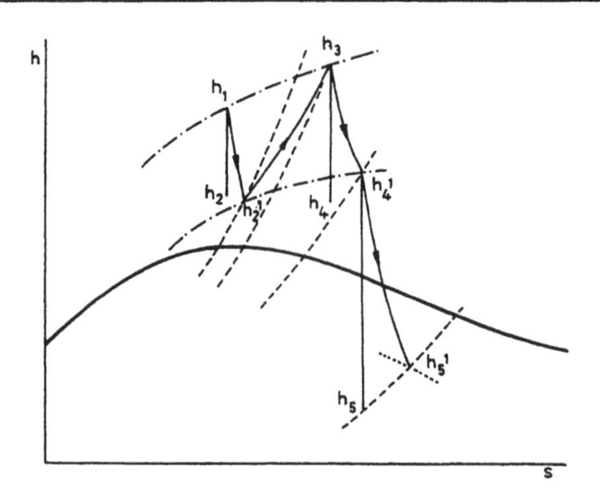

Stage 1. Isentropic efficiency $= \dfrac{h_1 - h_2{}^1}{h_1 - h_2}$

$= 0{\cdot}74$ or 74%

Stage 2. Isentropic efficiency $= \dfrac{h_3 - h_4{}^1}{h_3 - h_4}$

$= 0{\cdot}88$ or 88%

Stage 3. Isentropic efficiency $= \dfrac{h_4 - h_5{}^1}{h_4 - h_5}$

$= 0{\cdot}94$ or 94% Ans. (b)

Ratio of powers, $h_1 - h_2{}^1 : h_3 - h_4{}^1 : h_4 - h_5{}^1$

i.e. 310 : 345 : 715

i.e. 1 : 1·112 : 2·31 Ans. (c)

50.

$$\frac{T_1}{T_2} = \left(\frac{p_1}{p_2}\right)^{\frac{\gamma-1}{\gamma}}$$

2. $T_2 = 393 \times 10^{\frac{0{\cdot}4}{1{\cdot}4}} = 759\text{ K} = 486°\text{C}$

3. $T_3 = 800°\text{C}$

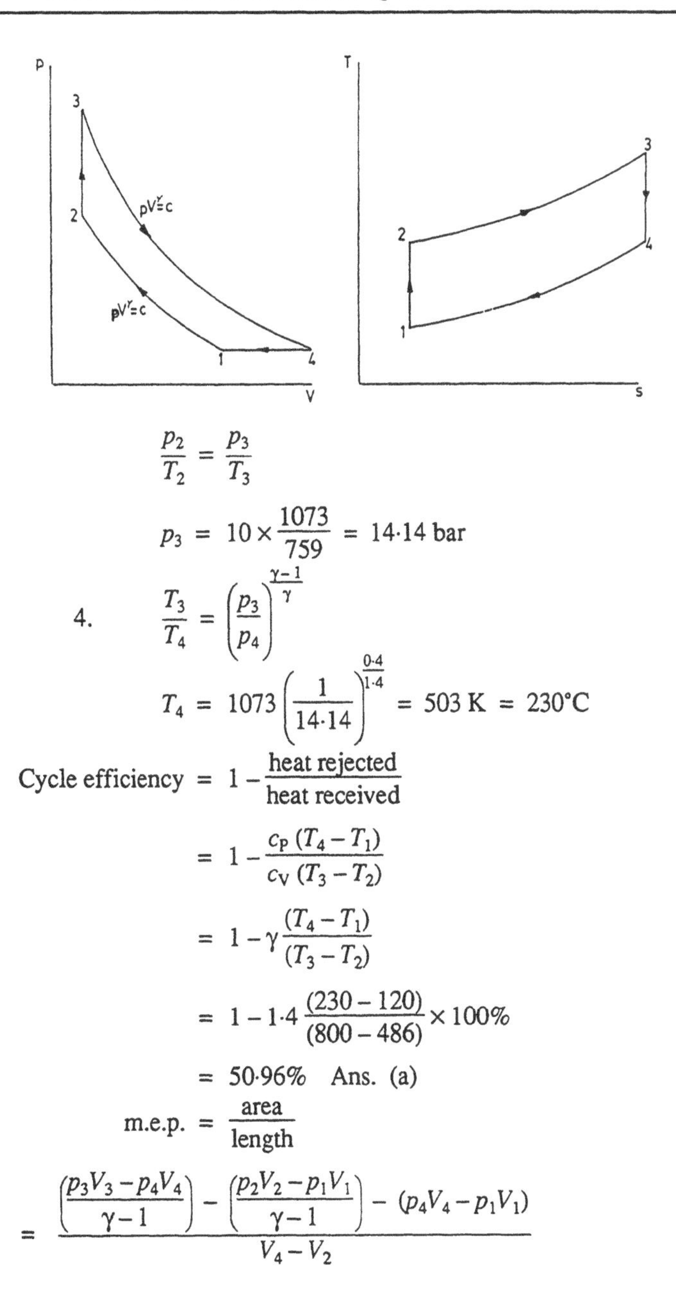

$$\frac{p_2}{T_2} = \frac{p_3}{T_3}$$

$$p_3 = 10 \times \frac{1073}{759} = 14{\cdot}14 \text{ bar}$$

4. $$\frac{T_3}{T_4} = \left(\frac{p_3}{p_4}\right)^{\frac{\gamma-1}{\gamma}}$$

$$T_4 = 1073\left(\frac{1}{14{\cdot}14}\right)^{\frac{0{\cdot}4}{1{\cdot}4}} = 503 \text{ K} = 230^\circ\text{C}$$

$$\text{Cycle efficiency} = 1 - \frac{\text{heat rejected}}{\text{heat received}}$$

$$= 1 - \frac{c_P\,(T_4 - T_1)}{c_V\,(T_3 - T_2)}$$

$$= 1 - \gamma\frac{(T_4 - T_1)}{(T_3 - T_2)}$$

$$= 1 - 1{\cdot}4\,\frac{(230 - 120)}{(800 - 486)} \times 100\%$$

$$= 50{\cdot}96\% \quad \text{Ans. (a)}$$

$$\text{m.e.p.} = \frac{\text{area}}{\text{length}}$$

$$= \frac{\left(\dfrac{p_3V_3 - p_4V_4}{\gamma - 1}\right) - \left(\dfrac{p_2V_2 - p_1V_1}{\gamma - 1}\right) - (p_4V_4 - p_1V_1)}{V_4 - V_2}$$

as $pV = mRT$

$$\text{m.e.p.} = \frac{\dfrac{(T_3 - T_4)}{\gamma - 1} - \dfrac{(T_2 - T_1)}{\gamma - 1} - (T_4 - T_1)}{\dfrac{T_4}{p_4} - \dfrac{T_2}{p_2}}$$

$$= \frac{\left(\dfrac{800 - 230}{0{\cdot}4}\right) - \left(\dfrac{486 - 120}{0{\cdot}4}\right) - (230 - 120)}{\dfrac{503}{10^5} - \dfrac{759}{10^6}}$$

$= 0{\cdot}937$ bar Ans. (b)

INDEX

REED'S MARINE ENGINEERING SERIES

Vol 1 Mathematics
Vol 2 Applied Mechanics
Vol 3 Applied Heat
Vol 4 Naval Architecture
Vol 5 Ship Construction
Vol 6 Basic Electrotechnology
Vol 7 Advanced Electrotechnology
Vol 8 General Engineering Knowledge
Vol 9 Steam Engineering Knowledge
Vol 10 Instrumentation and Control Systems
Vol 11 Engineering Drawing
Vol 12 Motor Engineering Knowledge

Reed's Engineering Knowledge for Deck Officers
Reed's Maths Tables and Engineering Formulae
Reed's Marine Distance Tables
Reed's Sextant Simplified
Reed's Skippers Handbook
Reed's Maritime Meteorology
Sea Transport – Operation and Economics

These books are obtainable from all good nautical booksellers or direct from:

Adlard Coles Nautical
A & C Black
P O Box 19
St Neots
Cambs PE19 8SF

Tel: 01480 212666
Fax: 01480 405014

Email: sales@acblack.com